AF356158

Micro/Nano Manufacturing Processes: Theories and Optimization Techniques

Micro/Nano Manufacturing Processes: Theories and Optimization Techniques

Editors

Zejia Zhao
Guoqing Zhang
Wai Sze YIP

Basel • Beijing • Wuhan • Barcelona • Belgrade • Novi Sad • Cluj • Manchester

Editors

Zejia Zhao
College of Mechatronics and
Control Engineering
Shenzhen University
Shenzhen
China

Guoqing Zhang
College of Mechatronics and
Control Engineering
Shenzhen University
Shenzhen
China

Wai Sze YIP
Department of Industrial and
Systems Engineering
The Hong Kong Polytechnic
University
Hong Kong
China

Editorial Office
MDPI AG
Grosspeteranlage 5
4052 Basel, Switzerland

This is a reprint of articles from the Special Issue published online in the open access journal *Processes* (ISSN 2227-9717) (available at: www.mdpi.com/journal/processes/special_issues/LN028R4O8S).

For citation purposes, cite each article independently as indicated on the article page online and as indicated below:

Lastname, A.A.; Lastname, B.B. Article Title. *Journal Name* **Year**, *Volume Number*, Page Range.

ISBN 978-3-7258-2034-4 (Hbk)
ISBN 978-3-7258-2033-7 (PDF)
doi.org/10.3390/books978-3-7258-2033-7

Contents

About the Editors

Zejia Zhao

Dr. Zhao Zejia is currently an Associate Professor in the College of Mechatronics and Control Engineering at Shenzhen University. He achieved his Master's degree in Materials Science at the Harbin Institute of Technology in 2012. He achieved his Ph.D. degree in Ultra-Precision Machining Technology at The Hong Kong Polytechnic University in 2019. His research interests mainly include ultra-precision machining, the cutting of difficult-to-cut and brittle materials, numerical modeling, and semiconductor manufacturing. As a Principal Investigator, he has been in charge of eight projects, such as those organized by the National Natural Science Foundation of China (NSFC), the Department of Education of Guangdong Province, and Shenzhen Stable Support. Dr. Zhao is also a Member of the Chinese Advanced Optical Manufacturing Youth Expert Committee and a Council Member of the Shenzhen Mechanical Engineering Advanced Manufacturing Branch. He has been awarded for his talent as part of the Shenzhen Peacock Project.

Guoqing Zhang

Prof. Dr. Zhang Guoqing is currently a Professor in the College of Mechatronics and Control Engineering at Shenzhen University. He achieved his Bachelor's degree in Mechanical Engineering at Northeast Petroleum University, China, in 2005. In 2009, he achieved his Master's degree in Mechanical Engineering at the Harbin Institute of Technology, China. Additionally, he achieved his Ph.D. degree at The Hong Kong Polytechnic University in 2014. He has been selected for the Shenzhen Peacock Program for Overseas/Professional High-Level Talents. He serves as a Deputy Director of Guangdong Key Laboratory of Electromagnetic Control and Intelligent Robots; Deputy Director of Shenzhen Key Laboratory of High-Performance Nontraditional Manufacturing; Committee Member of the Chinese Mechanical Engineering Society Design Branch; and a Member of the Advanced Optical Manufacturing Young Expert Committee of the Chinese Optical Engineering Society. His research interests include ultra-precision machining technology and equipment, robots and intelligent equipment, etc. To date, he has hosted more than 10 research projects, including 3 National Natural Science Foundation of China (NSFC) projects; he has published more than 80 peer reviewed papers, been granted 14 US patents and 28 CN patents, and has also co-edited 2 books. He now serves as an Academic Editor for the International journal Shock and Vibration and contributes as a reviewer to more than 10 other international journals. He has won more than five prizes, including the China Industry–University–Research Cooperation Promotion Award (individual) and a second prize in innovation achievement.

Wai Sze YIP

Dr. Yip Wai Sze is currently a Research Assistant Professor within the Department of Industrial and Systems Engineering of The Hong Kong Polytechnic University. She achieved her Ph.D. degree in Ultra-Precision Machining at The Hong Kong Polytechnic University in 2018. She achieved a double degree of a BBA in Marketing and a BEng in Industrial and Systems Engineering at The Hong Kong Polytechnic University in 2014, and a BEng degree in Electronic and Communication Engineering at the City University of Hong Kong in 2006. Before joining the ISE of PolyU, Dr Yip worked as a Research Fellow at the National University of Singapore. Her research focuses on the ultra-precision machining of difficult-to-cut materials, sustainable precision machining, and the sustainable development of precision manufacturing.

Dr Yip has published research in top-tier SCI journals such as *Energy*, the *Journal of Cleaner Production*, the *Journal of Alloy and Compounds*, and *IEEE Access*. In 2018, one of her papers in *Scientific Reports* was recognized as being one of the top 100 papers in materials science. In 2019, she received the Excellent Thesis Award, 9th Hiwin Doctoral Dissertation, from the Chinese Mechanical Engineering Society.

Preface

Micro/nano manufacturing has emerged as a cornerstone of modern engineering, enabling the production of highly precise and functional components that are integral to a wide array of industries, from electronics and biotechnology to aerospace and automotive sectors. At micro/nano scales, the fundamental theories in the manufacturing process are much different from those at the macro scale. Therefore, understanding the particular theories and proposing relevant optimization manufacturing methods are particularly essential to the micro/nano manufacturing processes.

This reprint, "Micro/Nano Manufacturing Processes: Theories and Optimization Techniques", aims to provide an exploration of the fundamental theories, cutting-edge techniques, and optimization strategies that underpin the micro/nano manufacturing processes. It mainly includes three parts, i.e., (1) fabrication of micro/nano-structured surfaces; (b) simulation and experimental investigation of laser manufacturing; and (c) optimization techniques in the microscale machining. The content of this reprint begins with a foundational overview of manufacturing micro/nano-structured surfaces, elucidating the primary fabrication methods such as lithography technology, LIGA technology, and ultra-precision machining technology to manufacture structured surfaces. Subsequent contents delve into simulation and optimization techniques of the microscale manufacturing, providing detailed insights into their mechanisms, advantages, and limitations.

This reprint is designed to serve as a guide for students, researchers, and professionals interested in understanding the fundamental principles of micro/nano manufacturing. By combining theoretical insights with practical optimization strategies, this reprint also aims to equip readers with the knowledge and tools necessary to navigate this rapidly evolving landscape.

We thank all the contributors and the editors (Mr. Scott Pan and Mr. Aleksandar Jovanović) for their enthusiastic support for this Special Issue, as well as the editorial staff of Processes for their efforts.

Zejia Zhao, Guoqing Zhang, and Wai Sze YIP
Editors

processes

MDPI

Editorial

Special Issue: "Micro/Nano Manufacturing Processes: Theories and Optimization Techniques"

Zejia Zhao [1], Guoqing Zhang [1],* and Wai Sze Yip [2]

[1] College of Mechatronics and Control Engineering, Shenzhen University, Shenzhen 518060, China; zhaozejia@szu.edu.cn
[2] Department of Industrial and Systems Engineering, The Hong Kong Polytechnic University, Hong Kong SAR, China; lenny.ws.yip@polyu.edu.hk
* Correspondence: zhanggq@szu.edu.cn

Citation: Zhao, Z.; Zhang, G.; Yip, W.S. Special Issue: "Micro/Nano Manufacturing Processes: Theories and Optimization Techniques". *Processes* **2024**, *12*, 1746. https://doi.org/10.3390/pr12081746

Received: 26 July 2024
Accepted: 19 August 2024
Published: 20 August 2024

Manufacturing at the micro/nano scale creates many opportunities to fabricate micro- and nanostructures or to manufacture high-precision components, which has attracted considerable attention in fields such as optics [1], electronics [2], precision instruments [3], the semiconductor industry [4], biomedical engineering [5], etc. Over the years, much effort has been devoted to investigating the micro/nano manufacturing processes, but there are still some challenges to fully understanding them since the deformation mechanism of materials at the micro/nano scales is quite different from the well-known deformation mechanisms at the macro scale [6,7]. This Special Issue is dedicated to the special theories and some optimization techniques of the micro/nano manufacturing processes. The Special Issue is available online at: https://www.mdpi.com/journal/processes/special_issues/LN028R4O8S.

Fabrication of micro/nano-structured surfaces

Material surfaces with micro-structured patterns have attracted much attention in communication, medical, and military fields due to their superior properties, such as light trapping, antibacterial, and self-cleaning [8]. Generally, the micro-structured functional surfaces could be achieved by methods of ultraprecision machining, photolithography, etching, ion beam machining, microforming, micromolding, etc. [9,10].

The paper by Ma et al. [11] gives a state-of-the-art in fabrication methods and assisted technologies for micro-structured surfaces. The primary methods include lithography technology, high-energy beam direct writing technology, special energy field machining technology, molding technology, LIGA (a combination of lithographie, galvanoformung, and abformung) technology, and ultra-precision machining technology. In particular, the authors compared the advantages and disadvantages of different ultra-precision machining technologies such as fast/slow tool servo machining, ultra-precision diamond scratching, ultra-precision fly cutting, and raster milling. They also give some perspectives about the future trend in fabricating micro-structured functional surfaces and suppose that a combination of different methods is of great significance for the fabrication of multi-layer and multi-scale micro-structures.

The paper by Wu et al. [12] proposed an efficient method to fabricate the mold cavity structure for a helical cylindrical pinion based on a plastic torsion-forming concept. A new method termed low-speed wire electrical discharge machining (LS-WEDM) was used to fabricate the spur gear cavity. In comparison to the multi-stage helical gear core electrode and the complex spiral EDM process, the proposed method provides an efficient and simple way to generate the structure of the helical gear cavity by twisting the spur gear cavity plastically around the central axis. Some theoretical models were also proposed to evaluate the precision of the helical gear geometry. Based on this theoretical model and experimental results, the proposed LS-WEDM method could precisely control the shape accuracy of a helical cylindrical structure by adjusting the machine torsion angle.

The paper by Du's team [13] successfully fabricates the micro/nanoscale hierarchical micropillar arrays on the thermoplastic polymer surface by an ultrasonic plasticizing and pressing (UPP) method. A high aspect ratio of up to 24.1 was achieved by the UPP methods since they could make full use of the ultrasonic vibration and avoid the violent friction between the raw material and template. In comparison to other template methods, the UPP has advantages in superior forming capability, a simple process, a short production cycle, and high cost-effectiveness. The authors also measured the wettability of the surface with micro/nanoscale hierarchical micropillar arrays and found that the micro-structured surface shows superhydrophobic and superoleophilic properties, which provide potential uses in both research and applications.

Simulation and experimental investigation of laser manufacturing

Laser manufacturing has been widely used in industry applications such as cutting, drilling, welding, scribing, and additive manufacturing due to its high precision, high speed, and versatility. The paper by Xu et al. [14] studied the femtosecond laser ablation of $Cu_{50}Zr_{50}$ metallic glass by molecular dynamics simulation based on a two-temperature model. They found that the maximum electron temperature at the same position on the target surface decreased as the pulse duration increased, but there was no significant change in the electron–lattice temperature coupling time. They also investigate the effect of absorbed fluence on the maximum electron temperature at the same position on the target surface and the electron–lattice temperature coupling time. Melting, spallation, and phase explosion caused by femtosecond laser irradiation give rise to the surface ablation of the target material.

The paper by Zhao et al. [15] investigated the high-speed laser-induced thermal crack propagation (LITP) dicing of a glass–silicon double-layer wafer. Both numerical simulations using the ABAQUS and experiments were conducted in this study. Their results indicate that the region near the upper surface of the glass layer cracked asynchronously with the silicon layer during the stable extension, and the crack propagation in the glass layer material is not synchronized with that in the silicon. In addition, a good surface roughness of 19 nm and 9 nm was achieved in the silicon and glass layers, respectively. In order to address the problem of trajectory deviation that usually occurs in actual glass cutting operations by the LITP, Zhao et al. [16] further proposed a low-temperature gas cooling trajectory deviation correction technique, which optimizes the temperature and stress distribution by spraying low-temperature gas onto the processing surface and maintaining a relative position with the laser. The proposed trajectory deviation correction benefits the application of LITP technology in practical glass cutting.

Optimization techniques in the microscale machining

The paper by Zhou et al. [17] presents a scheduling optimization approach for the micro-hole drilling production line of printed circuit boards (PCBs). A complex event model considered an emergency insertion event, a production line equipment operation event, and a tool failure event was proposed. A catastrophe genetic algorithm was also used to solve the initial scheduling scheme of the micro-hole drilling production line. Compared to the traditional genetic algorithm, the proposed catastrophic genetic algorithm is more accurate and effective in solving various production line scheduling problems. In addition, a dynamic scheduling of the production line was realized by considering complex events and scheduling optimization, which contributed to an improvement in the scheduling optimization rate to 25.1%. The paper by Liu et al. [18] proposed an improved prediction model to predict the remaining useful life (RUL) of the drill bit in micro-drilling of the packaging substrate. The experimental results indicate that the modified model, taking into account the degradation rate and offset coefficient, could provide more accurate and stable results in comparison to traditional models, which are supposedly a reliable theoretical foundation for the health state monitoring of drill bits in the micro-drilling of packaging substrates.

In their paper, Zhao et al. [19] propose a novel approach for the topology optimization of compliant mechanisms to solve design challenges including static strength, fatigue failure, and manufacturability. An improved solid isotropic material with penalization (SIMP) interpolation method is also used to derive the shape sensitivity of the optimization problem. The results indicate that the von Mises stresses in the force inverter and compliant gripper were below the materials strength limit of 275 MPa, and the fatigue-constrained topology optimization could reduce stress concentration. In order to meet the manufacturing process requirement, a three-field density projection approach was used to control the minimum size in the layout optimization. In addition, two numerical examples of an inverter and a gripper are given to illustrate the effectiveness of the proposed method.

The paper by Louis et al. [20] studied the performance improvement of milling tools by the chemical vapor deposition coating in the milling of zirconia ceramics (ZrO_2). Different diamond films with various diamond grain sizes, and film thicknesses were coated on the tungsten carbide milling tools by hot filament chemical vapor deposition (HFCVD). As the distances between substrate and filament decrease, the grain size and coating thickness of the diamond film on milling tools tend to decrease. In addition, the coating film with smaller grain sizes and thinner thicknesses could contribute to higher machining quality in terms of surface topology, surface roughness, and tool failure. Their findings provide useful instructions for parameter optimization in the production of coated tools by the HFCVD.

We thank all the contributors and the editors (Mr. Scott Pan and Mr. Aleksandar Jovanović) for their enthusiastic support for this Special Issue, as well as the editorial staff of *Processes* for their efforts.

Funding: This research received no external funding.

Conflicts of Interest: The authors declare no conflicts of interest.

References

1. Wang, W.; Qi, L. Light management with patterned micro-and nanostructure arrays for photocatalysis, photovoltaics, and optoelectronic and optical devices. *Adv. Funct. Mater.* **2019**, *29*, 1807275. [CrossRef]
2. Li, X.; Sun, W.; Fu, W.; Lv, H.; Zu, X.; Guo, Y.; Gibson, D.; Fu, Y.-Q. Advances in sensing mechanisms and micro/nanostructured sensing layers for surface acoustic wave-based gas sensors. *J. Mater. Chem. A* **2023**, *11*, 9216–9238. [CrossRef]
3. Zhou, T.; He, Y.; Wang, T.; Zhu, Z.; Xu, R.; Yu, Q.; Zhao, B.; Zhao, W.; Liu, P.; Wang, X. A review of the techniques for the mold manufacturing of micro/nanostructures for precision glass molding. *Int. J. Extrem. Manuf.* **2021**, *3*, 042002. [CrossRef]
4. Cho, D.; Park, J.; Kim, T.; Jeon, S. Recent advances in lithographic fabrication of micro-/nanostructured polydimethylsiloxanes and their soft electronic applications. *J. Semicond.* **2019**, *40*, 111605. [CrossRef]
5. Liu, Y.; He, X.; Yuan, C.; Cao, P.; Bai, X. Antifouling applications and fabrications of biomimetic micro-structured surfaces: A review. *J. Adv. Res.* **2023**, *59*, 201–221. [CrossRef]
6. Zhao, Z.; To, S.; Wang, J.; Zhang, G.; Weng, Z. A review of micro/nanostructure effects on the machining of metallic materials. *Mater. Des.* **2022**, *224*, 111315. [CrossRef]
7. Zhao, Z.; Fu, Y.; To, S.; Zhang, G.; Lin, J. Microstructural effects on the single crystal diamond tool wear in ultraprecision turning of Ti6Al4V alloys. *Int. J. Refract. Met. Hard Mater.* **2023**, *110*, 106038. [CrossRef]
8. Zhang, Y.; Jiao, Y.; Li, C.; Chen, C.; Li, J.; Hu, Y.; Wu, D.; Chu, J. Bioinspired micro/nanostructured surfaces prepared by femtosecond laser direct writing for multi-functional applications. *Int. J. Extrem. Manuf.* **2020**, *2*, 032002. [CrossRef]
9. Zhao, Z.; Yin, T.; To, S.; Zhu, Z.; Zhuang, Z. Material removal energy in ultraprecision machining of micro-lens arrays on single crystal silicon by slow tool servo. *J. Clean. Prod.* **2022**, *335*, 130295. [CrossRef]
10. Kodihalli Shivaprakash, N.; Banerjee, P.S.; Banerjee, S.S.; Barry, C.; Mead, J. Advanced polymer processing technologies for micro-and nanostructured surfaces: A review. *Polym. Eng. Sci.* **2023**, *63*, 1057–1081. [CrossRef]
11. Ma, Y.; Zhang, G.; Cao, S.; Huo, Z.; Han, J.; Ma, S.; Huang, Z. A review of advances in fabrication methods and assistive technologies of micro-structured surfaces. *Processes* **2023**, *11*, 1337. [CrossRef]
12. Wu, B.; Zhu, L.; Zhou, Z.; Guo, C.; Cheng, T.; Wu, X. An Efficient Method to Fabricate the Mold Cavity for a Helical Cylindrical Pinion. *Processes* **2023**, *11*, 2033. [CrossRef]
13. Wu, S.; Du, J.; Xu, S.; Lei, J.; Ma, J.; Zhu, L. Ultrasonic Plasticizing and Pressing of High-Aspect Ratio Micropillar Arrays with Superhydrophobic and Superoleophilic Properties. *Processes* **2024**, *12*, 856. [CrossRef]
14. Xu, J.; Xue, D.; Gaidai, O.; Wang, Y.; Xu, S. Molecular Dynamics Simulation of Femtosecond Laser Ablation of Cu50Zr50 Metallic Glass Based on Two-Temperature Model. *Processes* **2023**, *11*, 1704. [CrossRef]
15. Zhao, C.; Yang, Z.; Kang, S.; Qiu, X.; Xu, B. High-Speed Laser Cutting Silicon-Glass Double Layer Wafer with Laser-Induced Thermal-Crack Propagation. *Processes* **2023**, *11*, 1177. [CrossRef]
16. Zhao, C.; Yang, Z.; Qiu, X.; Sun, J.; Zhao, Z. Low-Temperature Gas Cooling Correction Trajectory Offset Technology of Laser-Induced Thermal Crack Propagation for Asymmetric Linear Cutting Glass. *Processes* **2023**, *11*, 1493. [CrossRef]

17. Zhou, Q.; Hu, X.; Peng, S.; Li, Y.; Zhu, T.; Shi, H. Scheduling Optimization of Printed Circuit Board Micro-Hole Drilling Production Line Based on Complex Events. *Processes* **2023**, *11*, 3073. [CrossRef]
18. Liu, X.; Tao, S.; Zhu, T.; Wang, Z.; Shi, H. Prediction Model of the Remaining Useful Life of the Drill Bit during Micro-Drilling of the Packaging Substrate. *Processes* **2023**, *11*, 2653. [CrossRef]
19. Zhao, D.; Wang, H. Topology Optimization of Compliant Mechanisms Considering Manufacturing Uncertainty, Fatigue, and Static Failure Constraints. *Processes* **2023**, *11*, 2914. [CrossRef]
20. Fan, L.L.; Yip, W.S.; Sun, Z.; Zhang, B.; To, S. Effect of Hot Filament Chemical Vapor Deposition Filament Distribution on Coated Tools Performance in Milling of Zirconia Ceramics. *Processes* **2023**, *11*, 2773. [CrossRef]

Review

A Review of Advances in Fabrication Methods and Assistive Technologies of Micro-Structured Surfaces

Yuting Ma, Guoqing Zhang *, Shuaikang Cao, Zexuan Huo, Junhong Han, Shuai Ma and Zejia Huang

Shenzhen Key Laboratory of High Performance Nontraditional Manufacturing, College of Mechatronics and Control Engineering, Shenzhen University, Nan-Hai Ave 3688, Shenzhen 518060, China
* Correspondence: zhanggq@szu.edu.cn; Tel.: +86-755-26536306; Fax: +86-755-26557471

Abstract: Micro-structured surfaces possess excellent properties of friction, lubrication, drag reduction, antibacterial, and self-cleaning, which have been widely applied in optical, medical, national defense, aerospace fields, etc. Therefore, it is requisite to study the fabrication methods of micro-structures to improve the accuracy and enhance the performance of micro-structures. At present, there are plenty of studies focusing on the preparation of micro-structures; therefore, systematic review of the technologies and developing trend on the fabrication of micro-structures are needed. In present review, the fabrication methods of various micro-structures are compared and summarized. Specially, the characteristics and applications of ultra-precision machining (UPM) technology in the fabrication of micro-structures are mainly discussed. Additionally, the assistive technologies applied into UPM, such as fast tool servo (FTS) technology and slow tool servo (STS) technology to fabricate micro-structures with different characteristics are summarized. Finally, the principal characteristics and applications of fly cutting technology in manufacturing special micro-structures are presented. From the review, it is found that by combining different machining methods to prepare the base layer surface first and then fabricate the sublayer surface, the advantages of different machining technologies can be greatly exerted, which is of great significance for the preparation of multi-layer and multi-scale micro-structures. Furthermore, the combination of ultra-precision fly cutting and FTS/STS possess advantages in realizing complex micro-structures with high aspect ratio and high resolution. However, residual tool marks and material recovery are still the key factors affecting the form accuracy of machined micro-structures. This review provides advances in fabrication methods and assistive technologies of micro-structured surfaces, which serves as the guidance for both fabrication and application of multi-layer and multi-scale micro-structures.

Keywords: micro-structures; ultra-precision machining; FTS/STS; fly cutting; raster milling

Citation: Ma, Y.; Zhang, G.; Cao, S.; Huo, Z.; Han, J.; Ma, S.; Huang, Z. A Review of Advances in Fabrication Methods and Assistive Technologies of Micro-Structured Surfaces. *Processes* **2023**, *11*, 1337. https://doi.org/10.3390/pr11051337

Academic Editor: Albert Renken

Received: 17 March 2023
Revised: 20 April 2023
Accepted: 21 April 2023
Published: 26 April 2023

1. Introduction

A micro-structured surface is small with special arranged topological structures [1,2]. The special arrangement of these microscale topological structures makes the micro-structured surface exhibit specific functions, such as superhydrophobicity, anti-fouling, drag reduction, structural color, etc., as shown in Figure 1 [3–6]. These specific functions are perfectly reflected in a range of plants and animals in nature.

In the 1980s, Barthlott et al. discovered that lotus leaf surfaces have a unique set of micro/nano structures, with thousands of microscale papillae (20–100 μm) and thousands of nanoscale wax filaments (100–500 nm) distributed on each papillae [7,8]. This unique surface result in superhydrophobicity and self-cleaning effect of the lotus leaf is known as the lotus effect. Similar to lotus leaves, rice leaves have superhydrophobicity and anisotropic wettability due to their specially arranged hierarchies [9]. These characteristics are also reflected in animals. For example, the rough nano-structures in shark skin endow it with anti-fouling and drag reduction properties [10]; the multi-scale structures on the butterfly wings can not only produce structural color but also have superhydrophobicity, self-cleaning, sensitive chemosensory abilities, and fluorescence emission functions [3,11]; insect compound eyes have high sensitivity and anti-reflex function [12]. These special micro-structured surfaces not only endow powerful functions to natural plants and animals, but also widely apply to optics, circuit systems, information and communication, precision engineering, biomedicine, and other fields [13]. For instance, micro-structured arrays have good optical properties, higher flexibility, and uniform shape, providing greater flexibility and innovation for optical design [14–17]. Moreover, the micro-structured surface is small in size and light in weight, which is easy to realize system integration and reduce the difficulty of assembly [18]. In addition to the above functions, the micro-structured surface also has great development potential in aerospace, navigation and guidance, and other military fields.

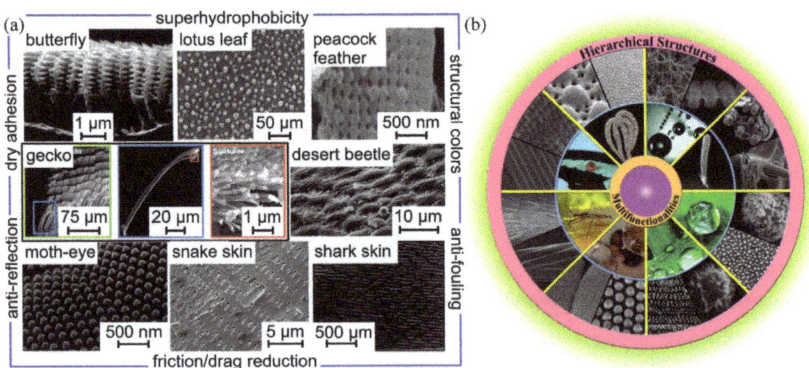

Figure 1. (**a**) Diversity in biological textures and their functions; and (**b**) hierarchical structures on the surfaces of various plants and animals. Reprinted with permission from Refs. [3,4].

However, due to the complexity of surface texture and the characteristics of high aspect ratio, the fabrication of micro-structures is very challenging. Typical machining techniques for a micro-structured surface mainly include lithography technology [19], high energy beam direct writing technology [20], special energy field machining technology [21], molding technology, LIGA technology [22], and ultra-precision machining (UPM) technology [23].

Lithography is suitable for fabricating two-dimensional (2D) and simple three-dimensional (3D) structures. For complex 3D structures, high energy beam direct writing technology and LIGA technology are more applicable, whereas the machining cost is high and it is difficult to fabricate large size micro-structures. Special energy field machining extends the machining range to difficult-to-cut materials, and the molding technology is suitable for the mass production of simple micro-structures. However, the above machining methods cannot meet all the requirements of high surface precision, complex 3D, and high aspect ratio micro-structures. UPM is an effective method to fabricate micro-structures, it has many advantages which could meet all these requirements.

UPM technology was developed in the 1960s to meet the manufacturing needs of nuclear power manufacturers, VLSI (Very Large Scale Integration), lasers, aircraft and other high-end products [24–27]. Production efficiency has been continuously improved and products have been gradually miniaturized, while higher requirements are put forward for micro-structures machining, the machining accuracy of UPM is constantly improved. The development of achievable machining accuracy is shown in Figure 2. UPM generally includes ultra-precision diamond turning, milling, scratching, ultra-precision grinding, and polishing, as shown in Figure 3. They could be employed for machining various freeform surfaces and complex micro-structures [28–34]. Ultra-precision diamond turning, milling, and scratching are usually employed with a natural single crystal diamond tool, this is called ultra-precision diamond cutting [35]. The surface roughness Ra can reach 1 nm and the form accuracy PV can reach 0.1 μm. Grinding is mainly used for machining difficult-to-cut materials, which is hard to do with diamond cutting. Polishing is a follow-up process that achieves higher surface quality.

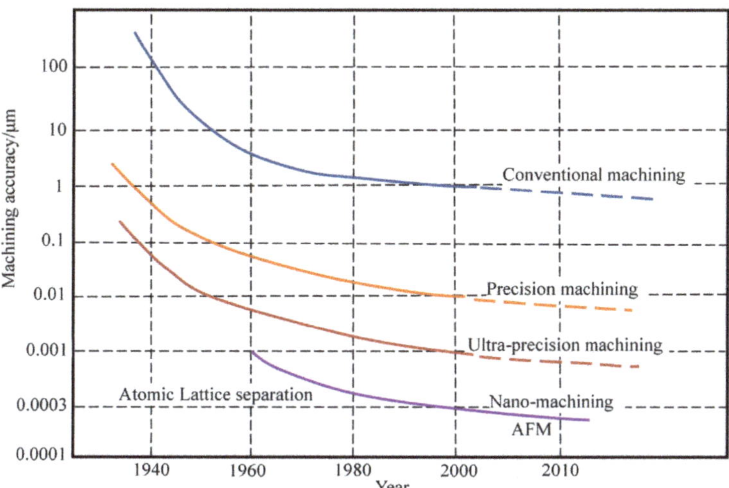

Figure 2. Achievable machining accuracy. Reprinted with permission from Ref. [36].

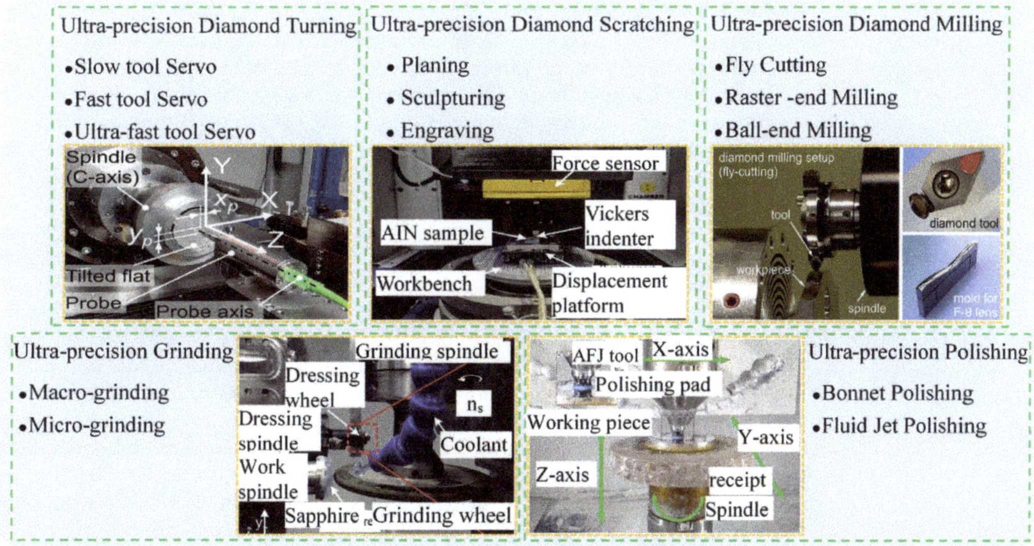

Figure 3. An overview of typical UPM methods. Reprinted with permission from Refs. [24,35,37–41].

Ultra-precision diamond turning, as a member of the UPM family, is generally employed for machining rotationally symmetric surfaces. To expand the application of turning, it is often combined with fast tool servo (FTS) technology or the slow tool servo (STS) system, where the feed depth can be dynamically adjusted and the positioning accuracy can be increased. As an assistive technology of turning, FTS system has high frequency response and high positioning accuracy. It has advantages for the surfaces with large change and complex structure machining, and the control system tracks the surface shape change in real time to control and adjust each servo axis during machining. However, the focus of research is how to make the FTS with high frequency overcome the defect of short stroke. Similar to FTS, the STS system also expands machining performance of turning, it enlarges the Z-axis stroke, and increases the high-precision multi-axis linkage characteristic of the lathe. It is suitable for off-axis surface, array structure, and freeform surface machining. Whereas, since the STS turning generates the target surface shape according to the 3D tool path and the three-axis linkage, whether the surface can be processed depends on the generation of the tool path. On the basis of turning, ultra-precision fly cutting is developed, it is a cutting technology with constant cutting speed and more flexible cutting trajectory, which is suitable for specific micro-structures machining [42–44]. Nowadays, the micro-structures fabricated by UPM are more diversified, such as spherical/aspherical lens [45–47], multi-focal lens, Fresnel lens [48–51], polygonal mirror [52], pyramid array [53,54], micro-structure array [55], anti-reflection channel, and V-groove [56,57], etc. [58].

In this paper, different machining methods of micro-structured surfaces are compared and their characteristics are discussed. According to their typical advanced application as vital results of their micro-structures fabrication, the main research achievements of various machining strategies are reported. As an essential method of fabricating micro-structures, the characteristics and advantages of UPM are reviewed, and the application of diamond tool fabrication and improvement by focused ion beam technology are introduced. The subsequent sections review the studies and applications of FTS/STS combined with UPM to fabricate complex micro-structures, and the research of fly cutting in the fabrication of micro-structures is emphasized. This paper will provide theoretical guidance and develop new ideas for the fabrication of different micro-structures.

2. Fabrication Methods of Micro-Structured Surfaces

2.1. Lithography Technology

Lithography is based on an optical projection printing system, in which the image on the mask is reduced and projected onto a photoresist coated substrate (such as a silicon wafer) through a high numerical aperture lens system, and the image is transferred to the surface of the substrate through an etching process. The general process of lithography includes substrate pretreatment, gluing, pre-baking, exposure, development, post-baking, etching, degluing, and engraving, etc.

Lithography includes many categories, such as electron beam lithography (EBL) [59], X-ray lithography [60], ion beam lithography (IBL) [19], grayscale lithography, and extreme ultraviolet (EUV) lithography [61]. EBL uses an electron beam to trace a pattern on the resist medium. When a very narrow electron beam passes through, it changes the physical properties of the resist layer, resulting in the appearance of submicron characteristics. EBL has been most widely implemented for patterning mesoscopic structures or systems with unique advantages of high resolution in feature size, high reliability in machining, and high flexibility in pattern replication [1,62]. X-ray lithography has the ability to penetrate thick resistance and produce high aspect ratio patterns to achieve side walls with optical quality. This fabrication technique has been used to fabricate micro- and nano-structures from materials such as methyl methacrylate [63,64].

IBL mask is a transmission/scattering two-phase mask made of Si material. Ion beam exposure has higher sensitivity and higher exposure rate. Compared with electron beam, at the same acceleration voltage, ion beam exposure resolution is higher due to the shorter ion wavelength. The ions are much more massive than the electrons, therefore, there is no proximity effect in ions. Common IBL mainly includes focused ion beam (FIB) lithography and ion projection lithography (IPL). FIB lithography uses physical interactions between ions to modify the surface layer of the substrate. Depending on the weight of the ions used (typically gallium, Ga^+), the working beam current, and the acceleration voltage, FIB technology can not only perform ion implantation, but also perform imaging, and addition and subtraction processes. It is notable for its ability to process any material by surface erosion and is widely used in micro-technology and metrology. FIB has a considerable advantage over EBL in terms of resolution when making high aspect ratio 3D structures. IPL is a mask process, it is suitable for mass production, mainly for repairing masks and writing directly to wafers [65,66].

To study the application of EBL and IBL, Palka et al. explored the applicability of As50Se50 thermal evaporation film in wet etching with amine solution in EBL and observed that the change of chemical resistance induced by light had the same trend as that induced by electron beam, and the chemical resistance increased significantly with the increase of irradiation dose. A diffraction grating with a period of 100 nm was prepared on As50Se50 film by EBL, as shown in Figure 4a [67]. Rius et al. studied the influence of electron beam and ion beam on CMOS circuit damage by local exposure of selected region and specific position around CMOS circuit. The importance of electron beam energy on exposure localization was studied, the optimal exposure conditions were determined, and the method of manufacturing monolithic nano-mechanical devices to CMOS circuit through EBL or IBL was presented, as shown in Figure 4b,c [68].

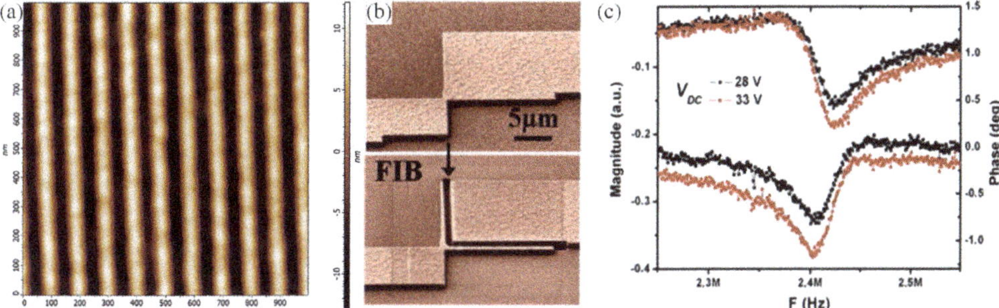

Figure 4. (**a**) Diffraction grating of 100 nm period prepared in $As_{50}Se_{50}$ thin film using EBL. (**b**) Nano-mechanical structures monolithically integrated into CMOS circuits have been fabricated with the combination of electron-and ion-beam lithography. (**c**) The compatibility of both fabrication processes is demonstrated from the electrical measurements of the nano-mechanical device in operation. Reprinted with permission from Refs. [67,68].

Micro-structures fabricated by lithography can be used in different fields. In terms of optical imaging, Bae et al. developed a new method for fabricating multi-focal micro-lens arrays with extended depth of field using multi-layer lithography and thermal reflux, providing a new approach for the development of various 3D imaging applications, such as light field cameras or 3D medical endoscopes [69]. Tong et al. used Monte Carlo simulation and EBL to study the grayscale lithography conditions systematically and copied resist as a template. The gold colored Kinoform lens B with a diameter of 200 μm and a height of 3.5 μm was successfully prepared by electroplating, as shown in Figure 5a. It has the advantages of high resolution lens and high focusing/imaging efficiency [70]. In terms of medicine, Au-Kathuria et al. fabricated polymer microneedle arrays by lithography, which provide potential applications in the delivery of low- and macro-molecular therapeutic drugs through the skin [71].

In order to solve and optimize the limitations and applications of EBL, different optimization methods have been proposed. EBL resolution below 10 nm is mainly limited by resistance contrast and proximity effect, Heusinger et al. used EBL to optimize the pseudo-stray light peak, also known as the "Rowland ghost", and developed a method to improve the stray light performance of binary spectrograph gratings [72]. Andrea et al. used the focused helium ion beam to expose resist, which could reduce the perfect development of dense lines by 20 nm, thus compensating for the proximity effect. An optimized reactive ion etching process was used to demonstrate the pattern transfer of 10 nm line with aspect ratio of 10 in silicon, its images of scanning electron microscope (SEM) as shown in Figure 5b [73].

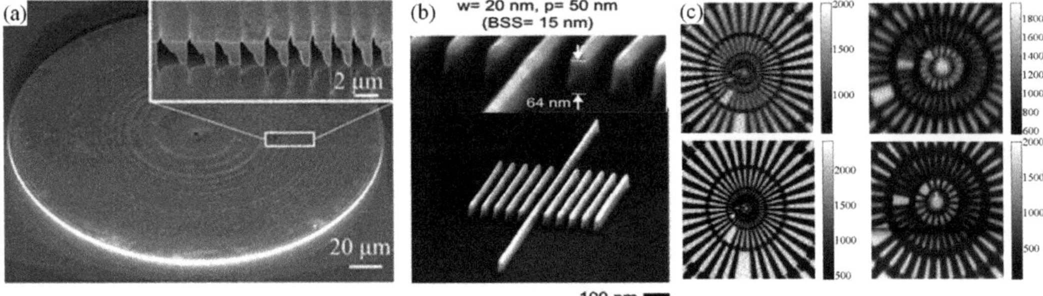

Figure 5. (a) A complete Au Kinoform lens and the insert gives the cross-sectional view of the plated zones. (b) SEM images of two examples of pattern transfer using the optimized reactive ion etching process at 80 V bias. (c) Images of 30 nm Siemens star by the 30 nm Fresnel zone plate in soft X-ray, taken with the scanning transmission X-ray microscope system. Reprinted with permission from Refs. [70,73,74].

When the width of the outermost region approaches 30 nm or less, electrodeposition of gold into EBL becomes increasingly difficult in the narrow grooves replicated in the resist. In this case, Zhu et al. studied the recent progress in the preparation of 30 nm Fresnel band sheets by EBL and controlled pulse voltage electroplating. Pulse electroplating instead of conventional direct-current has been attempted, and Siemens cluster imaging with the narrowest linewidth of 30 nm has been demonstrated in soft X-rays (706 eV) to verify system performance [74]. Figure 5c is a soft X-ray image of Siemens Star taken at 706 eV by self-made Fresnel zone plate.

Grayscale lithography displays 3D structures by modulation of ultraviolet (UV) exposure or control of UV dose through the mask to expand the lithography technology to 3D manufacturing [75]. However, it is only applicable to the manufacturing of 3D structures with small aspect ratio. For complex surface structures, it is not enough to achieve good machining accuracy [76]. The advantages of different technologies can be extended by combining different lithography and etching processes, Nachmias et al. described the use of grayscale lithography, reactive ion etching, or deep reactive ion etching to transfer patterns from grayscale resists to silicon substrates to fabricate Fresnel lenses efficiently [77].

EUV lithography uses EUV with a wavelength of 10–14 nm as a light source. EUV means that ultraviolet light is emitted from the K pole of the ultraviolet tube stimulated by electricity. It can extend lithography technology to less than 32 nm [78]. EUV is able to reflect optics, reflective masks, and vacuum environments, giving it an advantage over other lithography. EUV lithography is based on optical projection lithography, and the physical properties of 193 nm and 248 nm lithography can be directly applied to EUV. By employing an aggressive optical design with 0.45 numerical aperture and 0.32 numerical aperture, EUV can be extended to print half-pitch features of less than 22 nm. It is a reduced lithography technique, with a marking feature four times larger than the final pattern.

In addition, colloidal lithography is a very important technology for large surface area micro and nano fabrication; large area self-assembly of colloids without the need for expensive equipment can be achieved [79–82]. Liu et al. proposed an optimal demolding process to obtain high pattern transfer fidelity while avoiding distortion of soft imprinting molds and supports, and to effectively decrease the demolding force [83]. Olalla et al. proposed the use of colloidal lithography to map wavelength-sized pyramidal feature composition structures, and these top-coated structures as post-processing post-deposition solar cells to facilitate and expand their industrial applicability [84]. Centeno et al. proposed a simple, low temperature, low cost, and scalable colloidal lithography method for designing surfaces with effective light trapping and hydrophobic functions. The controllable nano/micro structure of its surface features also produce strong anti-reflection and light scattering effects, which increase the average daily energy generation by 35.2% [85].

In summary, although the lithography technology has high resolution, it has high cost of lithography equipment. It is usually suitable for the preparation of 2D structured surfaces, not applicable for the fabrication of 3D complex micro-structures, especially curved surface structures. The development of X-ray lithography and grayscale lithography allow lithography methods to be initially applied to the preparation of 3D continuous relief micro-structured surfaces. However, it needs masks with high production cost, the complexity of 3D surface is limited, and the surface controllability is low. The characteristics of lithography technologies are summarized in Figure 6.

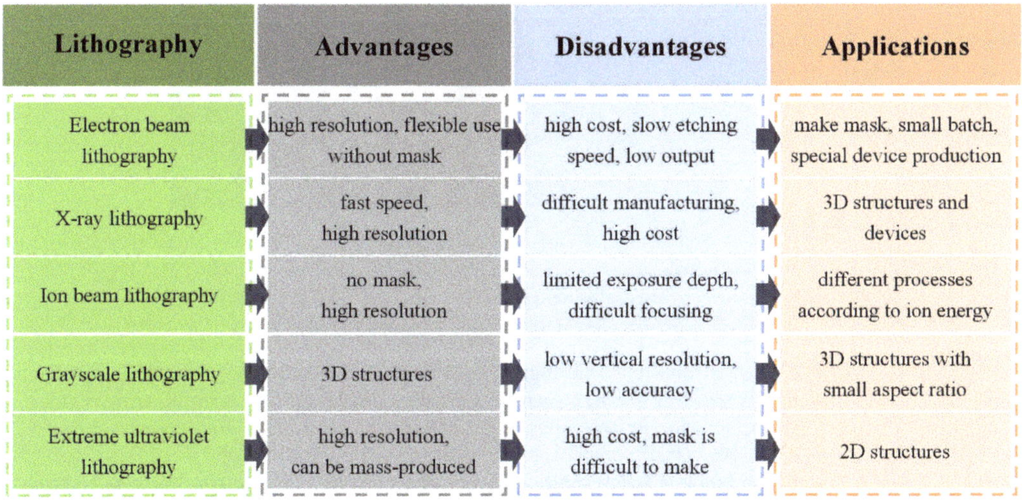

Lithography	Advantages	Disadvantages	Applications
Electron beam lithography	high resolution, flexible use without mask	high cost, slow etching speed, low output	make mask, small batch, special device production
X-ray lithography	fast speed, high resolution	difficult manufacturing, high cost	3D structures and devices
Ion beam lithography	no mask, high resolution	limited exposure depth, difficult focusing	different processes according to ion energy
Grayscale lithography	3D structures	low vertical resolution, low accuracy	3D structures with small aspect ratio
Extreme ultraviolet lithography	high resolution, can be mass-produced	high cost, mask is difficult to make	2D structures

Figure 6. The characteristics of lithography technologies.

2.2. High Energy Beam Direct Writing Technology

High energy beam direct writing technology is a non-contact machining technology, which changes the material state and property, realizing shape control and performance control through the interaction of high energy density beam and material. High energy beam refers to the directionally transmitted high density energy beam in free space, including laser beam, electron beam, ion beam, etc. It has controllable energy, beam density, time, and space. High energy beam direct writing can produce complex structures by non-contact and selective multi-scale control or material state change. Since time and space can be manipulated, it is possible to create complex 3D structures [20]. Among them, laser direct writing technology uses laser beam with variable focusing energy and focusing position to expose anticorrosive materials and form surface relief or 3D structures on them [15,86]. The basic working principle of the laser direct writing system is to use the computer to control the high precision focused beam to scan accurately and write the designed graphics directly on the photoresist [87,88]. As an important micro-machining technology, laser direct writing technology has the advantages of high precision, strong 3D machining ability, no mask, and high production efficiency, which is conducive to the production of high precision and complex micro-optical devices and meets the development requirements of micro-optical technology. Xu et al. used focused ion beam direct writing (FIBDW) method to prepare the star structures of spokes with a width of 25 nm~16 μm, as shown in Figure 7, which is applicable to the comparison of resolution measurement methods [89]. However, the machining efficiency of this method is low, which has advantages for small size precision micro-structures machining. Mizue et al. used femtosecond laser pulses to rapidly heat and reduce glyoxylic acid metal (Cu, Ni, and Cu/Ni mixed) complexes on glass substrates and directly write patterns of Cu, Ni, and Cu-Ni alloys without significant oxidation. This direct writing technique of pure metal and alloy allows various sensors to be printed in air [90].

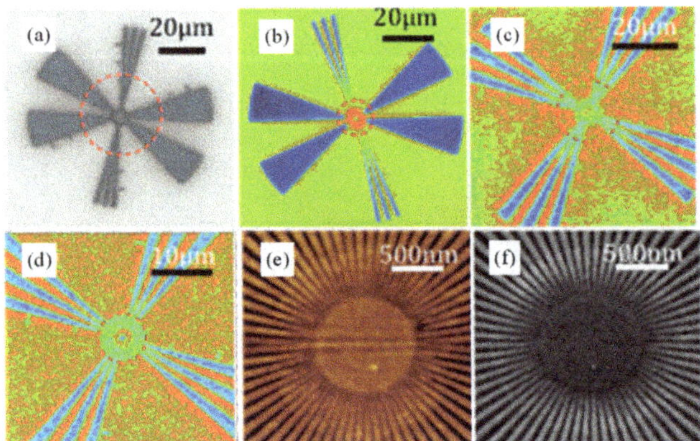

Figure 7. Star structures measured by various methods. (**a,b**) Star I measured by confocal laser scanning microscope with 427× and phase shifting interferometry with 503×, respectively; (**c,d**) Star II measured by vertical scanning interferometry with 201× and 503×, respectively. (**e,f**) Star center measured by SEM and AFM, respectively. Reprinted with permission from Ref. [89].

In summary, some high energy beam manufacturing technologies are capable of producing complex 3D structures with high resolution. For example, current electron beam manufacturing technologies have been able to produce smaller fine structures. However, high energy beam manufacturing is based on "point scanning" manufacturing technology, which has low machining efficiency and can only be used for the preparation of micro-structured surfaces with very small size.

2.3. Special Energy Field Machining Technology

Electromagnetic machining technology [21] and ultrasonic machining technology [91] can be classified as special energy field machining. Ultrasonic machining removes materials by means of ultrasonic frequency tools with small amplitude vibration in the abrasive liquid medium or dry abrasive, abrasive impact, polishing, and cavitation [92,93]. Along a certain direction of the tool or workpiece is subjected to ultrasonic vibration to carry out vibration machining, the workpiece surface materials to be processed mainly under the action of mechanical impact peeling, accompanied by polishing and super cavitation finally realize the forming process [94,95]. The workpiece is bonded to each other by ultrasonic vibration. The material removal rate is enhanced through the combination of ultrasonic vibrations and abrasive slurry actions [96,97]. It has a wide range of applications and is not limited by the electrical conductivity of materials. Special energy field machining has advantages of high strength and hardness, which can effectively solve the problem of difficult-to-cut materials [98,99]. It can obviously reduce the machining damage of cutting force, reduce tool wear, avoid surface microcrack, and improve surface machining quality and efficiency. It can machine not only hard alloy and other metal materials, but also non-metallic hard and brittle materials such as ceramics, glass, and gemstones. However, it is only suitable for the machining of simple micro-structures with low machining resolution [100]. Schematic diagram of the combined ultrasonic vibration assisted milling and minimum quantity lubrication methods is shown in Figure 8.

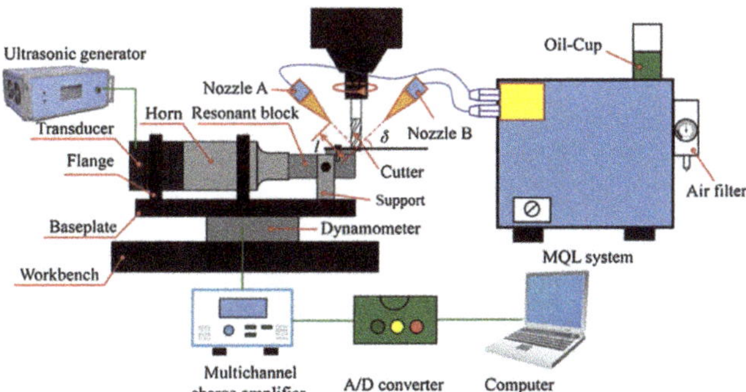

Figure 8. Schematic diagram of the combined ultrasonic vibration assisted milling and minimum quantity lubrication methods. Reprinted with permission from Ref. [101].

In conclusion, special energy field machining, including electromagnetic machining and ultrasonic machining, can be applied to ceramics, glass, and other brittle materials. However, it has low machining resolution. It can only be used for machining simple micro-structures such as micropores, but not for the fabrication of 3D complex micro-structured surfaces.

2.4. Molding Technology

Molding technology, hot pressing technology, and injection molding technology [102,103] can be unified as molding technology. Molding technology involves placing the plastic in the mold for heating to await plastic softening. Molding out of the workpiece is in line with the requirements of use under the external pressure. It is suitable for thin flat optical lenses, such as micro-lens arrays. The principle of molding equipment is simple and used for production of multiple varieties and small batches. Therefore, it has economic advantages in the manufacturing of experimental supplies and the research of optical plastic materials, and also shortens the experimental period. Hot pressing technology must go through several stages such as heating and heat preservation, hot pressing, slow cooling, and demolding. It has the advantages of high efficiency, is suitable for mass production of various optical devices, especially aspheric lens, lens array, and diffraction lens, and it is environmentally friendly and pollution-free. However, there is morphological deviation after machining. This kind of method has relatively simple equipment requirements, short machining cycle, and mass production, which is conducive to industrialization. It is widely used in the machining field of optical components with low precision requirements, such as freeform surface components for lighting. However, this method has strict requirements on technological parameters and process, and the accuracy of the product is related to the accuracy of the mold; whereas other machining methods are required to provide high-precision molds [104].

In order to optimize the method of machining high-precision products by injection molding technology, Guo et al. developed an online decision system consisting of a new reinforcement learning framework and a self-predictive artificial neural network model. The system has good convergence performance in lens production and the decision model has better robustness and effectiveness in the online production environment. Figure 9 presents the warpage results of different process conditions [105]. Warpage is one of the serious defects of thin-wall injection parts. Wang et al. used dynamic filling and filling process parameters as new design variables for the first time to optimize the warpage design. The ambiguous functional relationship between the target (maximum warpage) and the 12 process parameters was approximated by the Kriging proxy model. An efficient global optimization method (expected improvement method) is used to search for optimal solutions. Finally, a set of dynamic injection molding process parameters were given, through which the maximum warpage of plastic parts could be greatly reduced [106].

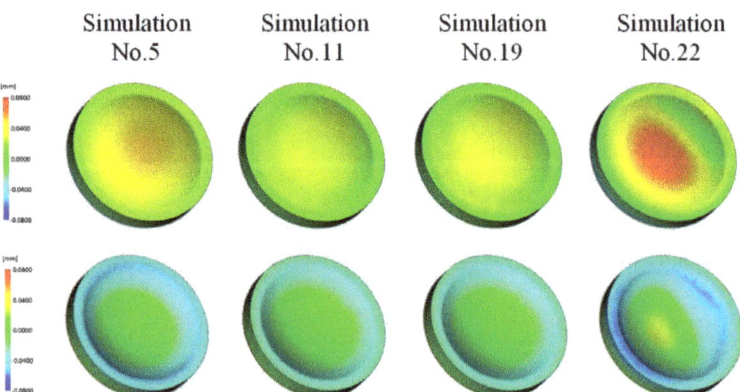

Figure 9. The warpage results of different process conditions. Reprinted with permission from Ref. [105].

2.5. LIGA Technology

LIGA technology [22,107] is a combination of lithography, molding, and injection molding technology. LIGA technology is short for German Lithographie Galvanformung and Abformug. It mainly uses X-ray deep exposure, micro-electroforming, micro-plastic forming, and other technologies to carry out micro-mechanical machining. LIGA technology can fabricate structures with large aspect ratios, up to submicron in width and hundreds of microns or even millimeters in depth, making it suitable for complex micromechanical structures [108,109]. It can be used in a wide range of materials, including metal, plastic, polymer materials, glass, ceramics, and they can also be combined. At the same time, the micro-structure obtained by LIGA technique has well-defined geometry and dimensions, straight and smooth sidewalls, and tight tolerances [110]. Polymer structures with high precision can be obtained through plastic casting after metal molds are obtained by X-ray deep exposure and micro-electroforming, which is suitable for large-scale production. It can be applied to the deep micro-structure of many research and development departments and industrial products [111]. However, LIGA technology requires expensive machining equipment, and it is difficult to fabricate micro-structures on curved substrates. Ma et al. fabricated multi-layer metal micro-structures with high precision and high quality by using ultraviolet (UV)-LIGA overlay processes, including mask manufacturing, substrate machining, and UV-LIGA overlay technology. The electroplating images of its multi-layer metal micro-structure are shown in Figure 10 [112].

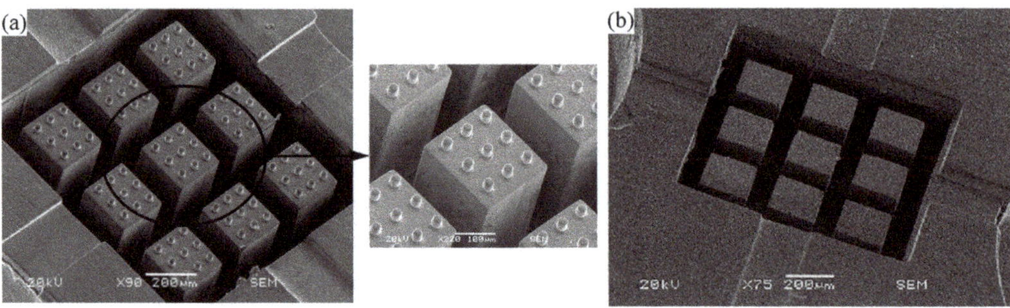

Figure 10. (**a**) SEM images of the multi-layer metal micro-structure was fabricated using UV-LIGA overlay technology; and (**b**) SEM image of the first layer micro-structures. Reprinted with permission from Ref. [112].

2.6. Ultra-Precision Machining Technology

2.6.1. Applications of Ultra-Precision Machining Technology

UPM includes ultra-precision diamond turning (turning classification as shown in Figure 11), scratching, milling, ultra-precision grinding, and polishing. The UPM technology is based on a diamond tool with a sharp cutting edge, high hardness, good wear resistance, ability to realize ultra-thin cutting thickness, and other characteristics. Diamond cutting is the most representative method of UPM.

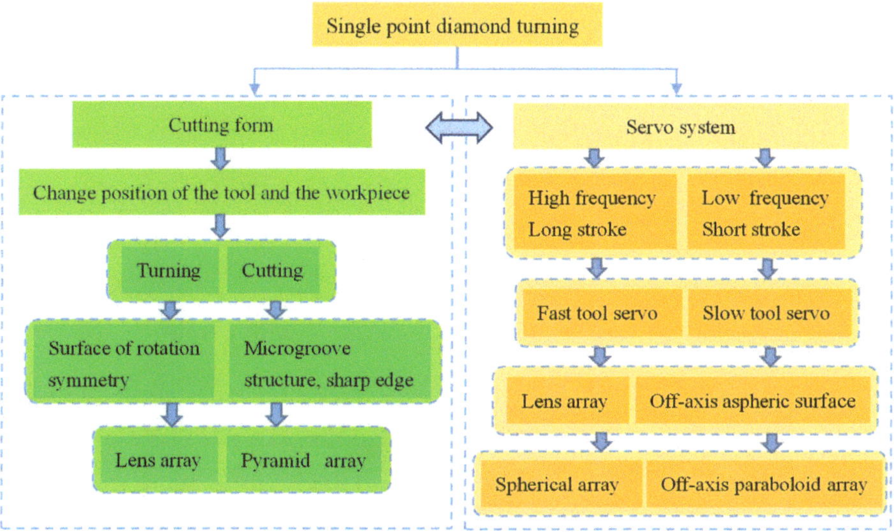

Figure 11. Classification of single point diamond turning.

Natural diamond is regarded as an ideal tool material because of its excellent properties, such as high hardness, high thermal conductivity, low coefficient of friction, high wear resistance, and low affinity with non-iron metals [16,113]. Due to these great tool cutting edge performances, ultra-precision diamond cutting can machine various freeform surfaces and complex optical components, and generate complex optical surfaces by manipulating the cutting depth within the micrometer range. Moreover, diamond tools have inherent tool nose radius, clearance angle, rake angle, and different tool geometry, which plays a key role in mathematical calculation and experimental preparation [114]. Moreover, ultra-precision diamond cutting has an effective tool path generation strategy for complex surface structures, which has great advantages in the machining of micro-structures [114–116]. Early diamond tools were limited to machining soft and malleable non-ferrous metal materials, such as aluminum and copper [117]. At present, ultra-precision diamond cutting has been extended to silicon, steel, and other difficult-to-cut materials [23], which have special functions and meet the needs of optical, semiconductor, mold, and other industries [118–120]. However, rapid tool wear is still a problem to be solved.

Ultra-precision diamond turning technology is generally termed single point diamond turning. The relative position of the tool and the workpiece is precisely controlled by the computer numerical control system of the lathe to turn, which can directly fabricate high-precision complex surface optical components. Turning technology is mainly used for machining infrared crystal, non-ferrous metal, and part of the laser crystal and optical materials such as plastic optical element. It can machine complex surface shape or special surface form of optical element, such as high order aspheric, diffraction optical element, the diffraction hybrid optical element such as rotational symmetry complex curved surface, and can also fabricate micro-lens array, precision die, micro-pyramid array, freeform surface, and other non-rotational symmetric surfaces [121].

The researchers developed different cutting methods based on turning. Shigeru et al. proposed a new method for single point diamond turning micro-cavity array based on fast tool servo (FTS). A computer program was developed to control the tool path, periodically arrange small quadrics on the large curved surface, and adjust the position, height, and size of each surface by using random values. Its dynamic range was preset to control spatial differences. The rotary steerable system was machined by continuous cutting method and subsection cutting method, respectively. The designed cutting method improved the edge accuracy and the form error was controlled below 10 nm level PV [122]. The basic shape of the quadric can be changed by changing the value of each coefficient of the quadric function, as shown in Table 1.

Table 1. The basic shape of a quadratic surface and typical shape changes obtained by varying the value of each coefficient of the quadratic surface function. Reprinted with permission from Ref. [122].

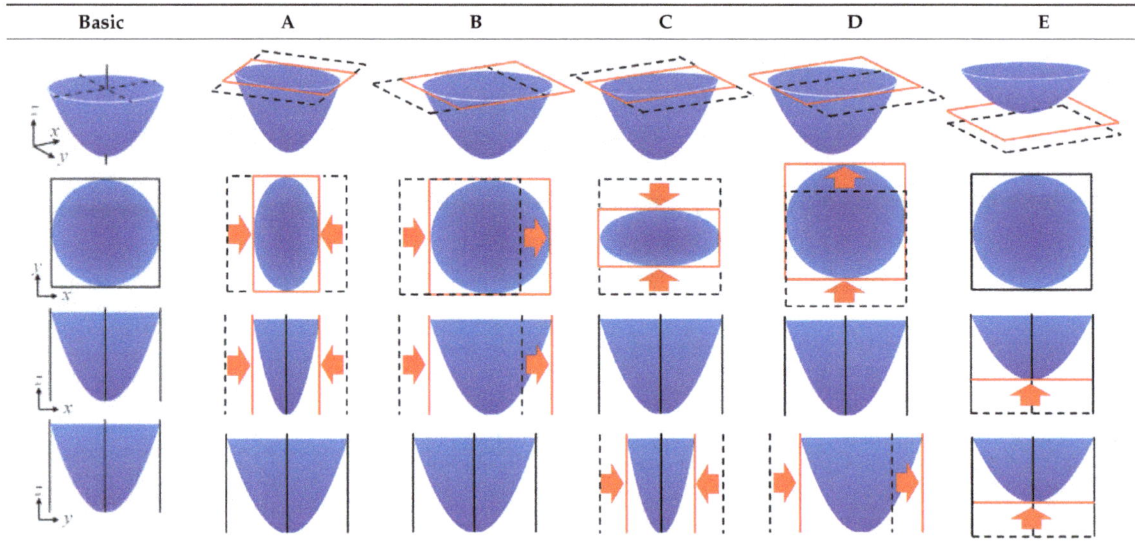

Ultra-precision diamond scratching is a fundamental approach to investigating surface integrity [123]. It can create the complex micro-structured surfaces such as three-focus Fresnel lens, micro cube corners, and freeform micro-lens array under the C-axis mode for re-orientating or re-positioning. In the machining, the cutting process is mainly controlled by linear slide servo motions and the C-axis is only used to orientate or position workpiece or diamond tools. Its material removal rate is also relatively low [35]. T. Moriya et al. fabricated curved V-shaped microgrooves with two flat-ends on a curved surface by using a six-axis controlled non-rotational cutting tool [124]. Y. Takeuchi et al. studied the method of designing and manufacturing micro Fresnel lenses with non-rotational diamond cutting, which can accurately and neatly produce multi-focus micro Fresnel lenses without any burrs [125]. Afterward, his team also developed a six-axis control machining CAM system using a non-rotational cutting tool. The system is used to generate microgrooves on sculpted surfaces [126].

Milling is a technology that uses workpiece fixation and tool rotation to obtain surface shape. Ultra-precision milling is generally used with a diameter of less than 50 μm milling cutter, tool materials are usually natural diamond, tungsten carbide, cubic boron nitride (CBN), etc. Machining workpiece materials are copper, aluminum, titanium, steel, and other non-ferrous metals. It is usually used for end milling on ultra-precision five-axis machine tools, which is suitable for machining freeform surfaces, but it is difficult to install and adjust the workpiece. The surface shape is affected by the size of the milling cutter. The machined surface has a large roughness value and a long machining cycle, thus expensive multi-axis ultra-precision machine tools are needed. To improve cutting performance, Chen et al. designed a miniature end milling cutter for machining GH4169, a nickel-base superalloy with high stiffness and sharpness. One-dimensional finite element method was used to determine the optimal geometric parameters of the cutter. Polycrystalline diamond (PCD) micro end milling cutter with better cutting performance was prepared by laser-induced graphitization assisted precision grinding method [101]. Owen et al. proposed an error reduction method based on manual, using spherical artifact to establish the tool error model, and applied it to the free surface machining [127].

Grinding is the process of removing excess material from the workpiece with abrasives and tools. The machining surfaces of the workpiece processed by diamond grinding show obvious peak-valley interphase structure and the surface does not appear the situation of tool copying. The workpiece surface integrity of grinding is good and the degree of subsurface damage is low. Therefore, grinding is suitable for some specific crystal materials. However, due to the wear and dressing error of grinding wheel, the machining accuracy is affected. Yu et al. established a three-axis linkage orthotic model of diamond grinding wheel and used true diamond grinding wheel to grind the aspherical structure array of tungsten carbide. An online shaping method of ultra-thin arc-shaped diamond grinding wheel was proposed, which can achieve the expected arc radius of grinding wheel, realize efficient and accurate grinding of non-binder carbide die, and obtain the surface of tungsten carbide aspherical micro-structure array with shape accuracy of 15 μm. The micro-structure array is shown in Figure 12 [55]. Luo et al. studied the wear process of metal-bonded diamond grinding wheel in the process of sapphire single crystal surface grinding to find the mechanism of the influence of grinding wheel wear on workpiece surface quality [128].

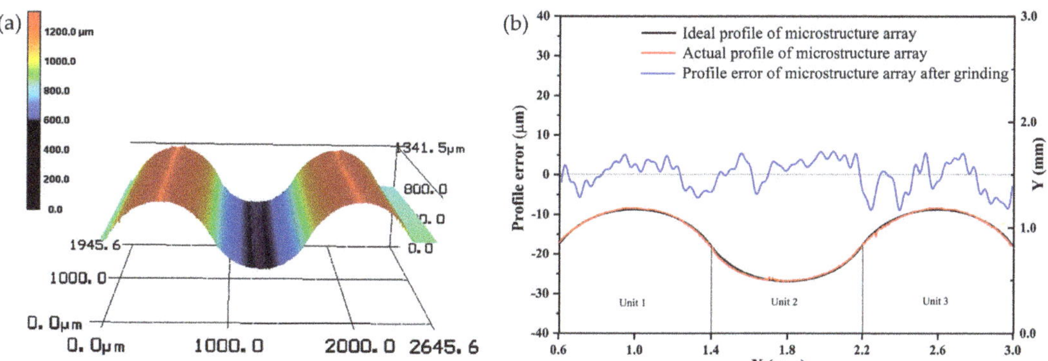

Figure 12. (**a**) Laser scanning microscope photos of the micro-structure array workpiece after fine machining; and (**b**) profile accuracy of the micro-structure array. Reprinted with permission from Ref. [55].

Polishing is usually one of the last processes of UPM, which can eliminate the surface and sub-surface damage of the machined parts and improve the surface shape. In the process of machining, it is inevitable to leave sharp points and stripes on the workpiece. These machining marks or chip residues must be removed by the polishing process [129]. Ultra-precision polishing is widely used in high-precision surface finish machining with high form accuracy and good surface roughness. Meng et al. proposed a new method for machining ultra-high precision textured surfaces from silicon carbide materials. The grooves and pits of two different sizes were compared. It is found that groove texture has a greater effect on polishing quality than pit texture at the same size [130]. Wang et al.

proposed an ultra-precision numerical control polishing method based on the principle of water dissolution, which uses small polishing tools to process large KDP surfaces [131]. Cheung et al. conducted theoretical and experimental studies on computer controlled ultra-precision polishing of the structure surface and established a model-based simulation system for structure surface generation, which was able to predict the shape error and pattern of 3D textures generated by computer controlled ultra-precision polishing [132]. Xu et al. proposed a novel computer controlled ultra-precision polishing hybrid manipulator, which is suitable for freeform surfaces polishing [129].

The characteristics of different micro-structures manufacturing methods are summarized in Table 2.

Table 2. Comparison of different fabrication methods for micro-structured surfaces. Reprinted with permission from Refs. [20,97,101,133–137].

Method	Classification	Advantage	Disadvantage	Schematic Diagram
Lithography	• electron beam • X-ray • ion beam • grayscale • extreme ultraviolet	• high resolution • mass produce	• high cost • difficult masks	
High energy beam direct writing	• laser beam • electron beam • ion beam	• high precision • 3D machining • no mask	• low efficiency • not for large size	
Special energy field machining technology	• electromagnetic • ultrasonic	• wide applications • no electrical conductivity • hard materials	• simple structure • low resolution	
Moldingtechnology	• molding • hot pressing • injection	• simple equipment • short cycle • mass produce	• strict requirements • difficult mold	
LIGA technology	• a collection technology	• high aspect ratio structures • wide available materials	• high cost • not for curved structures	
Ultra-precision machining technology	• turning • scratching • milling • grinding • polishing	• high precision • high efficiency • one-time molding • complex surface	• not for ferrous metal	

2.6.2. Studies of Diamond Tools by Focused Ion Beam Technology (FIB)

Due to the high precision of UPM, the machining quality is easily affected by various factors, such as precision machining lathes, cutting tools or grinding wheels, machining methods and parameters, machining objects, fixtures, and external environment. In ultra-precision cutting, the quality of cutting edge determines the machining quality as a tool that directly contacts the machined material. Natural single crystal diamond is generally selected as ultra-precision cutting tool material, mainly due to its own wonderful physical and chemical properties, such as high hardness, high wear resistance, good thermal conductivity, etc. FIB technology is an important method to prepare diamond tools. In recent years, the focused ion beam technology, which utilizes high-intensity focused ion beam to nano-machining materials, has become the main method for nanoscale analysis and manufacturing, combined with real-time observation by high-multiple electron microscopy such as SEM [71]. It has been widely used in diamond tool preparation, semiconductor integrated circuit modification, cutting, and fault analysis.

FIB system is a micro-cutting instrument that uses an electric lens to focus ion beam into a very small size, which can achieve material stripping, deposition, injection, cutting, and modification. At present, the ion beam of commercial systems is liquid metal ion source. The metal material is Ga because of its low melting point, low vapor pressure, and good oxidation resistance. Typical ion beam includes liquid metal ion source, lens, microscope scanning electrode, the secondary particle detector, 5–6 specimens of axial moving base, vacuum system, vibration resistance, magnetic devices, electronic control panel, and computer hardware equipment. Applying an electric field to the liquid metal ion source can make the liquid metal form a small tip, and the negative electric field is added to pull the metal or alloy at the tip, thus leading to the ion beam. The ion beam is focused through an electrostatic lens and passes through a series of automatic variable aperture (AVA) to determine its size. The desired ion species were screened by $E \times B$ mass analyzer. The ion beam is focused on the sample and scanned by an octupole deflector and an objective lens. The ion beam bombards the sample. The resulting secondary electrons and ions are collected and imaged using physical collisions to cut or grind.

There are several main functions in FIB technology. In the IC production process, if it is found that there are some errors in the etching of the micro-circuit, the original circuit can be cut by FIB, then the fixed area is sprayed with gold and connected to other circuits to achieve circuit modification, with the highest accuracy up to 5 nm. In addition, if there are micro and nano-level defects on the surface of the product, such as foreign bodies, corrosion, oxidation, and other problems, it is necessary to observe the interface between the defect and the substrate. Using FIB, the defect location section sample can be accurately positioned and cut, and the interface can be observed by SEM. For micron size samples, after surface treatment to form a film, it is necessary to observe the structure of the film and the degree of combination with the base material, using FIB cutting sample preparation, and then using SEM observation.

Different studies have been carried out to explore the characteristics of FIB technology. In order to obtain the specific tool geometry, diamond grinding wheel is an effective method for machining hard metals, with WC-Co carbide as the main material of the die. Yang et al. characterized the grinding damage of WC-Co class by FIB tomography, as shown in Figure 13a [138]. Rubanov et al. applied high pressure and high temperature annealing to graphitize the diamond implanted layer formed by a 30 keV Gat FIB. The implanted layer was studied by electron microscopy. The influence of ion implantation on diamond structure is studied by using electron microscope imaging and spectroscopy, as shown in Figure 13b,c [139]. Tong et al. established a large-scale multi-particle molecular dynamics simulation model to study the dynamic structure changes of single crystal diamond under 5 keV Gat irradiation, combined with transmission electron microscopy (TEM) experiments. Figure 13d shows the TEM section image of diamond sample irradiated by 5 keV Ga^+ [140].

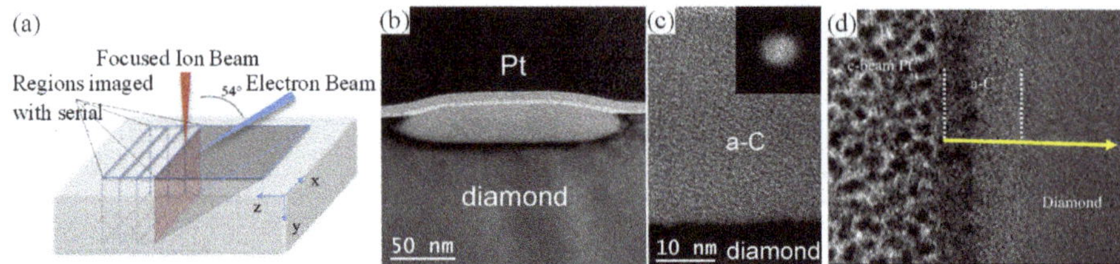

Figure 13. (**a**) Schematic of FIB serial sectioning process, (**b**) bright field image, (**c**) high resolution electron microscopy image with fast Fourier transformation on the inset of implanted layer with Ga fluence of 4×10^{15} ions/cm^2, and (**d**) TEM images of damage region after 5 keV Ga$^+$ irradiation with fluence of 1.0×10^{18} ions/cm^2. Reprinted with permission from Refs. [138–140].

FIB can be used with the preparation of anisotropic wetting surfaces; Wu et al. proposed a new method for high-throughput preparation of anisotropic wetting surfaces with good transparency and developed a multi-step FIB machining method to achieve accurate and reliable machining of non-conductive materials such as single crystal diamond tools. Figure 14 shows the evolution of tool tip microfeatures in FIB multi-step milling. The tool is used to process a variety of substrates, from metals to plastics, under a variety of conditions. All machined surfaces exhibit significant anisotropy at contact angles, being hydrophobic in one direction and highly hydrophobic in the other. It is found that anisotropic wettability depends only on micro-structure design and material properties. This method can produce anisotropic wetting more quickly than other methods [141].

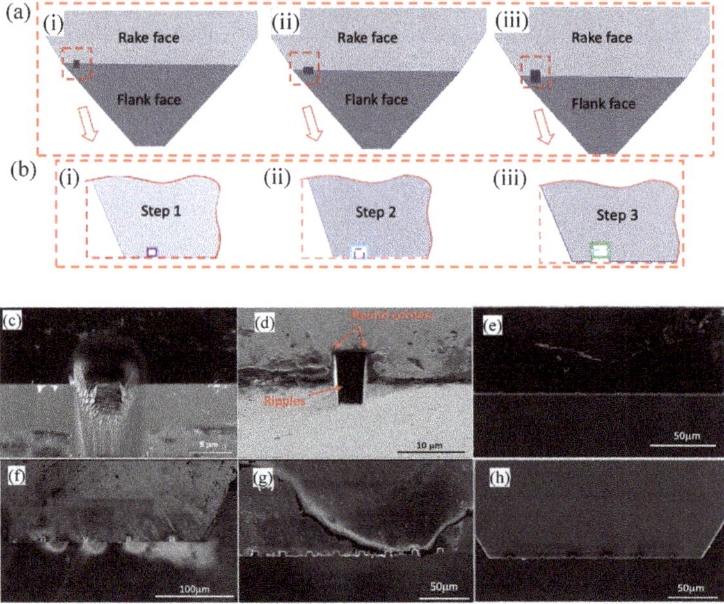

Figure 14. (**a**) Schematic of evolution of the micro-features on the tool tip during the multistep FIB milling process, (**b**) schematic of the corresponding top view of the three steps, (**c**) SEM image of diamond tool milled by FIB conventionally in a single step, (**d**) SEM images of diamond tool milling by FIB with multi-step with ripples and round corners. (**e**) Tool #1 with 1 µm × 1 µm grooves and 30 µm spacing, (**f**) tool #3 with 5 µm × 8 µm grooves and 50 µm spacing, (**g**) tool #4 with 5 µm × 5 µm grooves and 20 µm spacing, (**h**) tool #2 with 2 µm × 2 µm grooves and 30 µm spacing Reprinted with permission from Ref. [141].

FIB can be used to study diamond tool damage. Tong et al. studied the effect of FIB on the damaged layer of single crystal diamond tool under different FIB treatment voltages by TEM measurement and molecular dynamics simulation [142]. Takenori et al. studied the sharpening of the cutting edge of a single crystal diamond tool by an argon ion beam machine tool; the etching rate could be changed by changing the irradiation angle of the beam on the processed surface [143].

FIB can also be used to prepare and improve diamond tools with different requirements. Wei et al. can produce diamond tools with edge radius of nanometer by FIB technology, which can be used for UPM. The schematic diagram of tool edge radius is shown in Figure 15. In the nano-cutting process, the ratio of the minimum chip thickness to tool edge radius is about 0.3~0.4 [144].Wang et al. proposed a new cutting method based on force modulation method for multi-tip diamond tools to process micro-structured surfaces, its cutting schematic diagram is shown in Figure 16. A multi-tip diamond tool with periodic sinusoidal micro-structure was prepared by FIB technology [145].

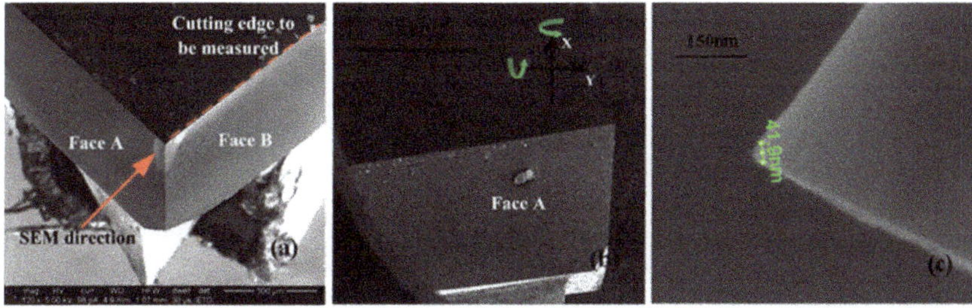

Figure 15. Field emission scanning electron microscopy measurement of the tool edge radius. (a) the cutting edge, (b) the error factors may come from the rotation of X and Y axes for the platform that fix the tool to be fabricated, (c) the measurement setup and result of a diamond tool developed by FIB with edge radius less than 20.95 nm. Reprinted with permission from Ref. [144].

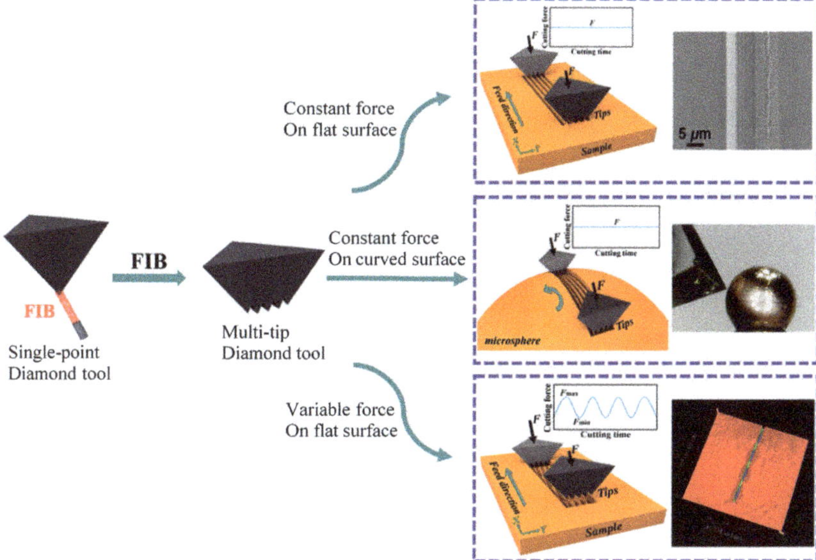

Figure 16. Schematic representation of multi-tip diamond tool cutting. Reprinted with permission from Ref. [145].

Kawasegi's research team had performed a lot of research on the preparation of diamond tools by FIB technology. Kawasegi et al. proposed a fabrication method for diamond tools with surface textures. In order to improve the machining performance of diamond tools, a texture with a depth of 43 nm and a width of 1.8 μm was prepared by FIB on the front surface of diamond tools, and heat treatment was carried out, as shown in Figure 17. Compared with femtosecond laser and direct FIB sputtering, this method has a better cutting effect on diamond tools. Due to the high resolution of FIB irradiation, FIB-induced non-diamond phase can be removed to avoid adverse effects on cutting performance [146]. Kawasegi et al. describe a technique to improve diamond tools for nanometer- and micrometer-scale machining and forming by FIB micromachining. Figure 18a,b shows NiP surface machining using FIB and processed tools [147]. Although the FIB technique is an effective means to fabricate nanometer- and micrometer-scale tool shapes, ion irradiation can lead to doping, defects, and reduce tool performance. Kawasegi et al. used 500°C heat treatment combined with aluminum deposition to remove Ga ions caused by ion irradiation in order to process FIB on a single crystal diamond tool without degrading the tool performance. The method was evaluated by machining aluminum alloy and NiP. The surface morphology of NiP is shown in Figure 18c–e [148].

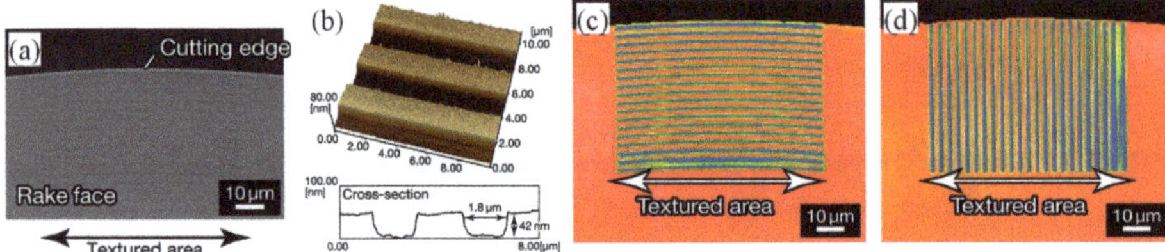

Figure 17. Textured diamond cutting tool fabricated by FIB irradiation and subsequent heat treatment. (**a**) SEM image of the rake face of the textured diamond cutting tool surface; (**b**) enlarged image of the texture measured by atomic force microscope. Surface topography of the rake faces textured in the; (**c**) perpendicular; and (**d**) parallel directions and measured with a coherence scanning interferometer. Reprinted with permission from Ref. [146].

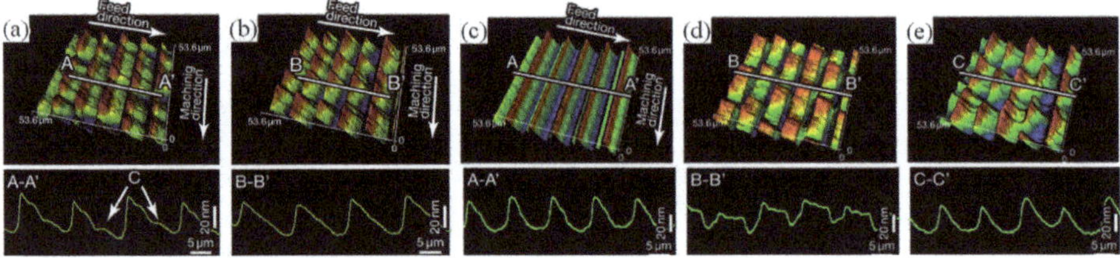

Figure 18. Cross-sectional images of the NiP surface following machining using the (**a**) FIB tool; (**b**) heat-treated tool, with a cutting distance of 19,040 m. Surface topography of the NiP surface after machining with; (**c**) non-FIB; (**d**) FIB; and (**e**) treated tools, measured using a white light interferometer. Reprinted with permission from Refs. [147,148].

3. Fast/Slow Tool Servo Technology

3.1. Fast Tool Servo (FTS)

With the continuous improvement of the requirements for micro-structured functional surfaces, traditional machining methods have been unable to meet the requirements of machining efficiency and accuracy. UPM technology combined with FTS has been developed and applied to the important machining methods of complex micro-structured functional surfaces [149].

FTS system is an independent closed-loop operating system, which is mainly composed of a FTS device and controller. In recent years, the guiding mechanism, actuator, and displacement sensor are the key components of FTS devices that have attracted the most attention. Various mechanical structures and actuators, high-performance controllers and trajectory tracking control algorithms are needed to control FTS devices to obtain high precision, high bandwidth, and high motion resolution. In addition, the tool path and machining parameters of a given freeform surface need to be calculated and optimized. The 3D surface topography of micro-structures such as lens array can be represented by cylindrical coordinates. The coordinates of each point on the surface are represented by three variables, spindle rotation angles, shaft feed, and tool feed. Before machining, preset the spindle speed and feed rate, then according to the tool path to discretization of micro-structure surface form, get the points on the tool path of cylindrical coordinates. According to the spindle speed and feed speed set before, the micro-structure machining is controlled synchronously by c-axis encoder cutter servo cutting motion and C-axis rotation. The FTS module and its control system are the core of the whole machining system. By collecting the angle signal of the rotary spindle, the control system can control the precision feeding movement of the diamond tool with high frequency and short stroke in real time, so as to complete the machining of the 3D profile of the workpiece.

The FTS system has the characteristics of high stiffness, high frequency, and high positioning accuracy, which is suitable for machining complex micro-structures with large surface changes or discontinuous parts and short stroke. The feed frequency of the FTS module can reach more than 1 KHz. It has gradually developed into one of the mainstream technologies of micro-structured surface machining from the error compensation of the original ultra-precision lathe.

Researchers are constantly developing and optimizing the FTS system to improve the machining performance. Tong et al. developed a freeform surface micro-groove machining system based on a FTS machining and measuring platform. It can synchronize and control all integrated sensor systems and CNC machine tool systems to achieve closed-loop process control for freeform surface manufacturing and metering. From improving tool alignment accuracy (<1 μm), fast data synchronization and conversion, integrated machine surface measurements and surface feature functions, including lens surface micro-grooves and Alvarez lens measurements, as shown in Figure 19 [150]. Zhong et al. developed a special controller with internal data machining algorithm, and successfully integrated online surface measurement into the ultra-precision FTS system by adjusting and synchronous data flow, which greatly improved the measurement efficiency and machining accuracy [151].

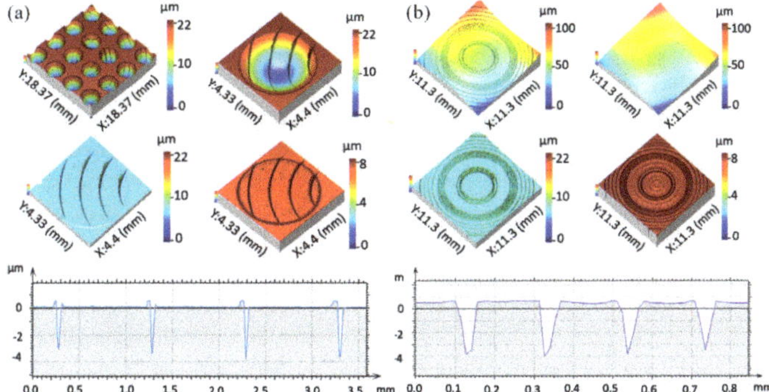

Figure 19. (**a**) Surface characterization of machined 4 × 4 lens array with micro grooves: overview of lens array with micro-grooves, a single lens with micro grooves, after form removal operation extracted micro-grooves after lens shape and measurement outlier removal, cross-sectional view micro-grooves. (**b**) Surface characterization of Alvarez lens with micro-grooves: on-machine surface measurement measured raw data, Alvarez lens shape calculated by 2nd order robust regression filter, after lens shape removal operation, extracted surface micro grooves after optical measurement outlier removal, cross-sectional view of micro-grooves. Reprinted with permission from Ref. [150].

3.1.1. Classified by Actuators

The main difference of FTS system lies in the actuator and mechanical structure. According to the different actuators, it could be divided into piezoelectric FTS (PZT-FTS), magnetostrictive FTS (MGS-FTS), Lorentz force FTS (LRI-FTS), and Maxwell normal stress FTS (MNM-FTS) [152]. The characteristics of the FTS system driven by various drivers are concluded in Table 3.

Table 3. The characteristics of the FTS system driven by various drivers.

Driver	Principle	Advantages	Limits	Applications
PZT-FTS	Piezoelectric effect	• compact structures • maximum response frequency • nanometric positioning accuracy • high dynamic stiffness	• hysteresis • heating problem	widest application
MGS-FTS	Magnetostrictive effect	• high output force • greater stiffness • simple structure	• temperature sensitive • hysteresis • heating problem	micro-displacement and precise positioning
LRT-FTS	Lorentz force	• large output • extra-long stroke • easy to control	• low response frequency • large mass of motion system • heating problem	large stroke with a small bandwidth
MNM-FTS	Maxwell normal stress	• ultra-high frequency • high force density	• non-linearity • positioning error	ultra-high frequency response and high acceleration potential

The piezoelectric actuator is designed by the inverse piezoelectric effect of piezoelectric materials (such as lead zirconate titanate series). They are energy transducers that convert electrical energy of voltage amplitude into mechanical energy of micro-displacement. It is a common FTS device with many advantages such as fast response speed, high acceleration, wide frequency response range—even up to several kilohertz, the closed-loop positioning accuracy can reach nanometer level and has a large output force density—high axial stiffness, simple structure, no transmission mechanism, and the open-loop system is stable and easy to control [153]. At the same time, there are some disadvantages: piezoelectric ceramic actuator requires high driving voltage, usually from several hundred volts, up to one thousand volts; its output displacement is small, and it cannot fabricate the topography with large surface drop; there are time delays and nonlinear phenomena between input voltage and output displacement; and the control algorithm is complicated. Therefore, FTS based on piezoelectric actuators is more suitable for compensating the motion errors of machine tools or machining micro-structural surfaces and small amplitude freeform surfaces.

To improve the tracking performance of traditional tool servo system, Zhao et al. developed a series two-stage FTS system driven by two piezoelectric actuators using flexible hinges for motion guidance [154]. Wang et al. developed a FTS mechanism based on piezoelectric actuator to add additional functions to the general CNC system to facilitate turning of middle-convex varying ellipse piston, which improved the requirements of cutting feed mechanism for strength, high stiffness, fast response, long stroke, and high precision [155], in order to overcome the defect of short servo stroke of PZT-FTS. Kim et al. described the optical surface of the second mirror, which is often needed in optical imaging system, in order to obtain large and high bandwidth tool motion, as shown in Figure 20.

A new long stroke fast tool servo system is proposed and installed on Z-axis of diamond lathe as an additional synchronous axis [44].

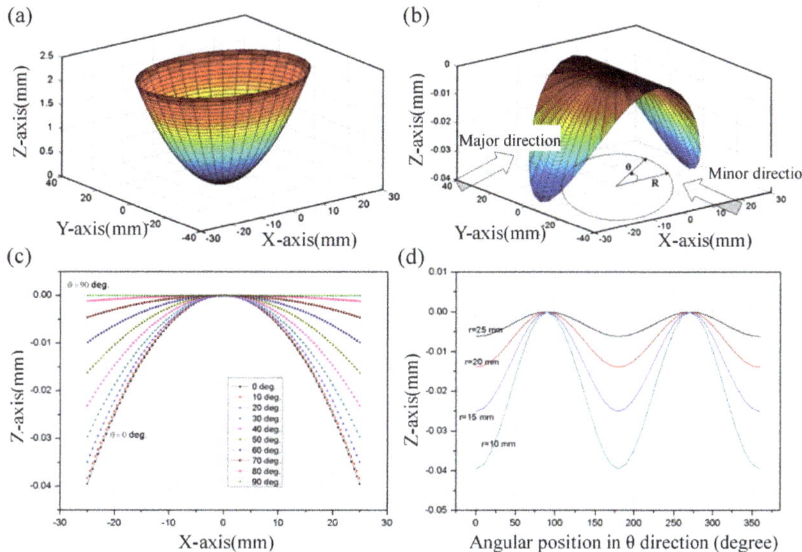

Figure 20. Freeform surface for second mirror in optical imaging system. (**a**) Rotationally symmetric component for the Z-axis; (**b**) non-rotationally symmetric components for the Long-stroke fast tool servo (LFTS); (**c**) the trajectories in the radial direction followed by the LFTS; and (**d**) the trajectories in the circumferential direction followed by the LFTS. Reprinted with permission from Ref. [44].

Magnetostrictive actuator is designed by using the properties of magnetostrictive material stretching or shortening along the direction of magnetization under the action of magnetic field [156,157]. The driving mode has the characteristics of large output force, fast frequency response, and high resolution. Compared with magnetostrictive materials, FTS driven by piezoelectric ceramics has a larger stroke, and magnetostrictive materials have inherent nonlinearity and magneto-hysteresis [158]. Hayato et al. proposed a FTS mechanism milling system based on giant magnetostrictive actuator. The system can control the axial movement of the milling cutter spindle when it is rotating in a non-contact state. FTS unit provides accurate machining movement through displacement feedback of coaxial capacitance sensor on the metering frame. The developed machining system combines the large feed movement of each straight axis of the workpiece shape with the FTS movement of the micro-geometry shape, which can efficiently process the surface with micro-geometry shape [159].

The Lorentz force driving method is suitable for long stroke FTS, which can be divided into three types, namely rotary motor, linear motor, and voice coil motor (VCM) [160]. The acceleration of linear motor and rotary motor is limited in a small range due to the limited current density and mass density of conductor material. The linear motor has a heating problem, which limits its working frequency. The rotary motor cannot directly achieve the linear requirements of FTS. It can indirectly achieve linear motion through the swing and obtain higher acceleration through the amplification of the swing radius. However, the machining accuracy and mechanism stiffness also decreased significantly.

VCM can achieve higher acceleration than linear motor. However, the accuracy of large stroke VCM is affected, with the increase of stroke brings cost increase, and it also has problems of high heat. VCM has a higher working frequency than linear motor, which is a long stroke linear FTS driving mode with a wide range of applications. The VCM adopts electric damping inside, and the flexible mechanism adopts two different viscoelastic damping materials. It is designed by using the galvanic conductor subjected to ampere force in the permanent magnet field. The VCM used in the drive of FTS system have many advantages such as the large stroke of millimeter level, the size of output force and wire length, current intensity and magnetic field intensity is linear, easy to control, etc. VCM usually has no hysteresis in small stroke and the relationship between current and force is nearly linear. Unlike piezoelectric actuators, voice coil actuators do not provide inherent system stiffness, and controller design becomes critical for adequate stiffness during the cutting process. The motion system of voice coil motor has large inertia, high energy consumption, and high heat generation, which limit its response frequency.

Different Lorentz force actuated of FTS systems are studied. Ding et al. revealed the influence of acceleration feedback control on the performance of LRI-FTS system by establishing the theoretical model of dynamic stiffness and error propagation [161]. Rakuff et al. developed a remote, precision FTS system based on voice coil actuator. It consists of a driven bending mechanism, a custom linear current amplifier, and a laser interferometer feedback system; the tool can be accurately converted on the lathe [162]. Tao et al. proposed a FTS system driven by VCM with self-sensing function of cutting force [163]. In FTS system, flexure hinges need to withstand alternating cutting forces, and large deformation that will reduce the stiffness, resulting in machining errors. Long stroke and high stiffness are two mutual restraint units of flexible hinge, which limit the application of long stroke FTS. Qiang et al. designed a cruciform flexure hinge structure with high stiffness and selected VCM to develop a long stroke flexure hinge [164]. Tian et al. presented a new voice coil motor driven high frequency response and long stroke remote FTS system. Experiments on micro-lens array fabrication were performed by the designed FTS to demonstrate its machining capability. Figure 21 is given as a diagram of the FTS structure and the processed micro-lens array [152].

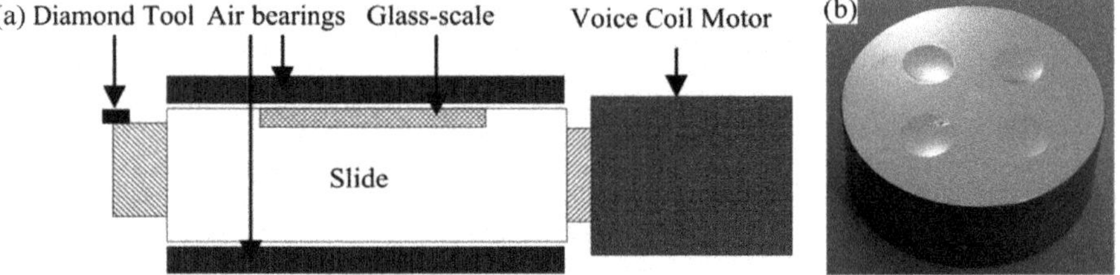

Figure 21. (**a**) The FTS structure diagram; and (**b**) the processed micro-lens array. Reprinted with permission from Ref. [165].

Maxwell normal stress actuated FTS has the characteristics of large driving force and small mass of system moving parts. The driving force of Maxwell force actuator has a linear relationship with excitation current and displacement. It has high force density, high acceleration, and low caloric value. It is a kind of FTS with the highest frequency response. Nie et al. designed and developed the FTS system driven by Maxwell Force to enable precise tool translation in a diamond turning lathe. The FTS system uses a bending-based mechanism to generate motion, a customized linear power amplifier to actuate the armature, and a capacitive sensor to gauge the precise distance. The designed FTS is capable of achieving a frequency response of 3 KHz [166].

3.1.2. Classification by Degree-of-Freedom (DOF)

Single DOF FTS is the most common type of FTS. The movement of single DOF FTS is to increase a high frequency linear motion along the spindle direction of the lathe. The projection of the diamond tip on the end face of the workpiece is an Archimedes spiral due to the joint action of the X- and Y-direction motion and the rotation of the spindle of the traditional lathe. If the Z-axis slide of the lathe moves slowly at the same time, the final workpiece shape is an axisymmetric surface. Single DOF FTS uses a high frequency

Z-direction DOF attached to the machine tool movement and carries out high frequency movement along the Z-direction in integer times of the spindle rotation angular frequency. The resulting surface will be non-axisymmetric.

Single DOF FTS are limited by stroke and frequency. In order to broaden the application scope of FTS and enhance the performance of FTS, some researchers began to look for new ways beyond the driving mode. FTS with multiple DOF began to emerge. Zhu et al. designed a piezoelectric driven 2-DOF FTS to assist diamond turning. Based on the bending deformation of Z-shaped flexure hinge beam, a new guiding bending mechanism based on Z-shaped flexure hinge beam was proposed to make the tool move in two directions with decoupled motion. A novel pseudo-random diamond turning (PRDT) method was also implemented to prepare micro-structure surfaces with scattering homogenization [167]. Liu et al. developed a piezo-actuated serial structure 2-DOF FTS system to obtain translational motions along with Z- and X-axis directions for UPM. A sinusoidal wavy surface is uniformly generated by the mechanism developed to demonstrate the effectiveness of the FTS system [168].

The cutting path types of single DOF FTS and 2-DOF FTS are limited to planar curves, which are difficult to meet 3D requirements. Therefore, the FTS mechanism with 3-DOF has high natural frequency and decoupling performance. It can construct complex 3D machining path. Han et al. designed a novel piezoelectric driven elliptic vibration assisted cutting system composed of flexible hinges. In addition, the developed elliptical vibration cutting system can not only be equipped with a variety of machine tools, but also can easily achieve arbitrary vibration in 3D space through two actuators [169]. Awtar et al. proposed a new constraint-based design of a parallel motion flexible mechanism that provides highly decoupled motion along the three translation directions (X, Y, Z) and high stiffness by rotating along the three directions [170]. Li et al. proposed a novel 3-DOF piezoelectric driven FTS with high natural frequency and decoupling characteristics, the tool path and cutting diagram are seen in Figure 22 [171].

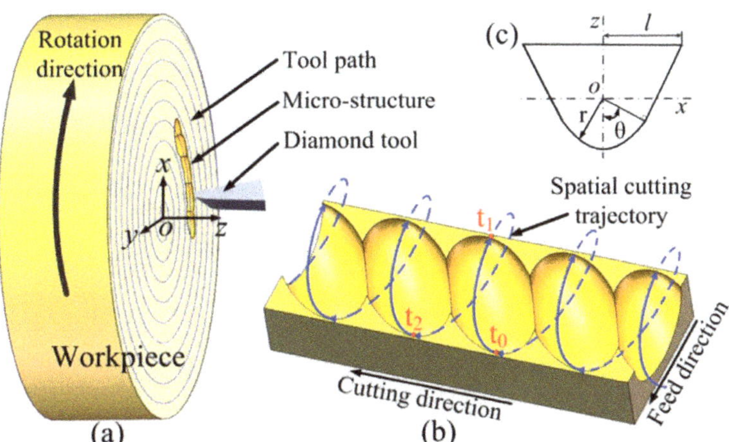

Figure 22. (**a**) Diagram of the designed 3-DOF FTS cutting path, (**b**) cutting path, and (**c**) tool tip shape. Reprinted with permission from Ref. [171].

3.2. Slow Too Servo (STS)

3.2.1. Principles of STS

STS technology changes the spindle of machine tool into a C-axis with controllable position, then transforms the 3D Cartesian coordinates of complex surface workpiece into polar coordinates through the numerical control system with high performance and high programming resolution. The interpolation feed instructions are sent to all motion axes. The system coordinates and controls the relative motion of the spindle and the tool precisely, so as to realize the diamond turning of the ultra-precision complex surface workpiece. In ultra-precision STS system, Z-axis follows X-axis and C-axis for sinusoidal reciprocating motion, which requires multi-axis precision interpolation linkage. The rotational motion of the spindle is both the main cutting motion and the feed motion. In order to ensure the normal turning motion conditions of the tool and meet the requirements of multi-axis linkage, STS has higher requirements on the dynamic characteristics of the feed axis than ordinary multi-axis linkage, especially for the Z-axis, which still needs to reciprocate with the ups and downs of the workpiece surface according to the position of C-axis even in the same radius. Therefore, high precision position servo control of spindle, high precision reciprocating motion, and high dynamic response of straight spindle are the key technical conditions necessary for ultra-precision STS system.

The Z-axis machining stroke of STS is large. Theoretically, the machining range can reach the whole z-axis stroke by using z-axis drag plate to drive the tool movement so as to process steeper and more complex surface; C, X, and Z axes are processed in linkage mode. Furthermore, the linkage of the three axes is completed by the same numerical control system, which can realize information sharing and simplify the control system structure. Moreover, the numerical control program is simplified and the tool locus calculated from the complex surface can be directly used in numerical control program.

Ultra-precision STS system is suitable for UPM industrialized production of small size optical freeform surface, and the structure of the array and off-axis surface. However, it is restricted by the structural characteristics of workpiece surface and dynamic response capability of feed shaft, some defects that exist in the actual application are summarized in Figure 23. Therefore, the selection of tool feed direction and tool contact should fully consider the workpiece surface characteristics. In order to improve the efficiency of ultra-precision production machining as much as possible, optimized machining parameters should be selected on the premise of ensuring the machining effect of the key structure.

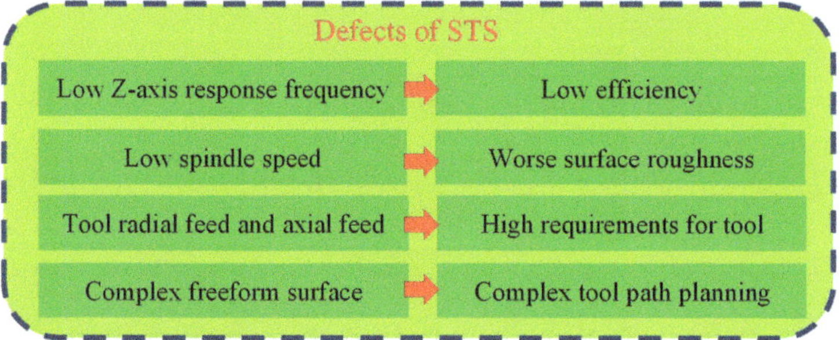

Figure 23. Some defects of STS.

3.2.2. Applications of STS

On the basis of STS technology, researchers continuously develop new methods and technologies to improve the machining performance of the system. In order to better predict the shape change trend under different tool centering errors and different cutting strategies, Yin et al. designed a method for cutting off-axis aspheric surface on an axis based on STS technology, which can achieve large stroke without additional devices [172]. In order to study the new technology of freeform surface optical high-precision manufacturing on hard and brittle materials Wang et al. proposed a new technique for optical surface generation of freeform surface based on STS of diamond grinding wheel, which improves the production efficiency and surface quality [31]. In order to accurately evaluate the contour error, Mishra et al. using STS technology to generate aspherical lens array and characterization of optical profilometer, introduced a STS machining method for machining aspherical lens array [173]. Huang et al. proposed a method of applying variable spindle speed to STS turning to reduce the turning shape error artificially [174].

In order to extend the machining capacity of existing machine tools, Kong et al. present a research method for machining wavy micro-structural patterns on precision rollers using orthogonal STS process. A tool path generator was developed for machining ripple structure patterns on roller surface. Different micro-structures, different wave patterns, and grooves were generated through modeling and simulation mode. A four-axis ultra-precision machine tool was proposed based on the initial experimental work of tool path generator. The generation of unique wave precision rollers in micro-structure mode is shown in Figure 24 [175]. Zhang et al. developed a rotating tool turning machining method based on STS for one-step machining of prisms [176].

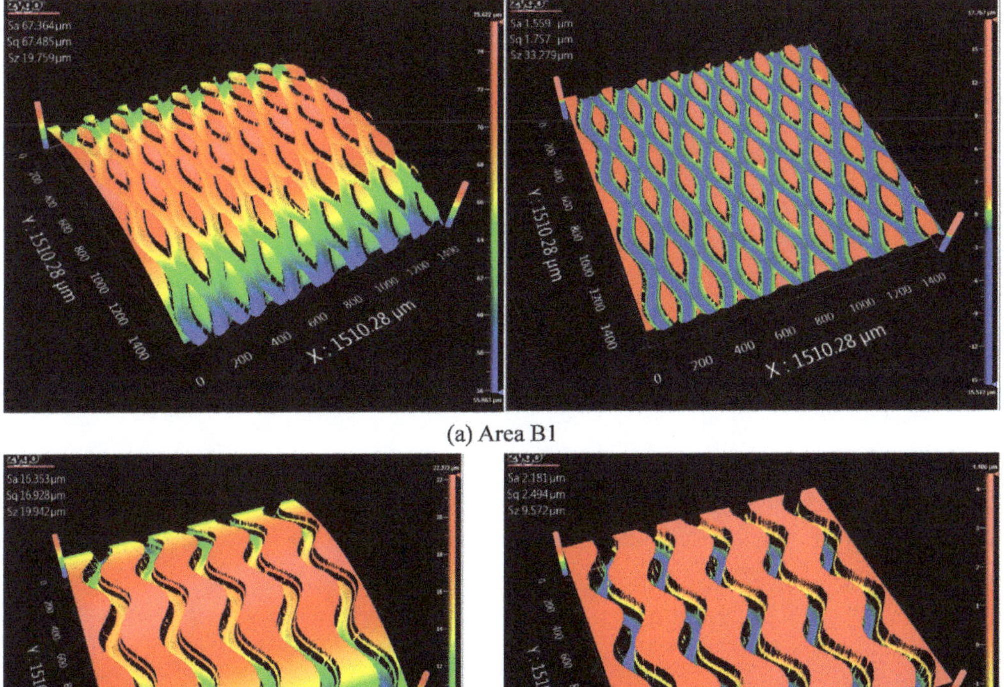

(a) Area B1

(b) Area B2

Figure 24. Three-dimensional topography on the roller surface and after cylindrical form removal (5.5X objective, 1X zoom). Reprinted with permission from Ref. [175].

Tool compensation is required to avoid overcutting, Li et al. conducted theoretical and experimental studies on surface topography generation in STS machining of freeform surfaces in order to realize nano-surface topography. A systematic tool path generation method was studied, including tool path planning, tool geometry selection, and tool radius compensation [177]. Peng et al., taking micro-lens array (MLA) as an example, analyzed the components of servo dynamics error in servo motion, including dynamic deformation, resonant vibration and trajectory tracking error, and established a dynamic surface generation model. The surface obtained by turning is shown in the Figure 25 [178].

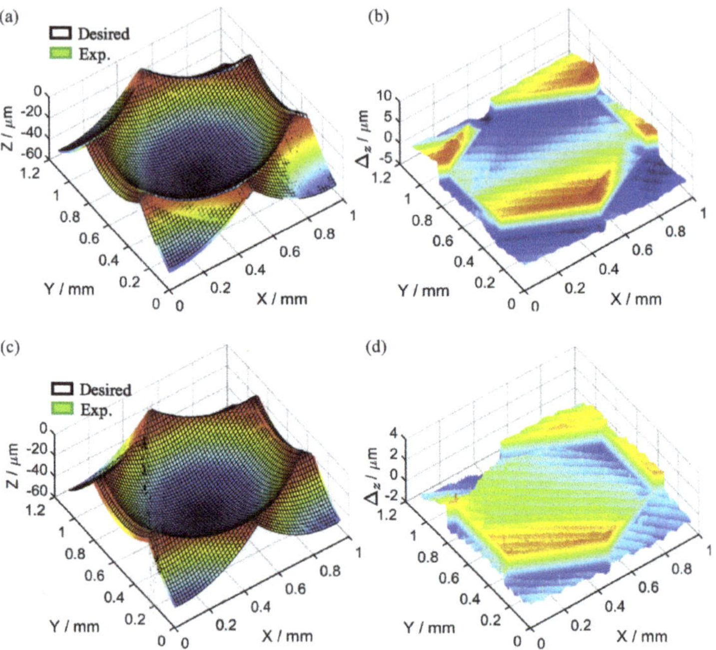

Figure 25. Characteristics of the surfaces obtained by practical turning: (**a**) The 3D profile; and (**b**) the resulting form error of the surface generated by STS turning; (**c**) the 3D profile; and (**d**) the resulting form error of the surface generated by cooperative tool servo turning. Reprinted with permission from Ref. [178].

Chen et al. established a 3D shape compensation model to solve the problem of tool path generation in the process of astigmatic contact lens asymmetric torus STS diamond turning. The ultra-high accuracy 3D profilometer with user-defined function was used to measure the shape accuracy of freeform surface [179]. Nagayama et al. proposed a deterministic machining process for low-speed servo turning manufacturing freeform surface optical components. The process comprehensively analyzed the main error factors before machining and carried out simulation and compensation based on feedforward method to accurately predict the workpiece shape error after compensation. In order to prove the effectiveness of the proposed method, a two-dimensional sinusoidal lattice cutting experiment was carried out on monocrystalline silicon (as shown in Figure 26) [180].

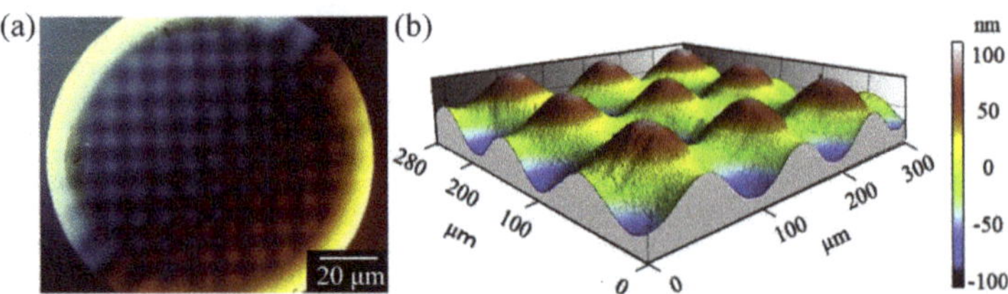

Figure 26. (**a**) Surface morphology of a 2D sinusoidal grid machined on a single crystal silicon workpiece; and (**b**) three-dimensional pictures of nanoscale sinusoidal lattices processed on single crystal silicon. Reprinted with permission from Refs. [179,180].

3.3. Comparison between FTS and STS

FTS machining and STS machining have similar machining action and machine tool structure, both need to obtain accurate spindle angle value and can realize the turning of free surface optical parts. Nevertheless, there are many differences, as can be summarized in Figure 27.

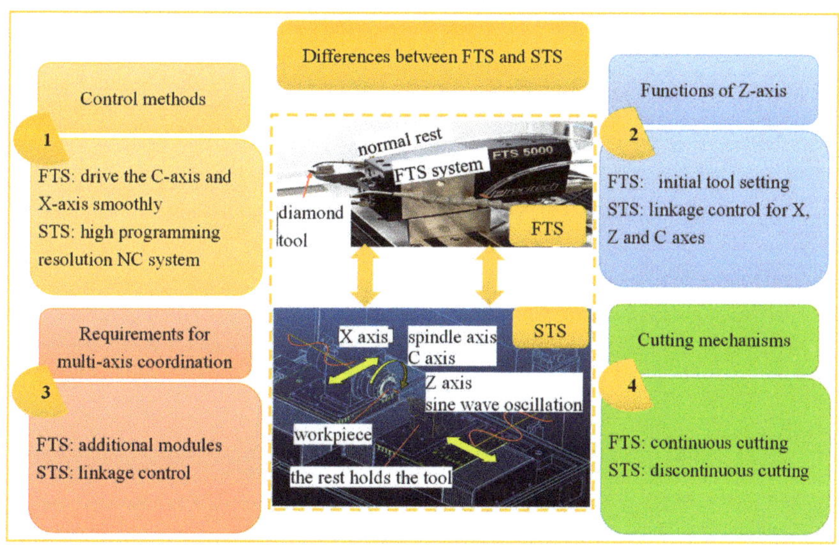

Figure 27. Differences between FTS and STS.

Having different control methods, STS requires a special high programming resolution numerical control system to carry out the three-axis interpolation algorithm including the spindle, and requires a special tool path generator and interpolation algorithm design. FTS does not necessarily need to be equipped with a special high-precision numerical control system. The task of the numerical control system is only to drive the C-axis and X-axis smoothly. Accurate position information is sensed by angle sensors and radial sensors and collected by FTS in real time [181].

Additionally, they have different functions of Z-axis. FTS module and its own control system are additional modules independent of ultra-precision machine tools. Z-direction feed is realized by FTS, and Z-axis is only used for initial tool setting. STS requires ultra-precise position servo linkage control for X, Z, and C axes, essentially three- or four-axis linkage machine tool. Therefore, the requirements on the sensors of each axis are extremely demanding, usually requiring the spindle encoder to reach hundreds of thousands of lines, and the grating ruler of X-axis and Z-axis has very high resolution, which increases the cost of the system to a certain extent.

Furthermore, they have different requirements for multi-axis coordination: The core of FTS machining is FTS servo module, and there is no linkage relationship between each axis. The sensors can meet the accuracy requirements of the final surface shape. STS needs to carry out multi-axis coordination on the surface shape of parts before machining, and then determine the tool path and tool compensation, so as to generate the optimal numerical control program.

In addition, they have different cutting mechanisms. Before FTS machining, the surface shape of parts should be accurately calculated to generate data files that can represent the surface shape of parts, and the precise cutting depends on the control performance of FTS. STS is continuous cutting, while FTS may contain discontinuous cutting at the abrupt change of profile.

Theoretically, STS can machine any complex surface shape and structures. For the small structure of surface shape mutation, it is necessary to reduce the spindle speed, and from the accuracy, efficiency and machining material consideration, the machining difficulty is greatly increased. It is not difficult to draw a conclusion that FTS and STS are completely different in control mode, complex surface shape generation, tool path planning, cutting mechanism, and process parameter selection due to the great differences in actual tool feeding mode, motion frequency response, stroke, and machining object. In a nutshell, FTS module has high motion frequency response (more than 100 Hz) and small stroke (less than 500 µm), which is more applicable for machining small structures with abrupt surface shape or discontinuous and limited stroke. STS motion has low frequency response (10 Hz) and large stroke (1–100 nm), which is applicable for machining complex surface parts with high surface roughness requirements, smooth surface, and large overall drop.

4. Ultra-Precision Fly Cutting

4.1. Fly Cutting

Ultra-precision fly cutting (UPFC) is a cutting process in which the single-crystal diamond cutter rotates simultaneously with the spindle of the lathe to cut the workpiece surface intermittently [182]. The fly cutting machining technology is a high-speed machining method with flexible trajectory and high machining efficiency. It can obtain micro-structures with nanometer surface roughness and submicron surface shape accuracy without post-polishing. In UPFC, cutting speed is constant and can deterministically generate special micro-structured surface. It is suitable for fabricating complex micro-structured surfaces, such as linear groove micro-structure, micro-groove array composed of multiple intersecting lines, pyramid matrix, F-Theta lens, repetitive prism matrix, and micro-structures applied to special reflective surface coatings, tapes and thin slices.

UPFC system installs the diamond tool on the circular fly tool disc and installs the corresponding counterweight on the fly tool disc, then install the fly tool disc on the lathe spindle, the diamond cutter rotates with the spindle at high speed. The workpiece is mounted on the workbench bracket, and the workpiece moves in a plane. UPFC is intermittent machining, and the diamond tool only contacts with the workpiece at a certain angle during each rotation, and the cutting depth changes with the rotation angle of the fly tool disc in each cutting process. For a specific micro-structure, UPFC can be divided into two machining methods from the perspective of the formation of the micro-structure morphology of the workpiece. One is the trajectory method; the shape of the processed micro-structure is formed by using the motion track of the diamond tool tip in the cutting movement. Since the workpiece surface is composed of the tool path, the tool path method can process micro-grooves of variable size without requiring accurate tool tip angle, as long as the tool does not interfere in the process of machining micro-structure. However, in this machining method, the tool trajectory is more complex, so the higher requirements for lathe, generally need multi-axis precision linkage. The other is the profile method, diamond tool geometry is consistent with the processed micro-structure. When machining, the positioning accuracy of the lathe is taken into account. The diamond tool is processed vertically in a micro-groove position, then the transverse feed is made after the ideal depth is processed. The tool is then processed vertically in the next micro-groove position until all the micro-groove matrix is processed. Compared with the trajectory method, this method has lower requirements on machine tool performance, simple tool path, easy machining, and better surface roughness. However, this machining method can only process the same sized micro-grooves, and the preparation of high requirements for cutting tools, which require precise geometric accuracy.

UPFC technology can be divided into two types according to the different tool installation direction: when the tool installation direction is parallel to the spindle axis, it is called end fly cutting [183,184]; and ultra-precision raster milling is when the tool direction is installed along the radial direction of the spindle [185,186]. Its schematic diagram is shown in Figure 28. Some micro-structures and SEM figures of machined structures fabricated by fly cutting are shown in Figure 29 and Table 4.

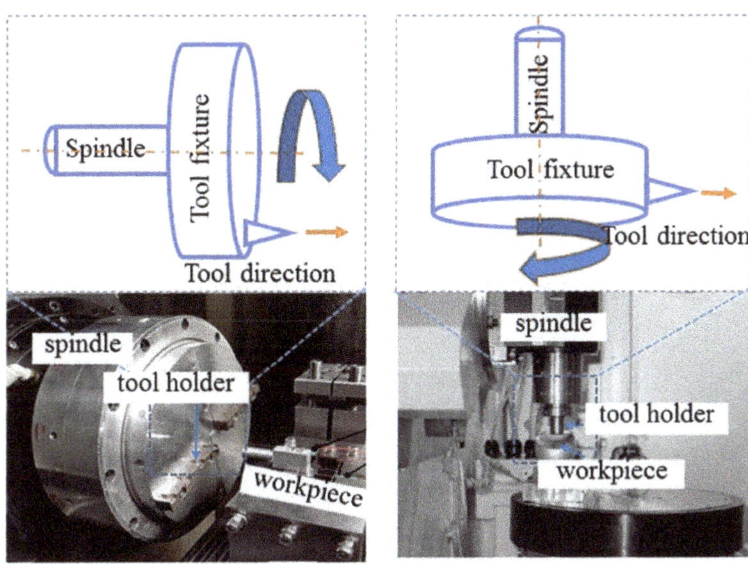

Figure 28. The two types of fly cutting. Reprinted with permission from Refs. [187,188].

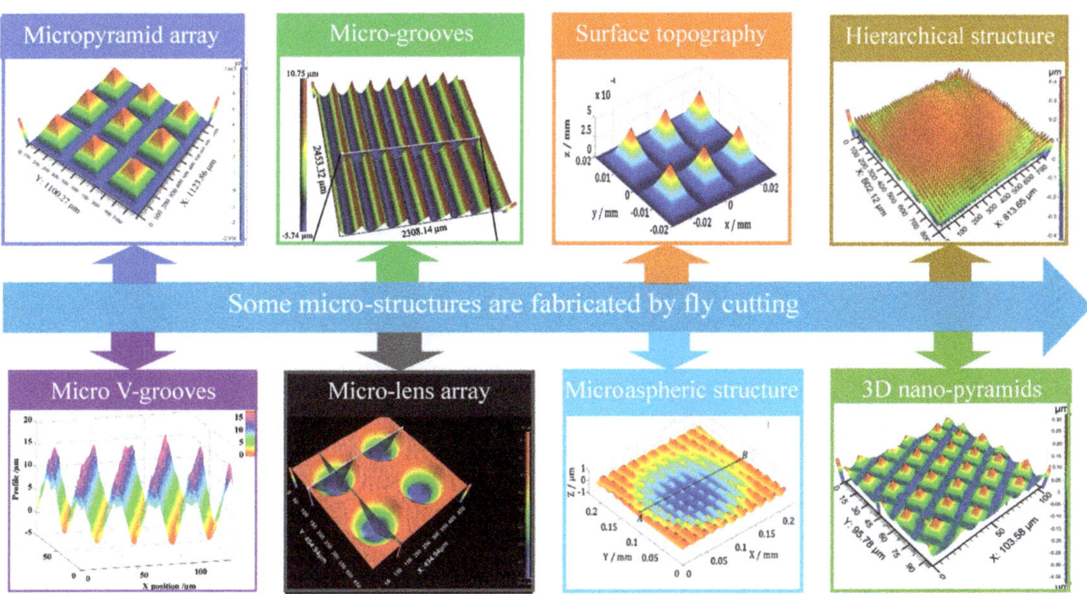

Figure 29. Some micro-structures are fabricated by fly cutting. Reprinted with permission from Refs. [119,177,184,189–191].

Table 4. SEM figures of machined structures by fly cutting [54,128,192,193].

Micro-Structure	Diagram	Material	Parameter	Surface Quality
Pyramid detail		Nickel phosphorus (Ni-P)	Microgroove width: 15 μm	Chipping eliminated
Micropyramid array		Nickel phosphorus (Ni-P)	Cutting depth: 10 μm	Maximum effective cutting thickness: 78.6 nm
Submicron grooves		Nickel phosphorus (Ni-P)	groove spacing: 500–800 nm	High-regularly arrayed and have a high parallelism
Triangular array		Brass	Cutting depth: 20 μm	Max deviation: 5.3 μm
Pyramid array		6061 aluminum	Cutting depth: 0.5 μm	Max error: 0.4 μm

4.2. End Fly Cutting

End fly cutting was originally used to fabricate large flat surfaces with a uniform surface quality. However, in recent years, fly cutting technology has been applied to UPM combined with FTS/STS to produce mixed structured surfaces such as micro-structured freeform surfaces [188]. In traditional FTS/STS, cutting is operated in a cylindrical coordinate system, where the cutting direction is always perpendicular to the polar axis of the workpiece and it is impossible to construct the intersection points of the cutting trajectory. For the end fly cutting, the diamond tool is mounted on the spindle and rotates with it, while the workpiece is sandwiched on the slide. By exchanging the position of the diamond tool and the workpiece, the operation of the end fly cutting system is transferred to the rectangular coordinate system. Due to the circumferential motion of the diamond tool, various cutting directions about the workpiece can be obtained. Translational servo motion along the Z-axis, like FTS or STS, acts on the workpiece and is responsible for determining the generation of complex micro-structures with complex shapes. Due to the unique advantages of end fly cutting in machining, it is suitable for machining complex specific micro/nano structures such as pyramid array, triangular pyramid array, and multi-layer complex micro-structures. The end fly cutting system configuration diagram is shown in Figure 30.

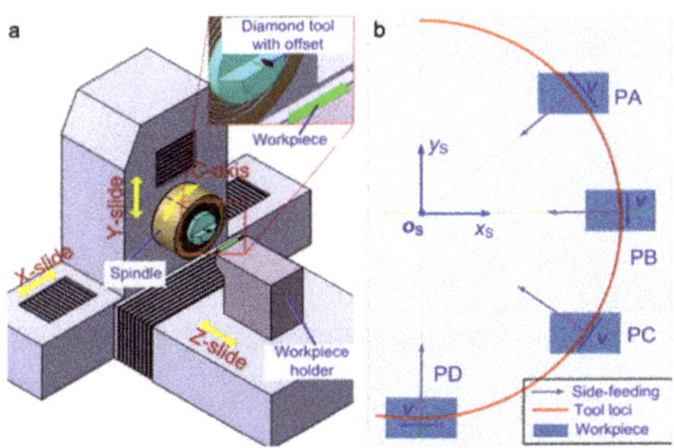

Figure 30. (**a**) End fly cutting system configuration diagram; (**b**) the induced cutting modes, where o_s-$x_s$$y_s$ denotes the coordinate system fixed on the spindle axis, and v denotes the relative cutting speed. Reprinted with permission from Refs. [177,184].

In some studies, improved methods were proposed for the preparation of large-aperture potassium dihydrogen phosphate (KDP) crystal by fly cutting. KDP is widely used in the laser path of inertial confinement fusion system. The most commonly used method to fabricate half meter KDP crystal is UPFC. When UPFC technology is used to process KDP crystal, the dynamic characteristics of the fly cutting lathe and the fluctuation of the fly cutting environment are transformed into surface errors in different spatial bands. To ensure that KDP crystal achieves the full band machining accuracy specified in the evaluation index, these machining errors should be effectively suppressed. Zhang et al. studied the anisotropic machinability of KDP crystal and the cause of typical surface errors in UPFC of materials. By analyzing the causes of machining errors in different frequency bands and measuring the brittle-ductile transition depth of the crystal, the structure, parameters and cutting environment of the fly cutting were optimized [194]. Tool tip is perpendicular to the workpiece surface due to the thermal expansion of the tool. To this end, Fu et al. established a surface cutting simulation model based on kinematic fly cutting and thermal models, then conducted UPFC tests on KDP crystals to explore the specific impact of the main cutting force on the thermal deformation of the workpiece, thus determining the relationship between the cutting parameters and the main cutting force. This model can be used to predict 3D surface topography after UPFC, and to evaluate surface corrugations, roughness, and other performance indicators [195]. In order to reveal the formation mechanism of high points on large optical surfaces during UPFC, Wang et al. established a thermodynamic model of the tool system during fly cutting. The temperature field distribution and axial displacement caused by thermal deformation of the tool system were solved and analyzed by finite element simulation [196].

To improve accuracy, researchers put forward improvement methods on fly cutting machine tools. Liang et al. proposed a mechanical structure-based design method to design and optimize UPFC machine tools. In order to study the effect of workpiece structure on machined surface roughness, an optimized spindle structure was designed to reduce workpiece roughness [197]. Lu et al. conducted dynamic modeling and simulation on the ultra-precision fly cutting machine tool to find out the relationship between structural parameters and the machined surface. The simulation surface and measurement surface of fly cutting machine tool are shown in Figure 31b,c [198]. The micro-structures fabricated by fly cutting also act as diffraction gratings, Du et al. used vibration-assisted diamond cutting process to generate structural colors on the surface of pure magnesium, and prepared periodic sawtooth micro-structures that acted as diffraction gratings to induce the generation of structural colors. The colors at different angles are shown in Figure 31e,f [199]. In order to suppress tool wear and improve cutting performance, Zhang et al. explored the tool wear by fly cutting experiments, found that increasing air pressure and the closer the nozzle position not only to improve the cutting performance, also reduced the tool wear [200].

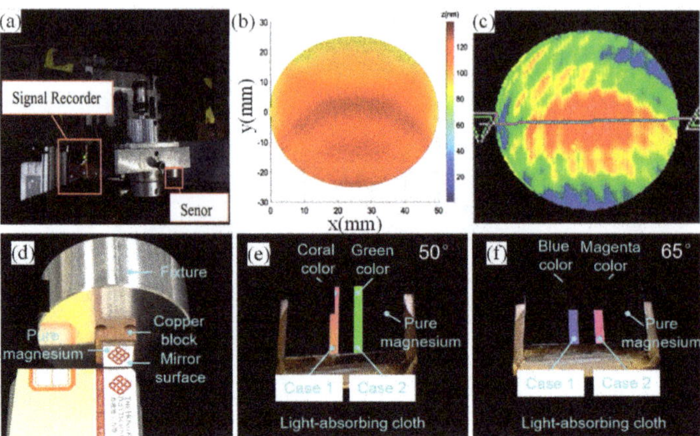

Figure 31. (**a**) Tool-tip acceleration test set up, ultra-precision fly-cutting machine tool; (**b**) simulated surface; (**c**) measured surface; (**d**) optical mirror surface; (**e**) coral color and green color at the viewing angle of 50°; and (**f**) blue color and magenta color at the viewing angle of 65°. Reprinted with permission from Refs. [198,199].

To overcome the generation of inherent residual marks (RTM) with specific patterns during cutting, Zhu et al. developed a novel biaxial servo assisted fly cutting (BSFC) method to achieve flexible control of RTM to generate functional freeform optical elements, which is difficult to achieve in FTS/STS diamond turning. By selecting the feed speed in each cutting rotation, the height and spacing of secondary micro/nano textures can be independently generated [189]. Furthermore, to overcome the problems of inconsistencies of cutting parameters induced by cutting speed and underutilization of vibration assistance caused by cutting direction in vibration assisted turning or milling of brittle materials. Zhu et al. introduced the machining method of rotary spatial vibration (RSV) in assisted diamond cutting. The feasibility and superiority of this process for brittle materials were demonstrated by manufacturing a group of annular micro-grooves in monocrystalline silicon with gradually changing cutting depth [201].

Different cutting strategies have been proposed to achieve complex micro/nano structures with high precision at multiple levels. Combining the concept of fast/slow tool servo system, Zhu et al. proposed an end fly cutting servo system (EFCS) with four-axis motion, which can be used for deterministic generation of layered micro/nano structures that are difficult to achieve by other methods. A nano-structured micro-aspheric array and nano-structured F-Theta freeform surface were successfully prepared, as shown in Figure 32a–c [177]. To et al. continuously develops the performance of EFCS and proposes a new machining method to generate hierarchical structure microstructure. By combining different cutting directions in the EFCS system, a variety of nano-polygons can be generated as secondary structures. Nano-pyramids for advanced optical applications are experimentally generated on the F-Theta freeform surface and high-precision micro-aspherical array. The fabricated nano-pyramid ellipsoidal aspherical array is shown in Figure 32d–f [184]. Zhu and co-researchers went further, they proposed a new mechanical micro/nano machining process that combines the rotary space vibration (RSV) of a diamond tool with the servo movement of the workpiece and applied it to the generation of multi-layer micro/nano structures. By combining non-resonant triaxial vibration with servo motion, a variety of micro/nano structures with complex shapes and flexible and adjustable feature sizes can be generated, as shown in Figure 32g–i [202]. In addition, Zhu et al. extended the application of FTS/STS diamond turning technology in end fly cutting because it could not achieve good large-scale machining of micro-lens array (MLA). Higher spindle speeds can also achieve higher cutting efficiency, and this method achieves uniform mass MLA in a large area, as shown in Figure 33 [190].

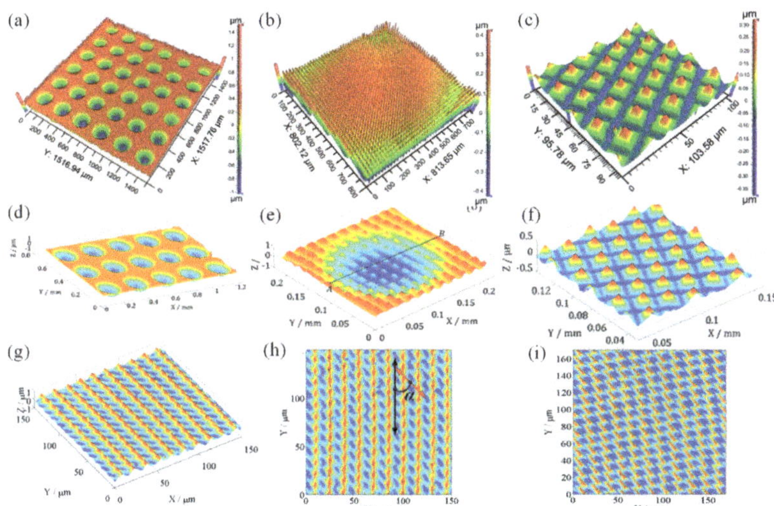

Figure 32. Surface characterizations of the micro-aspheric array with nano-pyramids: (**a**) large area structures; and (**b**) the extracted 3D nano-pyramids; (**c**) surface characterizations of the F-theta freeform surface with nano-pyramids, the 3D hierarchical structure. Characterizations of (**d**) the micro-aspheric array with nano-pyramids; (**e**) a single micro-aspheric structure with nano-pyramids; (**f**) the enlarged view of the secondary nano-pyramids. Characteristics of nano-structures fabricated by rotary vibration assistance with a single frequency: (**g**) the 3D, and (**h,i**) projected 2D nano-dimple array generated. Reprinted with permission from Refs. [177,184,202].

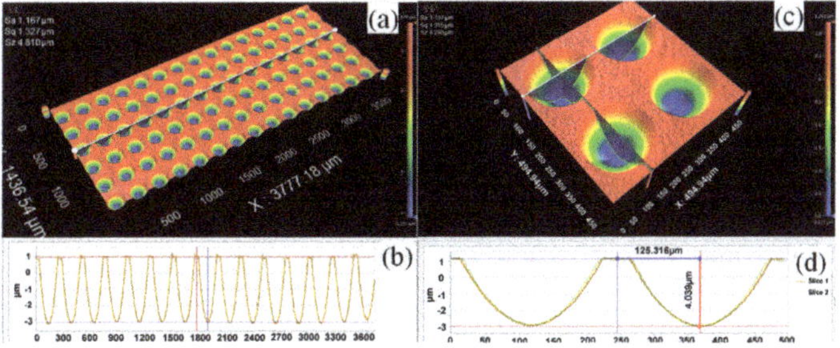

Figure 33. Characteristics of the machined micro-lens array: (**a**) the 3D structure of a large area; (**b**) the 2D profile of the cross-section; (**c**) an extracted square area; and (**d**) the corresponding profiles along the cross-hair directions of the micro-lens array. Reprinted with permission from Ref. [190].

To improve the shape accuracy, Huang et al. studied the influence of machine tool error on the shape accuracy of cone in fly cutting. The relationship between the minimum single cutting depth and the maximum tool displacement error is analyzed to avoid the size error of micro-cone [54]. Jiang et al. extended the application of fly cutting and proposed a new offset fly-cutting-servo system (OFCS) combining the concepts of fly cutting and STS, which can process straight slot microstructural arrays on commercial machine tools containing only X, Z, and C axes, such as pyramid array and triangular pyramid array. On the basis of this model, groove array, pyramid array, and triangular pyramid array are machined respectively, as shown in Figure 34 [128].

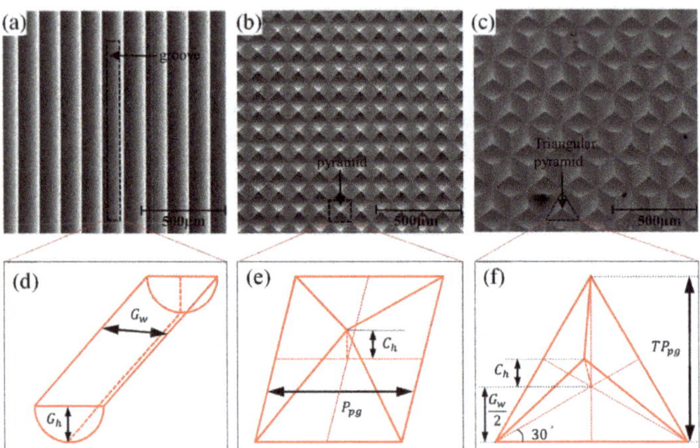

Figure 34. SEM figures of machined structure arrays: (**a**) micro groove array; (**b**) micro pyramid array; (**c**) micro triangular pyramid array; and (**d**–**f**) their geometric descriptions. Reprinted with permission from Ref. [128].

Single-crystal silicon is a brittle material widely used in infrared optics and optoelectronics industry. However, due to its extremely low fracture toughness, it is difficult to obtain ultra-smooth deep micro-structure on single-crystal silicon based on cutting, diamond milling and grinding. Sun et al. proposed a new UPFC ductile machining model to efficiently process deep micro-structure on silicon. The 3D topography of the optical micrograting composed of 15 μm deep curved micro-grooves prepared by UPFC on monocrystal silicon is shown in Figure 35a,b [119]. Furthermore, Sun et al. proposed a novel UPFC ductile machining model of silicon. By preparing two kinds of freeform surfaces on silicon, namely micro-grooves and F-Theta lenses; experiments were carried out to verify the superiority of UPFC in realizing deep ductile cutting regions. Figure 35c–e shows the surface characteristics of the F-Theta lens and its surface profile in the diagonal direction measured by the 3D optical surface profilometer. The surface roughness is up to nanometer level and the shape error is up to micron level without brittle fracture. Moreover, the model can be used to process freeform surfaces of other brittle materials under large cutting depth [120].

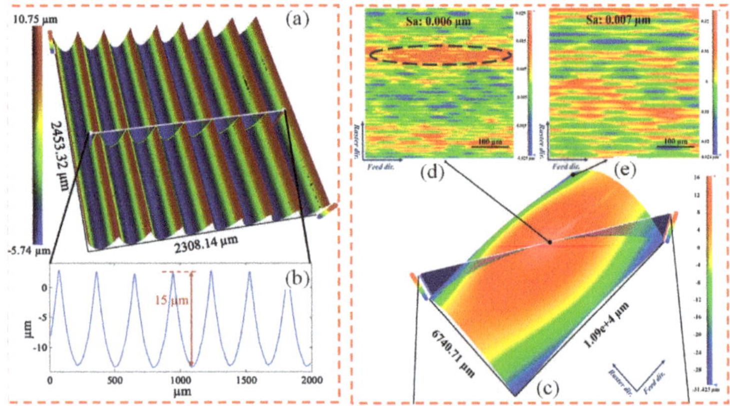

Figure 35. 15 μm deep micro-raster composed by arc-shaped micro-grooves in <100> direction: (**a**) 3D surface topography: (**b**) cross-sectional profile. Surface morphology of the F-theta lens surface machined by UPFC, (**c**) 3D morphology and (**d**) center area and (**e**) marginal area. Reprinted with permission from Refs. [119,120].

4.3. Raster Milling

In fly cutting, Ultra-precision Raster Milling (UPRM) is when the tool direction is perpendicular to the spindle. UPRM can be performed using horizontal and vertical cutting strategies and their geometry is shown in Figure 36. The diamond tool is fixed on the spindle and rotates with the spindle. The cutting process consists of two kinds of motion: feed motion and grating motion, determining the appropriate cutting strategy such as transverse cutting or longitudinal cutting. Horizontal cutting is performed by cutting the tool in a horizontal direction. After cutting a section, the diamond tool moves along the milling to process the entire workpiece surface. In vertical cutting, the cutter feeds vertically and moves horizontally. The surface roughness profile of the machined surface in both cutting methods is formed by the repeated cutting of the tool tip at intervals of the tool feed rate, by the tool moving a specified distance under ideal cutting conditions. The feed direction is perpendicular to the milling direction, and the two directions in horizontal cutting are opposite to those in vertical cutting [203].

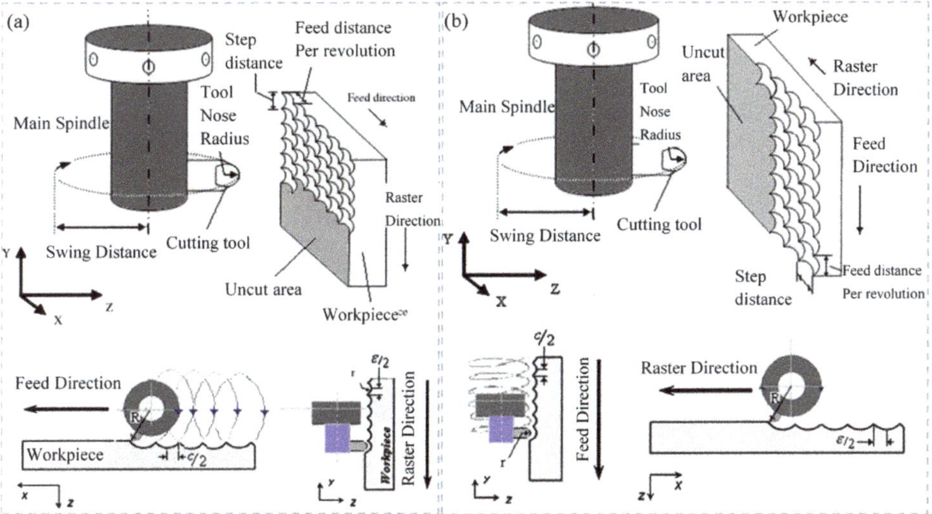

Figure 36. Cutting geometry for ultra-precision raster milling using: (**a**) horizontal cutting strategy; and (**b**) vertical cutting strategy. Reprinted with permission from Ref. [203].

To avoid a lot of trial-and-error experiments, the researchers proposed different simulation models to verify the cutting experiment. Cheung et al. established a simulation model to optimize the prediction mechanism of optical freeform surface generation during UPRM. It is not only possible to predict shape accuracy and simulate surface generation, but also to identify the optimal machining area to minimize shape error prior to actual production [185]. Cheng et al. conducted theoretical and experimental analysis on the generation of nano-surface in UPRM and established a theoretical model for surface roughness prediction [187].

Due to different cutting mechanisms, compared with ultra-precision diamond turning and conventional milling, the process factors that affect surface quality in UPRM are more complex. The process factors and cutting strategies have a certain influence on the surface quality. The influence of technological factors can be minimized or even eliminated by selecting reasonable cutting conditions and cutting strategies. Cheng et al. studied the process factors affecting the surface quality of UPRM and found that step spacing is one of the key factors affecting the surface quality of UPRM. The influence of process factors can be minimized through reasonable operation settings and control of dynamic characteristics of machine tools [52].

There are some studies on optimizing the precision of UPRM. Cheng et al. proposed an optimization system for surface roughness analysis UPRM. According to the minimum surface roughness criterion, the system studies the optimal cutting conditions of different cutting strategies within the preset process parameters [204]. Kong et al. studied the factors affecting the surface generation in UPRM. Compensation strategy was designed to improve surface generation and improve surface quality of UPRM [203]. The cutting force in milling process can be better studied by establishing motion model. Kong et al. has established a theoretical dynamic model for UPRM of optical freeform surfaces. The modification of machining parameters provides guidance for the control of cutting force excitation [205].

The special vibration of spindle has great influence on the surface formation of workpiece. Zhang et al. conducted relevant research on spindle vibration. Based on the 5-DOF dynamic model of the hydrostatic bearing spindle, a special model under the excitation of the batch cutting force was established, and the corresponding mathematical solution was derived. The surface topography is shown in Figure 37a,b [206]. In addition, Zhang et al. studied the surface generation problem under the excitation of spindle vibration in UPRM, proposed the nonlinear equation of spindle vibration in detail, and studied its special influence on the surface generation. The surface topography of spindle under horizontal cutting conditions is shown in Figure 37c,d [186].

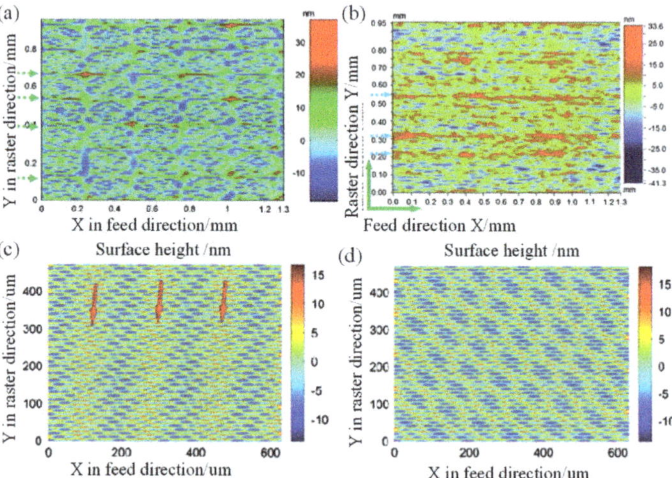

Figure 37. Surface generation under the cutting conditions: (**a**) simulated topography; (**b**) measured topography, simulated surface topographies with; (**c**) the uniform phase shift; and (**d**) the phase shift of 0.5. Reprinted with permission from Refs. [186,206].

To meet the requirements of precise forming control, the corresponding machining and measurement methods are proposed. Zhu et al. proposed a new method of fly cutting microstructured surface measurement based on the principle of scanning tunneling microscope. Figure 38a,b shows the surface morphology of unremoved machined tissue. It is suitable for the fly cutting process with frequent micro-V-shaped grooves, and significantly improves the surface form of rectangular pyramid array [191]. Cai et al. proposed a diamond fly cutting method based on tool path generation, which includes tool radius compensation and surface morphology simulation. The surface topography under different feed speeds and spindle speeds was simulated [207].

UPRM can also be applied to other aspects of optics. Submicron structure surface can produce view-angle dependent rainbow, which is widely used in multi-color printing, micro-display projection, invisibility cloak technology, and so on. He et al. designed several two-stage structures, including first-order microscopic geometric features corresponding to pattern shapes and second-order submicron grooves corresponding to diffraction gratings, which directly induced a variety of glow patterns according to their shape laws. The second-order submicron grooves at different feed rates are shown in Figure 39a–d [193]. The fabrication of micro-structures on brittle materials by grating milling is also the focus of research in recent years. Dong et al. analyzed a transverse planning method for machining NI-P die with micro-pyramid array. High-quality micron-scale micro-pyramid array was prepared on the phosphor nickel coating, as shown in Figure 39e,f [192]. To optimize the processing morphology of UPRM on brittle materials, Peng et al. introduced a simple model to analyze the mechanical properties around diamond tip. It is found that the stress state in the chip formation zone can be changed and the stress intensity factor at the crack tip can be reduced, which is beneficial to the plastic deformation of brittle materials [208]. Aramid fiber reinforced plastics (AFRP) is a typical hard cutting material due to burrs and cracks during machining. Bao et al. established a cutting mechanics model of AFRP to predict the influence of the angle between the feed direction and the fiber direction on tensile and shear behavior during cutting [209].The cutting chip, microwave, and burr produced in the machining process will greatly affect the machining accuracy. To reduce machining errors and obtain good surface roughness, researchers studied the error generation mechanism. Zhang et al. studied the influence of cutting chips and tool movement during cutting. By increasing spindle speed and decreasing feed speed, the probability of surface roughness patterns occurrence can be minimized [182]. Zhang et al. found that the microwaves on the UPRM surface were caused by the sliding of the material [210].

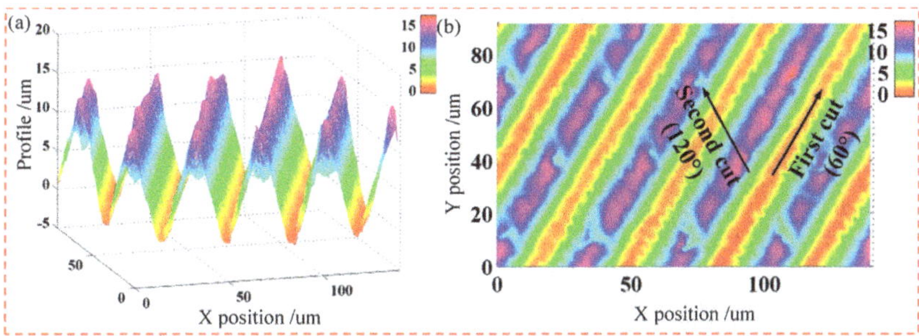

Figure 38. (**a**) Top view of on-machine measured surface of crossed micro-V-grooves; (**b**) top view of the measured surface. Reprinted with permission from Ref. [191].

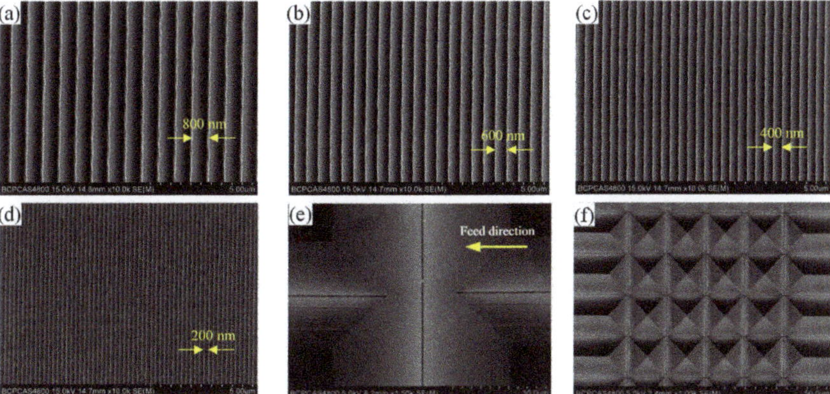

Figure 39. Second-order submicron grooves with different spacing: (**a**) 800 nm; (**b**) 600 nm; (**c**) 400 nm; (**d**) 200 nm; (**e**) High-quality exit and the entry part of the micro-pyramid; and (**f**) micro-pyramid array machined by fly cutting. Reprinted with permission from Refs. [192,193].

5. Summary and Outlook

5.1. Summary

Micro-structured surfaces not only show excellent performance in natural plants and animals, but also are widely used in many fields such as optical engineering, biomedicine etc. In this review, the advanced fabrication methods and assistive technologies for fabrication of micro-structures are reviewed. According to their specific applications, the main research achievements of various machining strategies are reported. From the review, typical machining technologies for micro-structured surface are mainly classified into lithography technology, high energy beam direct writing technology, special energy field machining technology, molding technology, LIGA technology, and UPM technology. However, these methods cannot meet all the requirements alone in the fabrication of high accuracy, complex 3D, and high aspect ratio micro-structure surface.

Among these methods, UPM exhibits unique advantages in obtaining micro-structures with certain geometry. As the key assistive technologies, FTS/STS system is combined with UPM to extend the machining technology to freeform surfaces and complex micro/nano structured surfaces. FTS system has high frequency response, which is applicable for machining the micro-structures with complex surface shape and small drop.

The different actuators in FTS device provide the possibility of machining a diversity of micro-structures. The piezoelectric actuated FTS system is suitable for machining small amplitude micro-structures such as micro-lens arrays with high frequency; magnetostrictive FTS system is usually used in micro-displacement and precise positioning; voice coil actuated FTS system is applicable for machining micro-structures with high aspect ratio; Maxwell actuated FTS system can achieve ultra-high frequency response and has great potential for micro-structures machining. FTS system can also be designed into different DOF according to machining requirements of complex structures. The versatility of FTS brings more possibilities for micro-structured surfaces machining. Different from FTS, STS system has low frequency response but large stroke for the feed shaft, which is applicable for machining micro-structures with high surface roughness, smooth surface shape and large overall drop. The different advantages of FTS/STS system broaden the machining performance of micro-structures.

Based on ultra-precision diamond turning, UPFC is a developed cutting technology. Under the turning mode, the cutting speed varies with the feed radius, which may lead to a uniformly surface quality. Compared with turning, though UPFC is discontinuous cutting with low cutting efficiency, its cutting speed is constant and it can generate special micro-structured surfaces deterministically. In UPFC, EFCS is applicable for machining freeform surfaces and periodic micro-structures, like pyramid array, triangular pyramid array, lens array, etc. And UPRM is applicable for machining V-grooves and large freeform surfaces. In the future, UPFC combined with FTS/STS system has great application potential in the preparation of special micro-structured surfaces.

5.2. Outlook

With the increasing demands of large size, high form accuracy and surface finish of micro-structures, UPM technology has developed rapidly. High precision, intelligence, automation, high efficiency, information, flexibility, and integration become the future development trend of UPM, as shown in Table 5 for details. Besides, the progress of micromachining trends in machining micro-structures is shown in Figure 40.

Specially, combining different machining methods together is of great significance for the fabrication of multi-layer and multi-scale micro-structures. For example, the compound eye structure can be obtained by fabricating planar substrate micro-lens using laser direct writing technology and fabricating sublayer surface using lithography and replication transfer technology. Higher flexibility and accuracy can be achieved by combination of various machining technologies, which is conducive to the realization of complex micro-structures with different characteristic sizes. It also can produce more diverse micro-structured surfaces and provide more possibilities for optical structures. In addition, UPM as a key method for machining micro-structures, combining with FTS/STS technology and FIB technology, will be fully exploited to their advantage. However, residual tool marks and material recovery are still the key factors affecting the micro-structured surface morphology, and more solutions are needed in the future.

Table 5. UPM future trends.

Trend	Content	Research
Precision	Improving the workpiece size, shape, position accuracy, surface roughness.	Aerospace, satellite, mobile phone, optical parts [17].
Intelligence	Replacing part of human mental work, independent analysis, reasoning, judgment, decision-making.	The dynamic adjustment model realizes the self-adjustment of the system's situational perception [211]. Intelligent monitoring [212,213].
Automation	Replacing part of human physical labor and realizing simple operation.	Quasi-online compensation method improves automatic adjustment of lathe accuracy [214–216]. Adjusting virtual model and controlling the process online [217].
Efficiency	Improving production efficiency and save time.	Efficient production [218].
Information	The collection, combination, input and machining of manufacturing information.	Real-time monitoring and data acquisition of digital twin in grinding machine system [219].
Flexibility	Computer numerical control machine tool-based manufacturing equipment to carry out large quantities, many varieties of production.	Dynamic clamping and positioning method of flexible machining system based on digital twin technology [220].
Integration	Multiple independently operating single modules are integrated into a coordinated and functional system.	Feedback-forward combined adaptive regulator of tool rest platform for active vibration control [221].

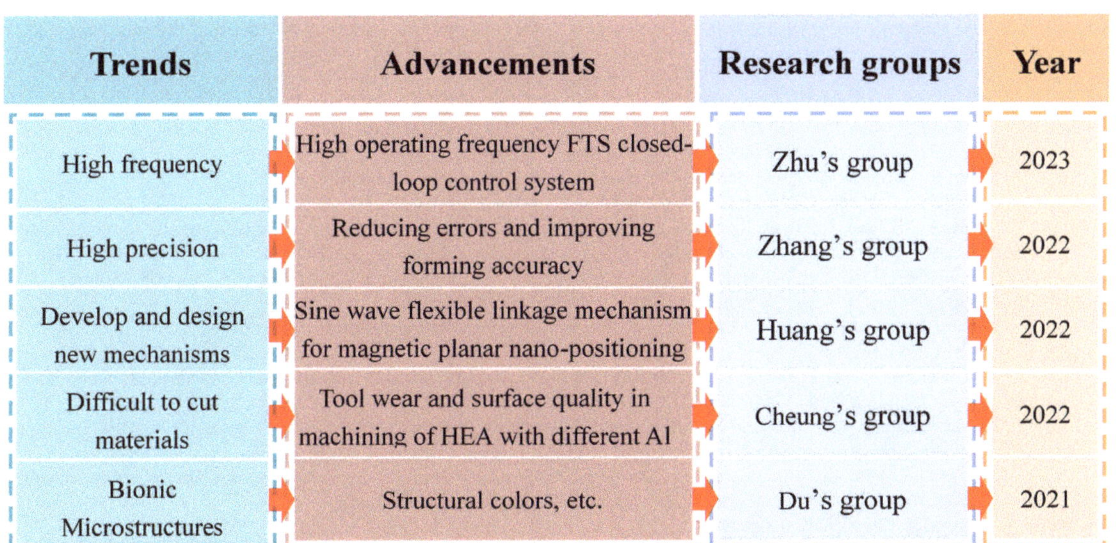

Figure 40. The progress of micromachining trends in machining micro-structures [199,222–225].

Author Contributions: Conceptualization, G.Z. and Y.M.; methodology, S.C.; validation, J.H., S.M. and Z.H. (Zexuan Huo).; resources, G.Z.; data curation, Y.M.; writing—original draft preparation, Y.M.; writing—review and editing, J.H., S.M. and Z.H. (Zejia Huang).; visualization, Y.M.; supervision, Y.M.; funding acquisition, G.Z. All authors have read and agreed to the published version of the manuscript.

Funding: The research proposed in this paper was supported by the National Natural Science Foundation of China (Grant No. 52275454, U2013603, 51827901) and the Shenzhen Natural Science Foundation (Grant No. JCYJ20220531103614032, 20200826160002001, JCYJ20220818102409021).

Data Availability Statement: Not applicable.

Conflicts of Interest: The authors declare no conflict of interest.

References

1. Tan, N.Y.J.; Zhang, X.; Neo, D.W.K.; Huang, R.; Liu, K.; Senthil Kumar, A. A review of recent advances in fabrication of optical Fresnel lenses. *J. Manuf. Process.* **2021**, *71*, 113–133. [CrossRef]
2. Kong, L.B.; Cheung, C.F. Modeling and characterization of surface generation in fast tool servo machining of microlens arrays. *Comput. Ind. Eng.* **2012**, *63*, 957–970. [CrossRef]
3. Liu, K.; Jiang, L. Bio-inspired design of multiscale structures for function integration. *Nano Today* **2011**, *6*, 155–175. [CrossRef]
4. Brinksmeier, E.; Karpuschewski, B.; Yan, J.; Schönemann, L. Manufacturing of multiscale structured surfaces. *CIRP Ann.* **2020**, *69*, 717–739. [CrossRef]
5. Fu, Y.F.; Yuan, C.Q.; Bai, X.Q. Marine drag reduction of shark skin inspired riblet surfaces. *Biosurf. Biotribol.* **2017**, *3*, 11–24. [CrossRef]
6. Wei, H.M.; Gong, H.B.; Chen, L.; Zi, M.; Cao, B.Q. Photovoltaic Efficiency Enhancement of Cu_2O Solar Cells Achieved by Controlling Homojunction Orientation and Surface Microstructure. *J. Phys. Chem. C* **2012**, *116*, 10510–10515. [CrossRef]
7. Zhong, M.; Huang, T. Laser hybrid fabrication of tunable micro- and nano-scale surface structures and their functionalization. In *Laser Surface Engineering*; Lawrence, J., Waugh, D.G., Eds.; Woodhead Publishing: Cambridge, UK, 2015; pp. 213–235, ISBN 978-1-78242-074-3.
8. Neinhuis, C.; Barthlott, W. Characterization and Distribution of Water-repellent, Self-cleaning Plant Surfaces. *Ann. Bot.* **1997**, *79*, 667–677. [CrossRef]
9. Zhao, W.; Wang, L.; Xue, Q. Fabrication of Low and High Adhesion Hydrophobic Au Surfaces with Micro/Nano-Biomimetic Structures. *J. Phys. Chem. C* **2010**, *114*, 11509–11514. [CrossRef]
10. Natarajan, E.; Inácio Freitas, L.; Rui Chang, G.; Abdulaziz Majeed Al-Talib, A.; Hassan, C.S.; Ramesh, S. The hydrodynamic behaviour of biologically inspired bristled shark skin vortex generator in submarine. *Mater. Today Proc.* **2021**, *46*, 3945–3950. [CrossRef]
11. Han, Z.; Fu, J.; Wang, Z.; Wang, Y.; Li, B.; Mu, Z.; Zhang, J.; Niu, S. Long-term durability of superhydrophobic properties of butterfly wing scales after continuous contact with water. *Colloids Surf. A* **2017**, *518*, 139–144. [CrossRef]
12. Gao, X.; Yan, X.; Yao, X.; Xu, L.; Zhang, K.; Zhang, J.; Yang, B.; Jiang, L. The Dry-Style Antifogging Properties of Mosquito Compound Eyes and Artificial Analogues Prepared by Soft Lithography. *Adv. Mater.* **2007**, *19*, 2213–2217. [CrossRef]
13. Malshe, A.P.; Bapat, S.; Rajurkar, K.P.; Haitjema, H. Bio-inspired textures for functional applications. *CIRP Ann.* **2018**, *67*, 627–650. [CrossRef]
14. Jung, H.; Jeong, K.-H. Monolithic polymer microlens arrays with high numerical aperture and high packing density. *ACS Appl. Mater. Interfaces* **2015**, *7*, 2160–2165. [CrossRef] [PubMed]
15. Zhou, W.; Li, R.; Li, M.; Tao, P.; Wang, X.; Dai, S.; Song, B.; Zhang, W.; Lin, C.; Shen, X.; et al. Fabrication of microlens array on chalcogenide glass by wet etching-assisted femtosecond laser direct writing. *Ceram. Int.* **2022**, *48*, 18983–18988. [CrossRef]
16. Zong, W.J.; Li, D.; Sun, T.; Cheng, K. Contact accuracy and orientations affecting the lapped tool sharpness of diamond cutting tools by mechanical lapping. *Diam. Relat. Mater.* **2006**, *15*, 1424–1433. [CrossRef]
17. Fang, F.Z.; Zhang, X.D.; Weckenmann, A.; Zhang, G.X.; Evans, C. Manufacturing and measurement of freeform optics. *CIRP Ann.* **2013**, *62*, 823–846. [CrossRef]
18. Luo, G.; Wu, D.; Zhou, Y.; Hu, Y.; Yao, Z. Laser printing of large-scale metal micro/nanoparticle array: Deposition behavior and microstructure. *Int. J. Mach. Tools Manuf.* **2022**, *173*, 103845. [CrossRef]
19. Lee, L.P.; Berger, S.A.; Liepmann, D.; Pruitt, L. High aspect ratio polymer microstructures and cantilevers for bioMEMS using low energy ion beam and photolithography. *Sens. Actuators A* **1998**, *71*, 144–149. [CrossRef]
20. Zhang, Z.; He, H.; Fu, W.; Ji, D.; Ramakrishna, S. Electro-Hydrodynamic Direct-Writing Technology toward Patterned Ultra-Thin Fibers: Advances, Materials and Applications. *Nano Today* **2020**, *35*, 100942. [CrossRef]
21. Biesuz, M.; Saunders, T.; Ke, D.; Reece, M.J.; Hu, C.; Grasso, S. A review of electromagnetic processing of materials (EPM): Heating, sintering, joining and forming. *J. Mater. Sci. Technol.* **2021**, *69*, 239–272. [CrossRef]
22. Stephens, L.S.; Siripuram, R.; Hayden, M.; McCartt, B. Deterministic Micro Asperities on Bearings and Seals Using a Modified LIGA Process. *J. Eng. Gas Turbines Power* **2004**, *126*, 147–154. [CrossRef]
23. Wang, Y.; Zhao, Q.; Shang, Y.; Lv, P.; Guo, B.; Zhao, L. Ultra-precision machining of Fresnel microstructure on die steel using single crystal diamond tool. *J. Mater. Process. Technol.* **2011**, *211*, 2152–2159. [CrossRef]
24. Wu, L.; Leng, J.; Ju, B. Digital twins-based smart design and control of ultra-precision machining: A Review. *Symmetry* **2021**, *13*, 1717. [CrossRef]
25. Joshi, S.S. Ultra-precision Machining. In *Encyclopedia of Nanotechnology*; Bhushan, B., Ed.; Springer: Dordrecht, The Netherlands, 2016; p. 4253, ISBN 978-94-017-9780-1.
26. Taniguchi, N. Current Status in, and Future Trends of, Ultraprecision Machining and Ultrafine Materials Processing. *CIRP Ann.* **1983**, *32*, 573–582. [CrossRef]
27. Byrne, G.; Dornfeld, D.; Denkena, B. Advancing Cutting Technology. *CIRP Ann.* **2003**, *52*, 483–507. [CrossRef]

28. Wang, Z.; Luo, X.; Sun, J.; Seib, P.; Phuagkhaopong, S.; Chang, W.; Gao, J.; Mir, A.; Cox, A. Investigation of chip formation mechanism in ultra-precision diamond turning of silk fibroin film. *J. Manuf. Process.* **2022**, *74*, 14–27. [CrossRef]
29. Foy, K.; Wei, Z.; Matsumura, T.; Huang, Y. Effect of tilt angle on cutting regime transition in glass micromilling. *Int. J. Mach. Tools Manuf.* **2009**, *49*, 315–324. [CrossRef]
30. Geng, R.; Yang, X.; Xie, Q.; Zhang, W.; Kang, J.; Liang, Y.; Li, R. Ultra-precision diamond turning of ZnSe ceramics: Surface integrity and ductile regime machining mechanism. *Infrared Phys. Technol.* **2021**, *115*, 103706. [CrossRef]
31. Wang, S.; Zhang, Q.; Zhao, Q.; Guo, B. Surface generation and materials removal mechanism in ultra-precision grinding of biconical optics based on slow tool servo with diamond grinding wheels. *J. Manuf. Process.* **2021**, *72*, 1–14. [CrossRef]
32. Zheng, Z.; Huang, K.; Lin, C.; Zhang, J.; Wang, K.; Sun, P.; Xu, J. An analytical force and energy model for ductile-brittle transition in ultra-precision grinding of brittle materials. *Int. J. Mech. Sci.* **2022**, *220*, 107107. [CrossRef]
33. Guo, J.; Goh, M.H.; Wang, P.; Huang, R.; Lee, X.; Wang, B.; Nai, S.M.L.; Wei, J. Investigation on surface integrity of electron beam melted Ti-6Al-4 V by precision grinding and electropolishing. *Chin. J. Aeronaut.* **2021**, *34*, 28–38. [CrossRef]
34. Arnold, T.; Pietag, F. Ion beam figuring machine for ultra-precision silicon spheres correction. *Precis. Eng.* **2015**, *41*, 119–125. [CrossRef]
35. Zhang, S.; Zhou, Y.; Zhang, H.; Xiong, Z.; To, S. Advances in ultra-precision machining of micro-structured functional surfaces and their typical applications. *Int. J. Mach. Tools Manuf.* **2019**, *142*, 16–41. [CrossRef]
36. Zhang, S.J.; To, S.; Wang, S.J.; Zhu, Z.W. A review of surface roughness generation in ultra-precision machining. *Int. J. Mach. Tools Manuf.* **2015**, *91*, 76–95. [CrossRef]
37. Buhmann, M.; Roth, R.; Liebrich, T.; Frick, K.; Carelli, E.; Marxer, M. New positioning procedure for optical probes integrated on ultra-precision diamond turning machines. *CIRP Ann.* **2020**, *69*, 473–476. [CrossRef]
38. Schönemann, L.; Riemer, O. Thermo-mechanical tool setting mechanism for ultra-precision milling with multiple cutting edges. *Precis. Eng.* **2019**, *55*, 171–178. [CrossRef]
39. Huang, X.-F.; Tang, P.; Yang, S.; Feng, J.-Y.; Wan, Z.-P. Investigation of AlN ceramic anisotropic deformation behavior during scratching. *J. Eur. Ceram. Soc.* **2022**, *42*, 2678–2690. [CrossRef]
40. Wang, S.; Zhao, Q.; Yang, X. Surface and subsurface microscopic characteristics in sapphire ultra-precision grinding. *Tribol. Int.* **2022**, *174*, 107710. [CrossRef]
41. Zhang, H.; Zhang, X.; Li, Z.; Wang, P.; Guo, Z. Removing single-point diamond turning marks using form-preserving active fluid jet polishing. *Precis. Eng.* **2022**, *76*, 237–254. [CrossRef]
42. Zhang, S.J.; To, S.; Zhang, G.Q.; Zhu, Z.W. A review of machine-tool vibration and its influence upon surface generation in ultra-precision machining. *Int. J. Mach. Tools Manuf.* **2015**, *91*, 34–42. [CrossRef]
43. Hong, D.; Zeng, W.; Yang, N.; Tang, B.; Liu, Q.-J. The micro-wear mechanism of diamond during diamond tool fly-cutting KDP (KH2PO4) from first principle calculations. *J. Mol. Model.* **2020**, *26*, 284. [CrossRef] [PubMed]
44. Kim, H.-S.; Lee, K.-I.; Lee, K.-M.; Bang, Y.-B. Fabrication of free-form surfaces using a long-stroke fast tool servo and corrective figuring with on-machine measurement. *Int. J. Mach. Tools Manuf.* **2009**, *49*, 991–997. [CrossRef]
45. Nakamura, M.; Mano, I.; Taniguchi, J. Fabrication of micro-lens array with antireflection structure. *Microelectron. Eng.* **2019**, *211*, 29–36. [CrossRef]
46. Zhang, X.; Chen, Z.; Chen, J.; Wang, Z.; Zhu, L. Fabrication of a microlens array featuring a high aspect ratio with a swinging diamond tool. *J. Manuf. Process.* **2022**, *75*, 485–496. [CrossRef]
47. Tang, L.; Zhou, T.; Zhou, J.; Liang, Z.; Wang, X. Research on single point diamond turning of chalcogenide glass aspheric lens. *Procedia CIRP* **2018**, *71*, 293–298. [CrossRef]
48. Meng, S.; Yin, Z.; Guo, Y.; Yao, J.; Chai, N. Ultra-precision machining of polygonal Fresnel lens on roller mold. *Int. J. Adv. Manuf. Technol.* **2020**, *108*, 2445–2452. [CrossRef]
49. Gao, X.; Li, Y.; Kong, Q.C.; Tan, Q.F.; Li, C.J. Uniform concentrating design and mold machining of Fresnel lens for photovoltaic systems. *Int. J. Adv. Manuf. Technol.* **2018**, *96*, 451–460. [CrossRef]
50. Huang, R.; Zhang, X.Q.; Rahman, M.; Kumar, A.S.; Liu, K. Ultra-precision machining of radial Fresnel lens on roller moulds. *Cirp Ann.-Manuf. Technol.* **2015**, *64*, 121–124. [CrossRef]
51. Meng, S.T.; Yin, Z.Q.; Xia, S.B.; Yao, J.H.; Zhang, J.W. Diamond micro-scraping for the fabrication of polygonal Fresnel lens structure array on roller molds. *Int. J. Adv. Manuf. Technol.* **2021**, *116*, 1951–1959. [CrossRef]
52. Cheng, M.N.; Cheung, C.F.; Lee, W.B.; To, S. A Study of Factors Affecting Surface Quality in Ultra-Precision Raster Milling. *Key Eng. Mater.* **2007**, *339*, 400–406. [CrossRef]
53. Xu, R.; Deng, Z.; Yue, Y.; Wang, S.; Li, X.; Ma, Z.; Jiang, Y.; Wang, L.; Du, C.; Jia, H.; et al. Fabrication of large-scale uniform submicron inverted pyramid pit arrays on silicon substrates by laser interference lithography. *Vacuum* **2019**, *165*, 1–6. [CrossRef]
54. Huang, Y.; Li, S.; Zhang, J.; Yang, C.; Liu, W. Research on shape error of microstructure array fabricated by fly cutting. *Optik* **2021**, *241*, 167031. [CrossRef]
55. Yu, S.; Yao, P.; Huang, C.; Chu, D.; Zhu, H.; Zou, B.; Liu, H. On-machine precision truing of ultrathin arc-shaped diamond wheels for grinding aspherical microstructure arrays. *Precis. Eng.* **2022**, *73*, 40–50. [CrossRef]
56. Wang, S.J.; To, S.; Chen, X.; Wang, H.; Xia, H.J. A study of the fabrication of v-groove structure in ultra-precision milling. *Int. J. Comput. Integr. Manuf.* **2014**, *27*, 986–996. [CrossRef]

57. Wacogne, B.; Sadani, Z.; Gharbi, T. Compensation structures for V-grooves connected to square apertures in KOH-etched (100) silicon: Theory, simulation and experimentation. *Sens. Actuators A-Phys.* **2004**, *112*, 328–339. [CrossRef]

58. Zhang, S.J.; To, S.; Zhang, G.Q. Diamond tool wear in ultra-precision machining. *Int. J. Adv. Manuf. Technol.* **2017**, *88*, 613–641. [CrossRef]

59. Mudgal, T.; Dwivedi, P.; Kulkarni, M. *Nano/Micro Structure Fabrication Using Combination of Self Organization, Electron Beam and Photolithography*; Centre For Nano-Sciences, Indian Institute of Technology (IIT): Kanpur, India, 2018.

60. Park, J.M.; Park, S.; Shin, D.S.; Kim, J.; Cho, H.; Yang, W.S.; Son, S.; Park, S.J. Microfabrication of Ni-Fe Mold Insert via Hard X-ray Lithography and Electroforming Process. *Metals* **2020**, *10*, 486. [CrossRef]

61. Finders, J.; de Winter, L.; Last, T. Mitigation of mask three-dimensional induced phase effects by absorber optimization in ArF i and extreme ultraviolet lithography. *J. Micro/Nanolithogr. MEMS MOEMS* **2016**, *15*, 021408. [CrossRef]

62. Chen, Y. Nanofabrication by electron beam lithography and its applications: A review. *Microelectron. Eng.* **2015**, *135*, 57–72. [CrossRef]

63. Andreev, M.; Marmiroli, B.; Schennach, R.; Amenitsch, H. Patterning a cellulose based dual-tone photoresist via deep X-ray lithography. *Microelectron. Eng.* **2022**, *256*, 111720. [CrossRef]

64. Marmiroli, B.; Amenitsch, H. X-ray lithography and small-angle X-ray scattering: A combination of techniques merging biology and materials science. *Eur. Biophys. J.* **2012**, *41*, 851–861. [CrossRef] [PubMed]

65. Watt, F.; Bettiol, A.A.; Van Kan, J.A.; Teo, E.J.; Breese, M.B.H. Ion beam lithography and nanofabrication: A reviw. *Int. J. Nanosci.* **2005**, *4*, 269–286. [CrossRef]

66. Reyntjens, S.; Puers, R. A review of focused ion beam applications in microsystem technology. *J. Micromech. Microeng.* **2001**, *11*, 287–300. [CrossRef]

67. Palka, K.; Kurka, M.; Slang, S.; Vlcek, M. Utilization of As50Se50 thin films in electron beam lithography. *Mater. Chem. Phys.* **2021**, *259*, 124052. [CrossRef]

68. Rius, G.; Llobet, J.; Arcamone, J.; Borrisé, X.; Pérez-Murano, F. Electron- and ion-beam lithography for the fabrication of nanomechanical devices integrated on CMOS circuits. *Microelectron. Eng.* **2009**, *86*, 1046–1049. [CrossRef]

69. Kathuria, H.; Kochhar, J.S.; Fong MH, M.; Hashimoto, M.; Iliescu, C.; Yu, H.; Kang, L. Microneedle Array Fabrication by Photolithography. *JoVE* **2015**, e52914. [CrossRef]

70. Tong, X.; Xu, C.; Zhu, J.; Mao, C.; Zhen, X.; Huang, W.; Chen, Y. A study of greyscale electron beam lithography for a 3D round shape Kinoform lens for hard X-ray optics. *Microelectron. Eng.* **2020**, *234*, 111435. [CrossRef]

71. Langford, R.M.; Nellen, P.M.; Gierak, J.; Fu, Y. Focused Ion Beam Micro- and Nanoengineering. *MRS Bull.* **2007**, *32*, 417–423. [CrossRef]

72. Heusinger, M.; Banasch, M.; Zeitner, U.D. Rowland ghost suppression in high efficiency spectrometer gratings fabricated by e-beam lithography. *Opt. Express* **2017**, *25*, 6182–6191. [CrossRef]

73. Cattoni, A.; Mailly, D.; Dalstein, O.; Faustini, M.; Seniutinas, G.; Rösner, B.; David, C. Sub-10nm electron and helium ion beam lithography using a recently developed alumina resist. *Microelectron. Eng.* **2018**, *193*, 18–22. [CrossRef]

74. Zhu, J.; Chen, Y.; Xie, S.; Zhang, L.; Wang, C.; Tai, R. Nanofabrication of 30 nm Au zone plates by e-beam lithography and pulse voltage electroplating for soft x-ray imaging. *Microelectron. Eng.* **2020**, *225*, 111254. [CrossRef]

75. Chung, J.; Hsu, W. Enhancement on forming complex three dimensional microstructures by a double-side multiple partial exposure method. *J. Vac. Sci. Technol. B* **2007**, *25*, 1671–1678. [CrossRef]

76. Totsu, K.; Fujishiro, K.; Tanaka, S.; Esashi, M. Fabrication of three-dimensional microstructure using maskless gray-scale lithography. *Sens. Actuators A* **2006**, *130*, 387–392. [CrossRef]

77. Nachmias, T.; Ohayon, A.; Meltzer, S.; Kabla, M.; Louzon, E.; Levy, U. Shallow Fresnel lens fabrication using grayscale lithography made by high energy beam sensitive mask (HEBS) technology and reactive ion etching. *SPIE* **2009**, *7205*, 68–79. [CrossRef]

78. Kuo, H.-F. Effect of Source Pupil Shape on Process Windows in EUV Lithography. *IEEE Trans. Nanotechnol.* **2014**, *13*, 136–142. [CrossRef]

79. Chen, H.L.; Chuang, S.Y.; Lin, C.H.; Lin, Y.H. Using colloidal lithography to fabricate and optimize sub-wavelength pyramidal and honeycomb structures in solar cells. *Opt. Express* **2007**, *15*, 14793–14803. [CrossRef]

80. Kosiorek, A.; Kandulski, W.; Glaczynska, H.; Giersig, M. Fabrication of Nanoscale Rings, Dots, and Rods by Combining Shadow Nanosphere Lithography and Annealed Polystyrene Nanosphere Masks. *Small* **2005**, *1*, 439–444. [CrossRef]

81. Vossen, D.L.J.; Fific, D.; Penninkhof, J.; van Dillen, T.; Polman, A.; van Blaaderen, A. Combined Optical Tweezers/Ion Beam Technique to Tune Colloidal Masks for Nanolithography. *Nano Lett.* **2005**, *5*, 1175–1179. [CrossRef]

82. Lee, J.-S.; Yu, J.-H.; Hwang, K.-H.; Nam, S.-H.; Boo, J.-H.; Yun, S.H. Patterning ITO by template-assisted colloidal-lithography for enhancing power conversion efficiency in organic photovoltaic. *J. Nanosci. Nanotechnol.* **2016**, *16*, 5024–5028. [CrossRef]

83. Hongzhong, L.; Yucheng, D.; Weitao, J.; Qin, L.; Lei, Y.; Yongsheng, S.; Bingheng, L. Novel imprint lithography process used in fabrication of micro/nanostructures in organic photovoltaic devices. *J. Micro/Nanolithography MEMS MOEMS* **2009**, *8*, 021170. [CrossRef]

84. Sanchez-Sobrado, O.; Mendes, M.J.; Haque, S.; Mateus, T.; Aguas, H.; Fortunato, E.; Martins, R. Lightwave trapping in thin film solar cells with improved photonic-structured front contacts. *J. Mater. Chem. C* **2019**, *7*, 6456–6464. [CrossRef]

85. Centeno, P.; Alexandre, M.F.; Chapa, M.; Pinto, J.V.; Deuermeier, J.; Mateus, T.; Fortunato, E.; Martins, R.; Águas, H.; Mendes, M.J. Self-Cleaned Photonic-Enhanced Solar Cells with Nanostructured Parylene-C. *Adv. Mater. Interfaces* **2020**, *7*, 2000264. [CrossRef]

86. Chen, J.; Liu, Z.; Su, F.; Wu, S.; Liang, C.; Liu, J. Surface modification of carbon fiber cloth by femtosecond laser direct writing technology. *Mater. Lett.* **2022**, *323*, 132483. [CrossRef]
87. Zhao, J.; Zhao, Y.; Peng, Y.; Lv, R.-q.; Zhao, Q. Review of femtosecond laser direct writing fiber-optic structures based on refractive index modification and their applications. *Opt. Laser Technol.* **2022**, *146*, 107473. [CrossRef]
88. Zhou, W.; Li, R.; Qi, Q.; Yang, Y.; Wang, X.; Dai, S.; Song, B.; Xu, T.; Zhang, P. Fabrication of Fresnel zone plate in chalcogenide glass and fiber end with femtosecond laser direct writing. *Infrared Phys. Technol.* **2022**, *120*, 104004. [CrossRef]
89. Xu, Z.; Fang, F.; Gao, H.; Zhu, Y.; Wu, W.; Weckenmann, A. Nano fabrication of star structure for precision metrology developed by focused ion beam direct writing. *CIRP Ann.* **2012**, *61*, 511–514. [CrossRef]
90. Mizoshiri, M.; Hayashi, T.; Narushima, J.; Ohishi, T. Femtosecond laser direct writing of Cu–Ni alloy patterns in ambient atmosphere using glyoxylic acid Cu/Ni mixed complexes. *Opt. Laser Technol.* **2021**, *144*, 107418. [CrossRef]
91. Egashira, K.; Kumagai, R.; Okina, R.; Yamaguchi, K.; Ota, M. Drilling of microholes down to 10μm in diameter using ultrasonic grinding. *Precis. Eng.* **2014**, *38*, 605–610. [CrossRef]
92. Banerjee, B.; Mondal, K.; Adhikary, S.; Nath Paul, S.; Pramanik, S.; Chatterjee, S. Optimization of process parameters in ultrasonic machining using integrated AHP-TOPSIS method. *Mater. Today Proc.* **2022**, *62*, 2857–2864. [CrossRef]
93. Airao, J.; Nirala, C.K. Machinability of Ti-6Al-4V and Nimonic-90 in ultrasonic-assisted turning under sustainable cutting fluid. *Mater. Today Proc.* **2022**, *62*, 7396–7400. [CrossRef]
94. Chen, F.; Bie, W.; Wang, X.; Zhao, B. Longitudinal-torsional coupled rotary ultrasonic machining of ZrO_2 ceramics: An experimental study. *Ceram. Int.* **2022**, *48*, 28154–28162. [CrossRef]
95. Chen, Y.; Hu, Z.; Yu, Y.; Lai, Z.; Zhu, J.; Xu, X.; Peng, Q. Processing and machining mechanism of ultrasonic vibration-assisted grinding on sapphire. *Mater. Sci. Semicond. Process.* **2022**, *142*, 106470. [CrossRef]
96. Sapkota, G.; Das, S.; Sharma, A.; Kumar Ghadai, R. Comparison of various multi-criteria decision methods for the selection of quality hole produced by ultrasonic machining process. *Mater. Today Proc.* **2022**, *58*, 702–708. [CrossRef]
97. Sabareesan, S.; Vasudevan, D.; Sridhar, S.; Kannan, R.; Sankar, V. Response analysis of ultrasonic machining process under different materials—Review. *Mater. Today Proc.* **2021**, *45*, 2340–2342. [CrossRef]
98. Juri, A.Z.; Nakanishi, Y.; Yin, L. Microstructural influence on damage-induced zirconia surface asperities produced by conventional and ultrasonic vibration-assisted diamond machining. *Ceram. Int.* **2021**, *47*, 25744–25754. [CrossRef]
99. Gao, H.; Ma, B.; Zhu, Y.; Yang, H. Enhancement of machinability and surface quality of Ti-6Al-4V by longitudinal ultrasonic vibration-assisted milling under dry conditions. *Measurement* **2022**, *187*, 110324. [CrossRef]
100. Xu, S.; Shimada, K.; Mizutani, M.; Kuriyagawa, T. Fabrication of hybrid micro/nano-textured surfaces using rotary ultrasonic machining with one-point diamond tool. *Int. J. Mach. Tools Manuf.* **2014**, *86*, 12–17. [CrossRef]
101. Ni, C.; Zhu, L. Investigation on machining characteristics of TC4 alloy by simultaneous application of ultrasonic vibration assisted milling (UVAM) and economical-environmental MQL technology. *J. Mater. Process. Technol.* **2020**, *278*, 116518. [CrossRef]
102. Liu, S.-J.; Chen, Y.-S. The manufacturing of thermoplastic composite parts by water-assisted injection-molding technology. *Compos. Pt. A-Appl. Sci. Manuf.* **2004**, *35*, 171–180. [CrossRef]
103. Michaeli, W.; Spennemann, A. A new injection molding technology for micro parts. *J. Polym. Eng.* **2001**, *21*, 87–98. [CrossRef]
104. Tsai, S.-W.; Chen, P.-Y.; Lee, Y.-C. Fabrication of a seamless roller mold with wavy microstructures using mask-less curved surface beam pen lithography. *J. Micromech. Microeng.* **2014**, *24*, 045022. [CrossRef]
105. Guo, F.; Zhou, X.; Liu, J.; Zhang, Y.; Li, D.; Zhou, H. A reinforcement learning decision model for online process parameters optimization from offline data in injection molding. *Appl. Soft Comput.* **2019**, *85*, 105828. [CrossRef]
106. Wang, X.; Gu, J.; Shen, C.; Wang, X. Warpage optimization with dynamic injection molding technology and sequential optimization method. *Int. J. Adv. Manuf. Technol.* **2015**, *78*, 177–187. [CrossRef]
107. Willke, T.; Gearhart, S. LIGA micromachined planar transmission lines and filters. *IEEE Trans. Microw. Theory Tech.* **1997**, *45*, 1681–1688. [CrossRef]
108. Us Sarwar, M.S.; Dahmardeh, M.; Nojeh, A.; Takahata, K. Batch-mode micropatterning of carbon nanotube forests using UV-LIGA assisted micro-electro-discharge machining. *J. Mater. Process. Technol.* **2014**, *214*, 2537–2544. [CrossRef]
109. Huang, M.-S.; Li, C.-J.; Yu, J.-C.; Huang, Y.-M.; Hsieh, L.-C. Robust parameter design of micro-injection molded gears using a LIGA-like fabricated mold insert. *J. Mater. Process. Technol.* **2009**, *209*, 5690–5701. [CrossRef]
110. Luo, S.Y.; Yu, T.H.; Liu, C.Y.; Chen, M.H. Grinding characteristics of micro-abrasive pellet tools fabricated by a LIGA-like process. *Int. J. Mach. Tools Manuf.* **2009**, *49*, 212–219. [CrossRef]
111. Malek, C.K.; Saile, V. Applications of LIGA technology to precision manufacturing of high-aspect-ratio micro-components and -systems: A review. *Microelectron. J.* **2004**, *35*, 131–143. [CrossRef]
112. Ma, Y.; Liu, W.; Liu, C. Research on the process of fabricating a multi-layer metal micro-structure based on UV-LIGA overlay technology. *Nanotechnol. Precis. Eng.* **2019**, *2*, 83–88. [CrossRef]
113. Zong, W.J.; Li, Z.Q.; Sun, T.; Cheng, K.; Li, D.; Dong, S. The basic issues in design and fabrication of diamond-cutting tools for ultra-precision and nanometric machining. *Int. J. Mach. Tools Manuf.* **2010**, *50*, 411–419. [CrossRef]
114. Yuan, Z.J.; Yao, Y.X.; Zhou, M.; Bal, Q.S. Lapping of Single Crystal Diamond Tools. *CIRP Ann.* **2003**, *52*, 285–288. [CrossRef]
115. Fujii, S.; Hayama, Y.; Imamura, K.; Kumazaki, H.; Kakinuma, Y.; Tanabe, T. All-precision-machining fabrication of ultrahigh-Q crystalline optical microresonators. *Optica* **2020**, *7*, 694–701. [CrossRef]

116. Vu, V.T.; Ui Hasan, S.A.; Youn, H.; Park, Y.; Lee, H. Imaging performance of an ultra-precision machining-based Fresnel lens in ophthalmic devices. *Opt. Express* **2021**, *29*, 32068–32080. [CrossRef] [PubMed]
117. Abou-El-Hossein, K.; Olufayo, O.; Mkoko, Z. Diamond tool wear during ultra-high precision machining of rapidly solidified aluminium RSA 905. *Wear* **2013**, *302*, 1105–1112. [CrossRef]
118. Peng, J.-W.; Zhang, F.-L.; Huang, Y.-J.; Liu, J.-M.; Li, K.-J.; Zhou, Y.-M.; Long, Y.; Wang, C.-Y.; Wu, S.-H.; Tang, H.-Q. Comparative study on NiAl and FeAl intermetallic-bonded diamond tools and grinding performance for Si_3N_4 ceramic. *Ceram. Int.* **2021**, *47*, 32736–32746. [CrossRef]
119. Sun, Z.; To, S.; Yu, K.M. Feasibility investigation on ductile machining of single-crystal silicon for deep micro-structures by ultra-precision fly cutting. *J. Manuf. Process.* **2019**, *45*, 176–187. [CrossRef]
120. Sun, Z.; To, S.; Zhang, S. A novel ductile machining model of single-crystal silicon for freeform surfaces with large azimuthal height variation by ultra-precision fly cutting. *Int. J. Mach. Tools Manuf.* **2018**, *135*, 1–11. [CrossRef]
121. Yuetian, H. Research on Single Point Diamond Turning Technology for Complex Surface. Ph.D. Thesis, University of Chinese Academy of Sciences, Beijing, China, 2019.
122. Tanikawa, S.; Yan, J. Fabrication of micro-structured surface with controllable randomness by using FTS-based diamond turning. *Precis. Eng.* **2022**, *73*, 363–376. [CrossRef]
123. Zhang, Z.; Wang, B.; Kang, R.; Zhang, B.; Guo, D. Changes in surface layer of silicon wafers from diamond scratching. *CIRP Ann.* **2015**, *64*, 349–352. [CrossRef]
124. Moriya, T.; Nakamoto, K.; Ishida, T.; Takeuchi, Y. Creation of V-shaped microgrooves with flat-ends by 6-axis control ultraprecision machining. *CIRP Ann.* **2010**, *59*, 61–66. [CrossRef]
125. Takeuchi, Y.; Maeda, S.; Kawai, T.; Sawada, K. Manufacture of Multiple-focus Micro Fresnel Lenses by Means of Nonrotational Diamond Grooving. *CIRP Ann.* **2002**, *51*, 343–346. [CrossRef]
126. Takeuchi, Y.; Yoneyama, Y.; Ishida, T.; Kawai, T. 6-Axis control ultraprecision microgrooving on sculptured surfaces with non-rotational cutting tool. *CIRP Ann.* **2009**, *58*, 53–56. [CrossRef]
127. Owen, J.D.; Shultz, J.A.; Suleski, T.J.; Davies, M.A. Error correction methodology for ultra-precision three-axis milling of freeform optics. *CIRP Ann.* **2017**, *66*, 97–100. [CrossRef]
128. Jiang, J.; Luo, T.; Zhang, G.; Dai, Y. Novel tool offset fly cutting straight-groove-type micro structure arrays. *J. Mater. Process. Technol.* **2021**, *288*, 116900. [CrossRef]
129. Xu, P.; Cheung, C.-F.; Li, B.; Ho, L.-T.; Zhang, J.-F. Kinematics analysis of a hybrid manipulator for computer controlled ultra-precision freeform polishing. *Rob. Comput. Integr. Manuf.* **2017**, *44*, 44–56. [CrossRef]
130. Meng, X.; Wu, W.; Liao, B.; Dai, H. Atomic simulation of textured silicon carbide surface ultra-precision polishing. *Ceram. Int.* **2022**, *48*, 17034–17045. [CrossRef]
131. Wang, X.; Gao, H.; Yuan, J. Experimental investigation and analytical modelling of the tool influence function of the ultra-precision numerical control polishing method based on the water dissolution principle for KDP crystals. *Precis. Eng.* **2020**, *65*, 185–196. [CrossRef]
132. Cheung, C.F.; Kong, L.B.; Ho, L.T.; To, S. Modelling and simulation of structure surface generation using computer controlled ultra-precision polishing. *Precis. Eng.* **2011**, *35*, 574–590. [CrossRef]
133. Maillard, D.; Benes, Z.; Piacentini, N.; Villanueva, L.G. Electron-beam lithography on M108Y and M35G chemically amplified DUV photoresists. *Micro Nano Eng.* **2021**, *13*, 100095. [CrossRef]
134. Senzaki, S.; Okabe, T.; Taniguchi, J. Fabrication of bifocal lenses using resin that can be processed by electron beam lithography after ultraviolet-nanoimprint lithography. *Microelectron. Eng.* **2022**, *258*, 111776. [CrossRef]
135. Janer, M.; López, T.; Plantà, X.; Riera, D. Ultrasonic nodal point, a new configuration for ultrasonic moulding technology. *Ultrasonics* **2021**, *114*, 106418. [CrossRef]
136. Song-Jo, C.; Herbert, H.; Juergen, M.; Franz Josef, P.; Joachim, S.; Ulrike, W. LIGA technology today and its industrial applications. *Proc.SPIE* **2000**, *4194*, 44–55. [CrossRef]
137. Ehrfeld, W.; Lehr, H. Synchrotron radiation and the LIGA technique. *Synchrotron Radiat. News* **1994**, *7*, 9–13. [CrossRef]
138. Yang, J.; Roa, J.J.; Odén, M.; Johansson-Jõesaar, M.P.; Llanes, L. 3D FIB/FESEM tomography of grinding-induced damage in WC-Co cemented carbides. *Procedia CIRP* **2020**, *87*, 385–390. [CrossRef]
139. Rubanov, S.; Suvorova, A.; Popov, V.P.; Kalinin, A.A.; Pal'yanov, Y.N. Fabrication of graphitic layers in diamond using FIB implantation and high pressure high temperature annealing. *Diam. Relat. Mater.* **2016**, *63*, 143–147. [CrossRef]
140. Tong, Z.; Xu, Z.; Wu, W.; Luo, X. Molecular dynamic simulation of low-energy FIB irradiation induced damage in diamond. *Nucl. Instrum. Methods Phys. Res. Sect. B* **2015**, *358*, 38–44. [CrossRef]
141. Wu, R.; Tauhiduzzaman, M.; Ravi Selvaganapathy, P. Anisotropic wetting surfaces machined by diamond tool with tips microstructured by focused ion beam. *Mater. Des.* **2021**, *210*, 110014. [CrossRef]
142. Tong, Z.; Luo, X. Investigation of focused ion beam induced damage in single crystal diamond tools. *Appl. Surf. Sci.* **2015**, *347*, 727–735. [CrossRef]
143. Ono, T. Sharpening of the diamond tool edge by the Ar ion beam machine tool. *Procedia Manuf.* **2021**, *53*, 246–250. [CrossRef]
144. Wu, W.; Xu, Z.; Fang, F.; Liu, B.; Xiao, Y.; Chen, J.; Wang, X.; Liu, H. Decrease of FIB-induced lateral damage for diamond tool used in nano cutting. *Nucl. Instrum. Methods Phys. Res. Sect. B* **2014**, *330*, 91–98. [CrossRef]

145. Wang, Y.; Fan, P.; Luo, X.; Geng, Y.; Goel, S.; Wu, W.; Li, G.; Yan, Y. Fabrication of three-dimensional sin-shaped ripples using a multi-tip diamond tool based on the force modulation approach. *J. Manuf. Process.* **2021**, *72*, 262–273. [CrossRef]
146. Kawasegi, N.; Ozaki, K.; Morita, N.; Nishimura, K.; Yamaguchi, M. Development and machining performance of a textured diamond cutting tool fabricated with a focused ion beam and heat treatment. *Precis. Eng.* **2017**, *47*, 311–320. [CrossRef]
147. Kawasegi, N.; Ozaki, K.; Morita, N.; Nishimura, K.; Sasaoka, H. Single-crystal diamond tools formed using a focused ion beam: Tool life enhancement via heat treatment. *Diam. Relat. Mater.* **2014**, *49*, 14–18. [CrossRef]
148. Kawasegi, N.; Niwata, T.; Morita, N.; Nishimura, K.; Sasaoka, H. Improving machining performance of single-crystal diamond tools irradiated by a focused ion beam. *Precis. Eng.* **2014**, *38*, 174–182. [CrossRef]
149. Gong, Z.; Huo, D.; Niu, Z.; Chen, W.; Shyha, I. Investigation of control algorithm for long-stroke fast tool servo system. *Precis. Eng.* **2022**, *75*, 12–23. [CrossRef]
150. Tong, Z.; Zhong, W.; To, S.; Zeng, W. Fast-tool-servo micro-grooving freeform surfaces with embedded metrology. *CIRP Ann.* **2020**, *69*, 505–508. [CrossRef]
151. Zhong, W.; Tong, Z.; Jiang, X. Integration of On-machine Surface Measurement into Fast Tool Servo Machining. *Procedia CIRP* **2021**, *101*, 238–241. [CrossRef]
152. Gong, Z.; Huo, D.; Niu, Z.; Chen, W.; Cheng, K. A novel long-stroke fast tool servo system with counterbalance and its application to the ultra-precision machining of microstructured surfaces. *Mech. Syst. Sig. Process.* **2022**, *173*, 109063. [CrossRef]
153. Chang, K.-M.; Cheng, J.-L.; Liu, Y.-T. Machining control of non-axisymmetric aspheric surface based on piezoelectric fast tool servo system. *Precis. Eng.* **2022**, *76*, 160–172. [CrossRef]
154. Zhao, D.; Zhu, Z.; Huang, P.; Guo, P.; Zhu, L.; Zhu, Z. Development of a piezoelectrically actuated dual-stage fast tool servo. *Mech. Syst. Sig. Process.* **2020**, *144*, 106873. [CrossRef]
155. Wang, H.; Yang, S. Design and control of a fast tool servo used in noncircular piston turning process. *Mech. Syst. Sig. Process.* **2013**, *36*, 87–94. [CrossRef]
156. Karunanidhi, S.; Singaperumal, M. Design, analysis and simulation of magnetostrictive actuator and its application to high dynamic servo valve. *Sens. Actuators A* **2010**, *157*, 185–197. [CrossRef]
157. Xiaobo, T.; Baras, J.S. Modeling and control of a magnetostrictive actuator. In Proceedings of the 41st IEEE Conference on Decision and Control, Las Vegas, NV, USA, 10–13 December 2002; Volume 1, pp. 866–872. [CrossRef]
158. Park, G.; Bement, M.T.; Hartman, D.A.; Smith, R.E.; Farrar, C.R. The use of active materials for machining processes: A review. *Int. J. Mach. Tools Manuf.* **2007**, *47*, 2189–2206. [CrossRef]
159. Yoshioka, H.; Kojima, K.; Toyota, D. Micro patterning on curved surface with a fast tool servo system for micro milling process. *CIRP Ann.* **2020**, *69*, 325–328. [CrossRef]
160. Chen, Y.-L.; Tao, Y.; Hu, P.; Wu, L.; Ju, B.-F. Self-sensing of cutting forces in diamond cutting by utilizing a voice coil motor-driven fast tool servo. *Precis. Eng.* **2021**, *71*, 178–186. [CrossRef]
161. Ding, F.; Luo, X.; Cai, Y.; Chang, W. Acceleration feedback control for enhancing dynamic stiffness of fast tool servo system considering the sensor imperfections. *Mech. Syst. Sig. Process.* **2020**, *141*, 106429. [CrossRef]
162. Rakuff, S.; Cuttino, J.F. Design and testing of a long-range, precision fast tool servo system for diamond turning. *Precis. Eng.* **2009**, *33*, 18–25. [CrossRef]
163. Tao, Y.; Chen, Y.-L.; Hu, P.; Ju, B.-F.; Du, H. Development of a voice coil motor based fast tool servo with a function of self-sensing of cutting forces. *Precis. Eng.* **2020**, *65*, 130–137. [CrossRef]
164. Liu, Q.; Zhou, X.; Liu, Z.; Lin, C.; Ma, L. Long-stroke fast tool servo and a tool setting method for freeform optics fabrication. *Opt. Eng.* **2014**, *53*, 092005. [CrossRef]
165. Tian, F.; Yin, Z.; Li, S. A novel long range fast tool servo for diamond turning. *Int. J. Adv. Manuf. Technol.* **2016**, *86*, 1227–1234. [CrossRef]
166. Nie, Y.H.; Fang, F.Z.; Zhang, X.D. System design of Maxwell force driving fast tool servos based on model analysis. *Int. J. Adv. Manuf. Technol.* **2014**, *72*, 25–32. [CrossRef]
167. Zhu, Z.; Zhou, X.; Liu, Z.; Wang, R.; Zhu, L. Development of a piezoelectrically actuated two-degree-of-freedom fast tool servo with decoupled motions for micro-/nanomachining. *Precis. Eng.* **2014**, *38*, 809–820. [CrossRef]
168. Liu, Y.; Zheng, Y.; Gu, Y.; Lin, J.; Lu, M.; Xu, Z.; Fu, B. Development of Piezo-Actuated Two-Degree-of-Freedom Fast Tool Servo System. *Micromachines* **2019**, *10*, 337. [CrossRef] [PubMed]
169. Han, J.; Lin, J.; Li, Z.; Lu, M.; Zhang, J. Design and Computational Optimization of Elliptical Vibration-Assisted Cutting System With a Novel Flexure Structure. *IEEE Trans. Ind. Electron.* **2019**, *66*, 1151–1161. [CrossRef]
170. Awtar, S.; Ustick, J.; Sen, S. An XYZ Parallel-Kinematic Flexure Mechanism With Geometrically Decoupled Degrees of Freedom. *J. Mech. Rob.* **2012**, *5*, 015001. [CrossRef]
171. Li, H.; Tang, H.; Li, J.; Chen, X. Design, fabrication, and testing of a 3-DOF piezo fast tool servo for microstructure machining. *Precis. Eng.* **2021**, *72*, 756–768. [CrossRef]
172. Yin, Z.Q.; Dai, Y.F.; Li, S.Y.; Guan, C.L.; Tie, G.P. Fabrication of off-axis aspheric surfaces using a slow tool servo. *Int. J. Mach. Tools Manuf.* **2011**, *51*, 404–410. [CrossRef]
173. Mishra, V.; Kumar, N.; Sharma, R.; Garg, H.; Karara, V. Development of Aspheric Lenslet Array by Slow Tool Servo Machining. *Mater. Today Proc.* **2020**, *24*, 1602–1607. [CrossRef]

174. Huang, W.; Yu, D.; Chen, D.; Zhang, M.; Liu, J.; Yao, J. Investigation of variable spindle speed in slow tool servo-based turning of noncircular optical components. *Proc. SPIE* **2016**, *9683*, 17–25. [CrossRef]
175. Kong, L.B.; Cheung, C.F.; Lee, W.B. A theoretical and experimental investigation of orthogonal slow tool servo machining of wavy microstructured patterns on precision rollers. *Precis. Eng.* **2016**, *43*, 315–327. [CrossRef]
176. Zhang, X.; Gao, H.; Guo, Y.; Zhang, G. Machining of optical freeform prisms by rotating tools turning. *CIRP Ann.* **2012**, *61*, 519–522. [CrossRef]
177. Li, D.; Qiao, Z.; Walton, K.; Liu, Y.; Xue, J.; Wang, B.; Jiang, X. Theoretical and Experimental Investigation of Surface Topography Generation in Slow Tool Servo Ultra-Precision Machining of Freeform Surfaces. *Materials* **2018**, *11*, 2566. [CrossRef] [PubMed]
178. Huang, P.; Wu, X.; To, S.; Zhu, L.; Zhu, Z. Deterioration of form accuracy induced by servo dynamics errors and real-time compensation for slow tool servo diamond turning of complex-shaped optics. *Int. J. Mach. Tools Manuf.* **2020**, *154*, 103556. [CrossRef]
179. Chen, C.C.; Cheng, Y.C.; Hsu, W.Y.; Chou, H.Y.; Wang, P.J.; Tsai, D.P. Slow tool servo diamond turning of optical freeform surface for astigmatic contact lens. *Proc. SPIE Opt. Manuf. Test. IX* **2011**, *8126*, 358–366. [CrossRef]
180. Nagayama, K.; Yan, J. Deterministic error compensation for slow tool servo-driven diamond turning of freeform surface with nanometric form accuracy. *J. Manuf. Process.* **2021**, *64*, 45–57. [CrossRef]
181. Yu, S.; Wang, Y.; Yao, P.; Wang, H.J.; Liu, Z.L.; Jin, X.Y.; Liu, C.H.; Bao, X.Y.; Chu, D.K.; Qu, S.S.; et al. A novel machining approach of freeform multi-mirror mold via normal swing cutting. *J. Manuf. Process.* **2023**, *94*, 316–327. [CrossRef]
182. Zhang, G.; Ran, J.; To, S.; Wu, X.; Huang, P.; Kuz'min, M.P. Size effect on surface generation of multiphase alloys in ultra-precision fly cutting. *J. Manuf. Process.* **2020**, *60*, 23–36. [CrossRef]
183. Zhu, Z.; To, S.; Zhang, S. Theoretical and experimental investigation on the novel end-fly-cutting-servo diamond machining of hierarchical micro-nanostructures. *Int. J. Mach. Tools Manuf.* **2015**, *94*, 15–25. [CrossRef]
184. To, S.; Zhu, Z.; Zeng, W. Novel end-fly-cutting-servo system for deterministic generation of hierarchical micro–nanostructures. *CIRP Ann.* **2015**, *64*, 133–136. [CrossRef]
185. Cheung, C.F.; Kong, L.B.; Lee, W.B.; To, S. Modelling and simulation of freeform surface generation in ultra-precision raster milling. *Proc. Inst. Mech. Eng. Part B J. Eng. Manuf.* **2006**, *220*, 1787–1801. [CrossRef]
186. Zhang, S.J.; To, S. A theoretical and experimental study of surface generation under spindle vibration in ultra-precision raster milling. *Int. J. Mach. Tools Manuf.* **2013**, *75*, 36–45. [CrossRef]
187. Cheng, M.N.; Cheung, C.F.; Lee, W.B.; To, S.; Kong, L.B. Theoretical and experimental analysis of nano-surface generation in ultra-precision raster milling. *Int. J. Mach. Tools Manuf.* **2008**, *48*, 1090–1102. [CrossRef]
188. Zhang, S.J.; To, S.; Zhu, Z.W.; Zhang, G.Q. A review of fly cutting applied to surface generation in ultra-precision machining. *Int. J. Mach. Tools Manuf.* **2016**, *103*, 13–27. [CrossRef]
189. Zhu, Z.; To, S.; Zhang, S. Active control of residual tool marks for freeform optics functionalization by novel biaxial servo assisted fly cutting. *Optica* **2015**, *54*, 7656–7662. [CrossRef]
190. Zhu, Z.; To, S.; Zhang, S. Large-scale fabrication of micro-lens array by novel end-fly-cutting-servo diamond machining. *Opt. Express* **2015**, *23*, 20593–20604. [CrossRef]
191. Zhu, W.-L.; Yang, S.; Ju, B.-F.; Jiang, J.; Sun, A. Scanning tunneling microscopy-based on-machine measurement for diamond fly cutting of micro-structured surfaces. *Precis. Eng.* **2016**, *43*, 308–314. [CrossRef]
192. Dong, X.; Zhou, T.; Pang, S.; Liang, Z.; Yu, Q.; Ruan, B.; Wang, X. Comparison of fly cutting and transverse planing for micropyramid array machining on nickel phosphorus plating. *Int. J. Adv. Manuf. Technol.* **2019**, *102*, 2481–2489. [CrossRef]
193. He, Y.; Zhou, T.; Dong, X.; Liu, P.; Zhao, W.; Wang, X.; Hu, Y.; Yan, J. Generation of high-saturation two-level iridescent structures by vibration-assisted fly cutting. *Mater. Des.* **2020**, *193*, 108839. [CrossRef]
194. Zhang, F.H.; Wang, S.F.; An, C.H.; Wang, J.; Xu, Q. Full-band error control and crack-free surface fabrication techniques for ultra-precision fly cutting of large-aperture KDP crystals. *Front. Mech. Eng.* **2017**, *12*, 193–202. [CrossRef]
195. Fu, P.; Xue, J.; Zhou, L.; Wang, Y.; Lan, Z.; Zhan, L.; Zhang, F. Influence of the heat deformation of ultra-precision fly cutting tools on KDP crystal surface microstructure. *Int. J. Adv. Manuf. Technol.* **2019**, *103*, 1009–1018. [CrossRef]
196. Wang, M.; Zhang, Y.; Guo, S.; An, C.; Zhang, F. Understanding the high-point formation mechanism on large optical surface during ultra precision fly cutting process. *Int. J. Adv. Manuf. Technol.* **2018**, *95*, 2521–2528. [CrossRef]
197. Liang, Y.; Chen, W.; Sun, Y.; Luo, X.; Lu, L.; Liu, H. A mechanical structure-based design method and its implementation on a fly-cutting machine tool design. *Int. J. Adv. Manuf. Technol.* **2014**, *70*, 1915–1921. [CrossRef]
198. Lu, H.; Ding, Y.; Chang, Y.; Chen, G.; Rui, X. Dynamics Modelling and Simulating of Ultra-precision Fly-Cutting Machine Tool. *Int. J. Precis. Eng. Manuf.* **2020**, *21*, 189–202. [CrossRef]
199. Du, H.; Yin, T.; Yip, W.S.; Zhu, Z.; To, S. Generation of structural colors on pure magnesium surface using the vibration-assisted diamond cutting. *Mater. Lett.* **2021**, *299*, 130041. [CrossRef]
200. Zhang, X.; Huang, R.; Wang, Y.; Liu, K.; Deng, H.; Neo, D.W.K. Suppression of diamond tool wear with sub-millisecond oxidation in ultrasonic vibration cutting of steel. *J. Mater. Process. Technol.* **2022**, *299*, 117320. [CrossRef]
201. Zhu, Z.; To, S.; Xiao, G.; Ehmann, K.F.; Zhang, G. Rotary spatial vibration-assisted diamond cutting of brittle materials. *Precis. Eng.* **2016**, *44*, 211–219. [CrossRef]
202. Zhu, Z.; To, S.; Ehmann, K.F.; Xiao, G.; Zhu, W. A novel diamond micro-/nano-machining process for the generation of hierarchical micro-/nano-structures. *J. Micromech. Microeng.* **2016**, *26*, 035009. [CrossRef]

203. Kong, L.B.; Cheung, C.F.; To, S.; Lee, W.B. An investigation into surface generation in ultra-precision raster milling. *J. Mater. Process. Technol.* **2009**, *209*, 4178–4185. [CrossRef]
204. Cheng, M.N.; Cheung, C.F.; Lee, W.B.; To, S. Optimization of Surface Finish in Ultra-precision Raster Milling (Ultra-precision machining). In Proceedings of the International Conference on Leading Edge Manufacturing in 21st Century: LEM21 2005.3, Nagoya, Japan, 19–22 October 2005; The Japan Society of Mechanical Engineers: Tokyo, Japan, 2005; pp. 1019–1024. [CrossRef]
205. Kong, L.B.; Cheung, C.F.; To, S.; Lee, W.B. Development of a Dynamic Model for Ultra-Precision Raster Milling. *Key Eng. Mater.* **2008**, *364–366*, 58–63. [CrossRef]
206. Zhang, S.J.; To, S. The effects of spindle vibration on surface generation in ultra-precision raster milling. *Int. J. Mach. Tools Manuf.* **2013**, *71*, 52–56. [CrossRef]
207. Cai, H.; Shi, G. Ultra-Precision Machining for Micro Structural Functional Surfaces. In Proceedings of the 2018 3rd International Conference on Mechanical, Control and Computer Engineering (ICMCCE), Huhhot, China, 14–16 September 2018; pp. 277–280. [CrossRef]
208. Peng, Y.; Jiang, T.; Ehmann, K.F. Research on single-point diamond fly-grooving of brittle materials. *Int. J. Adv. Manuf. Technol.* **2014**, *75*, 1577–1586. [CrossRef]
209. Bao, Y.-J.; Hao, W.; Wang, Y.-Q.; Gao, H.; Liu, X.-S. Formation mechanism of burr defect in aramid fiber composites based on fly-cutting test. *Int. J. Adv. Manuf. Technol.* **2019**, *104*, 1531–1540. [CrossRef]
210. Zhang, G.; To, S.; Zhang, S.; Zhu, Z. Case study of surface micro-waves in ultra-precision raster fly cutting. *Precis. Eng.* **2016**, *46*, 393–398. [CrossRef]
211. Zhao, R.; Yan, D.; Liu, Q.; Leng, J.; Wan, J.; Chen, X.; Zhang, X. Digital Twin-Driven Cyber-Physical System for Autonomously Controlling of Micro Punching System. *IEEE Access* **2019**, *7*, 9459–9469. [CrossRef]
212. Selvaraj, V.; Xu, Z.; Min, S. Intelligent Operation Monitoring of an Ultra-Precision CNC Machine Tool Using Energy Data. *Int. J. Precis Eng Manuf.-Green Technol.* **2022**, *10*, 59–69. [CrossRef]
213. Cai, W.; Zhang, W.; Hu, X.; Liu, Y. A hybrid information model based on long short-term memory network for tool condition monitoring. *J. Intell. Manuf.* **2020**, *31*, 1497–1510. [CrossRef]
214. Li, D.; Wang, B.; Tong, Z.; Blunt, L.; Jiang, X. On-machine surface measurement and applications for ultra-precision machining: A state-of-the-art review. *Int. J. Adv. Manuf. Technol.* **2019**, *104*, 831–847. [CrossRef]
215. Jeong, S.-H.; Park, J.-A. System identification and admittance model-based nanodynamic control of ultra-precision cutting process. *KSME Int. J.* **1997**, *11*, 620–628. [CrossRef]
216. Tao, H.; Chen, R.; Xuan, J.; Xia, Q.; Yang, Z.; Zhang, X.; He, S.; Shi, T. A new approach to identify geometric errors directly from the surface topography of workpiece in ultra-precision machining. *Int. J. Adv. Manuf. Technol.* **2020**, *106*, 5159–5173. [CrossRef]
217. Altintas, Y.; Aslan, D. Integration of virtual and on-line machining process control and monitoring. *CIRP Ann.* **2017**, *66*, 349–352. [CrossRef]
218. Lin, C.-J.; Lin, C.-H.; Wang, S.-H. Using Fuzzy Control for Feedrate Scheduling of Computer Numerical Control Machine Tools. *Appl. Sci.* **2021**, *11*, 4701. [CrossRef]
219. Qi, B.; Park, H.S. Data-driven digital twin model for predicting grinding force. *IOP Conf. Ser. Mater. Sci. Eng.* **2020**, *916*, 012092. [CrossRef]
220. Liu, J.; Du, X.; Zhou, H.; Liu, X.; ei Li, L.; Feng, F. A digital twin-based approach for dynamic clamping and positioning of the flexible tooling system. *Procedia CIRP* **2019**, *80*, 746–749. [CrossRef]
221. Aggogeri, F.; Merlo, A.; Pellegrini, N. Active vibration control development in ultra-precision machining. *J. Vib. Control* **2020**, *27*, 107754632093347. [CrossRef]
222. Zhao, D.; Du, H.; Wang, H.; Zhu, Z. Development of a novel fast tool servo using topology optimization. *Int. J. Mech. Sci.* **2023**, *250*, 108283. [CrossRef]
223. Guo, P.; Li, Z.; Xiong, Z.; Zhang, S. A theoretical and experimental investigation into tool setting induced form error in diamond turning of micro-lens array. *Int. J. Adv. Manuf. Technol.* **2023**, *124*, 2515–2525. [CrossRef]
224. Niu, Y.; Chen, X.; Chen, L.; Zhu, Z.; Huang, P. Development of a Sinusoidal Corrugated Dual-Axial Flexure Mechanism for Planar Nanopositioning. *Actuators* **2022**, *11*, 276. [CrossRef]
225. Liang, X.; Zhang, C.; Wang, C.; Li, K.; Loh, Y.M.; Cheung, C.F. Tool wear mechanisms and surface quality assessment during micro-milling of high entropy alloy FeCoNiCrAlx. *Tribol. Int.* **2023**, *178*, 108053. [CrossRef]

Article

An Efficient Method to Fabricate the Mold Cavity for a Helical Cylindrical Pinion

Bo Wu [1], Likuan Zhu [2,*], Zhiwen Zhou [2], Cheng Guo [2], Tao Cheng [1] and Xiaoyu Wu [2]

[1] Guangdong University Engineering Technology Research Center for Precision Components of Intelligent Terminal of Transportation Tools, College of Urban Transportation and Logistics, Shenzhen Technology University, Shenzhen 518118, China; wubo@sztu.edu.cn (B.W.); chengtao@sztu.edu.cn (T.C.)

[2] Guangdong Provincial Key Laboratory of Micro/Nano Optomechatronics Engineering, College of Mechatronics and Control Engineering, Shenzhen University, Shenzhen 518060, China; zwzhou5431@163.com (Z.Z.); cheng.guo@szu.edu.cn (C.G.); wuxy@szu.edu.cn (X.W.)

* Correspondence: zhulikuan@yeah.net; Tel.: +86-185-8896-0496

Abstract: An efficient method was proposed to fabricate the mold cavity for a helical cylindrical pinion based on a plastic torsion forming concept. The structure of the spur gear cavity with the same profile as the end face of the target helical gear cavity was first fabricated by low-speed wire electrical discharge machining (LS-WEDM). Then, the structure of the helical gear cavity could be obtained by twisting the spur gear cavity plastically around the central axis. In this way, the fabrication process of a helical cylindrical gear cavity could be greatly simplified, compared to the fabrication of a multi-stage helical gear core electrode and the highly difficult and complex spiral EDM process in the current gear manufacturing method. Moreover, several experiments were conducted to verify this novel processing concept, and a theoretical model was established to show the relationship between the machine torsion angle and the helical angle of a helical gear. Based on this theoretical model, the experimental results showed that it is feasible to precisely control the shape accuracy of a helical cylindrical pinion mold cavity by adjusting the machine torsion angle.

Keywords: injection mold; helical cylindrical pinion; plastic torsion forming; LS-WEDM

Citation: Wu, B.; Zhu, L.; Zhou, Z.; Guo, C.; Cheng, T.; Wu, X. An Efficient Method to Fabricate the Mold Cavity for a Helical Cylindrical Pinion. *Processes* **2023**, *11*, 2033. https://doi.org/10.3390/pr11072033

Academic Editor: Jun Zhang

Received: 5 June 2023
Revised: 1 July 2023
Accepted: 4 July 2023
Published: 7 July 2023

1. Introduction

With the rapid development of some emerging fields of science and technology such as electronic information, the trend of product miniaturization is becoming increasingly popular. Thus, small modulus gears are now widely used in aircraft, communication facilities, traffic vehicles, intelligent instruments, appliances, etc. In particular, plastic gears are getting more and more attention in recent years due to their light weight, low noise, and inherent lubricity [1,2]. For plastic gear manufacturing, injection molding is the main method and the mold cavity is crucial for the gear injection molding process [3]. In order to reduce micro-cell formation in the gear and produce more accurate plastic gears, Yoon et al. proposed a new injection molding process, the pressurized mold, which is designed to suppress the nucleation of micro-voids [4]. Ni et al. used micro powder injection molding (μPIM) to fabricate the micro gear cavity. The results showed that the micro gears can be successfully fabricated under an injection pressure of 70 MPa and 60% injection speed [5]. An effective method to control the non-linear shrinkage of micro-injection molded small module plastic gears by combining multi-objective optimization with Moldflow simulation is proposed. The accuracy of the simulation model was verified in a micro-injection molding experiment using reference process parameters [6]. In order to reduce the excess costs of the molds used to produce parts in injection molding and the problems of wastes that occur during production in hobbing, Tunalioglu et al. analyzed the wear resistance of plastic spur gears produced by the Fused Deposition Modeling (FDM) method [7]. In order to determine the service life of gears, wear tests were carried out in the Forschungsstelle fur Zahnrader und Getriebebau (FZG) type test device at the same load and rotational speeds. Polylactic acid (PLA), acrylonitrile butadiene styrene (ABS), and polyethylene terephthalate (PETG) thermoplastic polymer materials were used in the production of gears. Singh et al. produced ABS, HDPE, and POM gears by the injection molding method, and their thermal and wear behaviors were investigated [8]. While ABS and HDPE gears completed 0.5 and 1.1 million cycles, respectively, before failure, they claimed that the POM gear completed 2 million cycles without any signs of failure. Duzcukoglu and Imrek increased the tooth width in PA 66 gears, delaying the occurrence of thermal damage in the single-tooth region [9,10]. Experiments show that the appearance of heat damage is delayed for width-modified gear teeth compared with unmodified gear teeth. Kalin et al. investigated the tribological properties of POM gears at different temperatures and torques [11]. In a word, the machining method for a spur gear mold cavity is quite stable and reliable, but fabricating a mold cavity for a helical gear with a high-precision is still a big challenge.

LS-WEDM is a general method for machining micro spur gears or their mold cavities, and its machining quality can reach Grade 5 for the DIN standard. Unfortunately, there is no evidence that this method can be used for machining micro helical mold cavities [12]. Limited by the complex enclosed space, the helical cylindrical pinion cavity cannot be fabricated by LS-WEDM directly like the micro spur gears and its mold cavities. The common process is to first fabricate the helical gear core into several rough and fine processing sections by hobbing, milling, and other cutting methods (Figure 1a) and then the helical cylindrical pinion mold cavity can be fabricated by the spiral EDM using the obtained helical gear core as the tool electrode [13]. Considering that the discharging gap will change its size at the rough and finish machining steps in the spiral EDM process, it is quite difficult to accurately compensate for the machining error in this process. Moreover, chip removal will become more difficult in the spiral discharge machining process, which can cause a more serious electrode loss for the multi-stage helical gear core. It should also be mentioned that the discharge machining process for a helical cylindrical cavity is quite time-consuming and more and smaller metal debris would be generated, which is detrimental to maintaining a clean processing environment. Wang et al. forward the forging process of spiral bevel gear with large modulus, including blanking, uptuning-punching compound, rolling, and final forging. It was found that the tooth shape of the lower die produces uneven deformation under the action of thermal coupling, and as the height of the tooth shape of the lower die decreases, the spiral angle of the center point of the helix decreases, and the tooth profile becomes narrower [14]. Qi et al. proposed a forming method for straight bevel gear, which used a special mold with a flash edge and a convex to manufacture straight bevel gear, and used deform-3D V6.1 to optimize the forming load and die wear [15]. Xia et al. proposed using a positive extrusion process to form a four-leaf aluminum alloy spiral surface rotor and combined deform-3D V6.1 with numerical simulation to study the influence of extrusion temperature, extrusion speed, extrusion ratio, and other parameters on the extrusion process [16]. Li et al. put forward a combination process including hot forging, upsetting finishing, and radial extrusion of gears and conducted simulation and experimental research to analyze the flow law of metals in the extrusion process and the influence of finishing parameters on surface quality and the stress–strain distribution in the extrusion process [17]. Moreover, it is still a challenge to predict the machining error in the fabricating process of gear structure. Ma et al. built a prediction model of free tooth splitting error in the axial rolling of a straight gear, and characterized the mathematical relationship between the tooth spacing error of a free tooth splitting axial rolling workpiece and the initial diameter of the workpiece, the initial phase of the rolling gear teeth and the contact degree. Based on the theory of gear meshing, the assumption of plane strain, and the principle of equal volume, the estimation model of workpiece tooth forming height at any time is constructed. The geometric design method of the axial rolling wheel is also proposed. The structure of the axial rolling wheel is divided into the cutting section, finishing section, and exiting section. The cutting section, finishing section, and exiting section are designed from the angles of the force of rolling wheel teeth, root stress, slip rate, and tooth surface scratches [18,19].

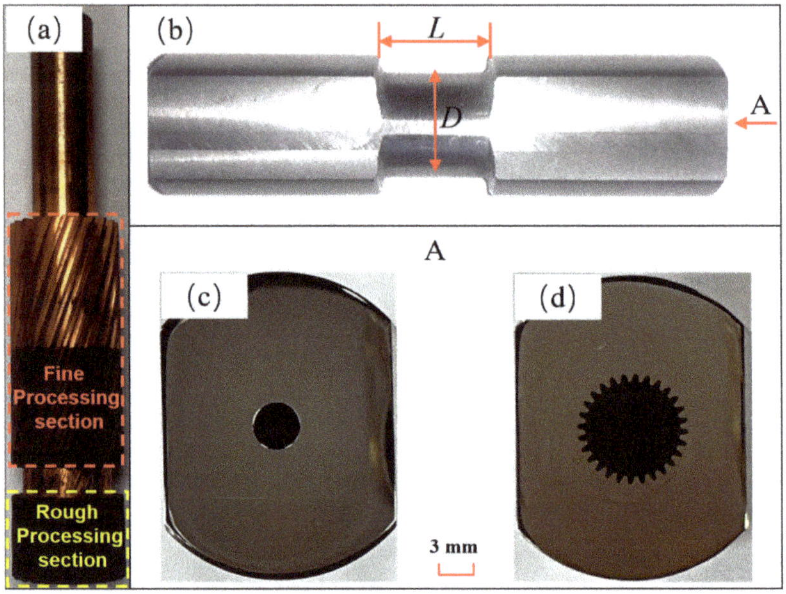

Figure 1. Conventional machined electrode and plastic torsion samples: (**a**) a helical gear core electrode; (**b**) the size view of a torsion sample; (**c**) the torsion sample with a through-hole; (**d**) the torsion sample with a spur tooth cavity.

In a word, there is a lack of a more effective way to fabricate the cavity of the helical cylindrical pinion mold by the current machining methods. Therefore, it is urgent for the industry to develop a new approach to fabricate the structure of the helical cylindrical pinion cavity.

In this paper, an efficient method was proposed to fabricate the mold cavity for a helical cylindrical pinion based on a plastic torsion forming concept. A mathematical model was established for the relationship between the machine torsion angle and the helix angle on the gear cavity reference circle. By analyzing and comparing the torsion experiment and calculation results at different helix angles, the established mathematical model was found valid.

2. Research Methodology and Experiment Details

2.1. Proposed Concept

According to the profile data of the helical gear cavity end face, its spur gear cavity was first machined by LS-WEDM. The spur gear cavity was then twisted plastically by a torsion machine to obtain the targeted helical gear cavity. Compared with the highly difficult and complex steps for the gear cavity machining in the spiral EDM process, this novel method was based on a more simple and more reliable 2D machining concept, which greatly improved the machining reliability and efficiency. It should be mentioned that the high-precision LS-WEDM was also adopted in this novel method to obtain the spur gear mold cavity, which is crucial for ensuring the forming accuracy of its helical cylindrical pinion cavity. Thus, this novel fabricating method for a helical gear cavity shows its promising application in the involved industry.

2.2. Theoretical Background

As shown in Figure 1b, the torsion sample could be divided into three sections. The two ends were the clamping sections to facilitate clamping the sample piece on the torsion machine chuck, and the middle was the torsion section. A wire-cutting threaded hole (Figure 1c) was initially processed on the axis of the torsion sample. According to the end surface profile data of the helical gear cavity, the spur gear mold cavity (Figure 1d) was first fabricated by LS-WEDM, and then the helical gear cavity could be obtained from the torsion section by twisting the spur gear mold cavity slowly. As shown in Figure 2a, the torsion sample could be further divided into three different deformation zones after the torsion based on its plastic deformation characteristics: (1) Helical gear zone (uniform deformation zone): l_m was the length, and β_m was the reference circle helix angle, which was found constant in this zone, which indicated that the materials twisted and deformed uniformly in this zone. A helical cylindrical gear mold cavity could be obtained from this part. (2) Spur gear zone: there was no plastic deformation in this zone, and the internal cavity structure still maintained its complete spur gear cavity. (3) Transition gear zone: the internal cavity shape contained both spur and helical gears in this zone. This zone deformation played a transitional role between the helical gear zone and the spur gear zone to guarantee the continuity of the material deformation in the whole torsion sample. l_{em} represented the length of the intersection zone between the transition gear zone and torsion section and l_{eo} meant the length of the intersection zone between the transition gear zone and clamping section. l_e was the total length of the transition gear zone ($l_e = l_{em} + l_{eo}$). β_e was the reference circle helix angle of the transition gear zone, which changed continuously in the range of $0 \sim \beta_m$ along the axial direction of the torsion sample.

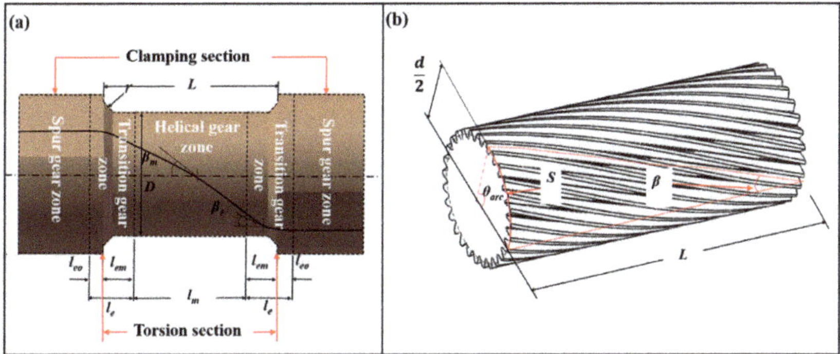

Figure 2. Analysis of the torsion forming for the gear cavity: (**a**) different deforming zones; (**b**) the geometrical relationship in the torsion section.

In order to analyze the deformation law of the helical gear cavity in the torsion section, the torsion section part was taken for analysis (Figure 2b). d was the diameter of the reference circle on the end face of the helical gear cavity and L was the length of the torsion section. Considering the torsion section theoretically included both the helical gear zone (uniform deformation) and the transition gear zone (non-uniform deformation), β represented the approximate helix angle in the torsion section. Based on the forming characteristics of the spiral involute surface, the relationship between the circumferential travel S of the tooth line on the reference circle of the torsion section, the machine torsion radian θ_{arc}, and the reference circle helix angle β was as follows:

$$S = \theta_{arc}\frac{d}{2} = Ltg\beta \tag{1}$$

After converting radian units (θ_{arc}) into angle units (θ), the machine torsion angle θ could be calculated as follows:

$$\theta = \frac{2Ltg\beta}{d} \times \frac{180}{\pi} \tag{2}$$

From Equation (2), the reference circle helix angle β of the helical gear cavity could be also calculated from the machine torsion angle θ directly in the torsion section.

Considering a part of the transition gear zone might be located in the torsion section, the helix angle β calculated by Equation (2) would not be completely consistent with the helix angle β_m in the helical gear zone with a uniform deformation, which indicated that there might be some deviation in controlling the helix angle β_m directly by the machine torsion angle θ. However, Equation (2) still provided a kind of theoretical relationship between the operating parameters of the torsion machine and the characteristic parameters of the helical gear cavity. If it could be proved that the error between the predicted β and the β_m is acceptable or controllable, this new method proposed would be quite promising for fabricating the helical gear cavity in the industry.

2.3. Experiment Setup

The torsion sample material was S136 steel, the length of the torsion section of the sample L = 15 mm, and the diameter of the torsion section D = 15 mm. An LS-WEDM machine (AP250L, Sodick Co., Ltd., Kanagawa, Japan) was used to fabricate the spur gear cavity with the highest machining accuracy of 0.001 μm and the best surface roughness of Ra0.1. A pulse width of 0.5 μs, a pulse interval of 15 μs, a peak current of 0.1 A, and a wire speed of 2 mm/min were used in the LS-WEDM process. An electronic torsion testing machine (CTT1000, Xinsansi Co., Ltd., Hangzhou, China) was adopted for the plastic torsion forming to obtain the helical gear cavity. Based on the helical gear cavity obtained, a series of injection molding experiments were conducted to verify the novel fabrication method for a helical cylindrical pinion. First, the POM granules were dried at 100 °C for 3 h. An injection molding machine (HA1700, ShengDa machinery equipment company, Wenzhou, China) was used and the injection parameters were as follows: Injection temperature was 200 °C, injection pressure was 110 MPa, and pressure holding time was 8 s. A small gear testing machine (GTR-4LS, Osaka Precision Co., Ltd., Osaka, Japan) was used to determine the quality of the helical cylindrical pinions.

The design parameters of the helical cylindrical gear cavity were chosen as follows: the number of teeth z = 28, the normal module m_n = 0.2, the pressure angle α = 20°, and the helix angle β = 20°. The diameter of the end reference circle d = 5.959 mm, and the diameter of the end addendum circle d_a = 6.359 mm. According to the end face profile data of the helical gear cavity, several spur gear cavity samples were machined by LS-WEDM, and then they were twisted around the central axis with the machine torsion angles (θ) of 20°, 40°, 60°, 80°, 100°, 120°, and 140°, respectively. After the torsion experiments, each sample was cut into five sections and each cutting surface was located 3 mm away from the nearest interface between the clamping section and the torsion section as shown in Figure 3a. From the experimental observations, this cutting method could guarantee that each cutting section included only one tooth shape feature: helical gear shape, spur gear shape, or transition gear shape.

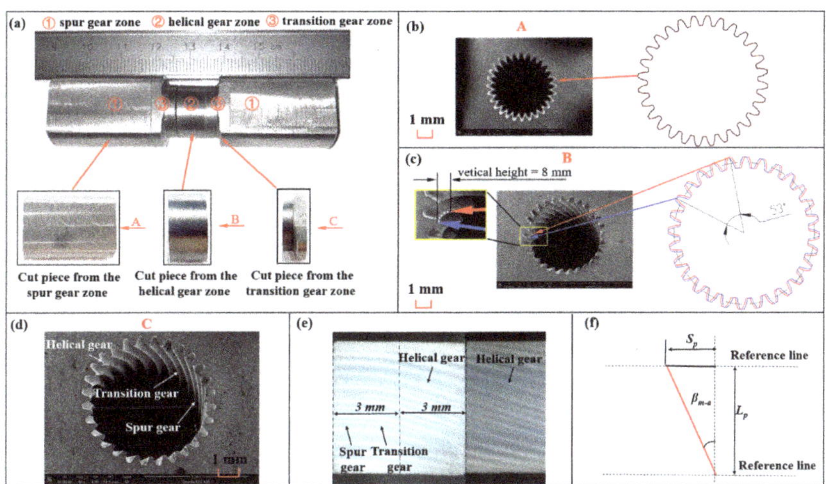

Figure 3. Torsion forming result for the gear cavity and its silicone copy: (**a**) different cut pieces from the torsion sample; (**b**) a cut piece from the spur gear zone; (**c**) a cut piece from the helical gear zone; (**d**) a cut piece from the transition gear zone; (**e**) a silicone copy; (**f**) the spiral tooth line measurement.

3. Result and Discussion

Figure 3b,c showed the cutting cavities in the spur gear zone without any plastic deformation and the helical gear zone with a large plastic deformation with the machine torsion angle θ of 100°. As shown in those two figures (Figure 3b,c), the end face contours were in good consistency. Figure 3d showed the cutting cavity in the transition gear zone, and its shape was represented by filling the cavity with silicone resin (Figure 3e left). Using the same method, the cavity shape of the helical gear zone could be also obtained as shown in the right of Figure 3e. It could be clearly seen from the above figures (Figure 3b–e) that the torsion sample really could be divided into three different zones: the helical gear zone with a uniform deformation, the spur gear zone without any deformation, and the transition gear zone.

In order to analyze the evolution law of the spiral tooth line from the cavity, one of the tooth surfaces from the silicone gear was painted using black ink and then the silicone gear was rolled along a reference line on paper. The axial travel of the tooth line (L_p) and the circumferential travel of the tooth line (S_p) could be measured, and the torsional angle (θ_p) of the tooth profile in the circumferential direction could be calculated as follows:

$$\theta_p = 2\frac{S_p}{d_a} \times \frac{180}{\pi} \tag{3}$$

where d_a was the diameter of the addendum circle.

In fact, θ_p was also the torsion angle for the cut piece from the helical gear zone. At the assumption that the entire torsion section was located at the helical gear zone, the equivalent torsion angle θ_e for the entire torsion section could be determined by the following equation:

$$\theta_e = \frac{L}{L_p}\theta_p \tag{4}$$

The helix angle on the addendum circle $\beta_{m\text{-}a}$ in the helical gear zone could be calculated as:

$$\beta_{m-a} = arctg\left(\frac{S_p}{L_p}\right) \times \frac{180}{\pi} \tag{5}$$

From Equation (2) above, the helix angle on the reference circle (β_m) in the helical gear zone could be determined:

$$\beta_m = arctg\left(\frac{\theta_e d}{2L} \times \frac{\pi}{180}\right) \times \frac{180}{\pi} \tag{6}$$

Based on the above theoretical formulas and the experiment data, the calculated and measured values for the helical angles with different machine torsion angles are shown in Table 1. The results showed that the machine torsion angle θ and equivalent torsion angle θ_e were quite close, which suggested that the proportion of the transition gear zone probably should be very limited. This could also be verified from the silicone copy obtained from the cut piece from the transition gear zone (Figure 3e). In the silicone copy of the transition gear zone with a length of about 6 mm, the part with the uniform helical gear cavity occupied almost half of this zone (Figure 3e), which indicated that there was no transition zone in the torsion section actually ($l_{em} \approx 0$ or $l_e \approx l_{eo}$) (Figure 2a). That is, the uniform deformation occurred in almost the entire torsion section, which suggested that this section could be used as the helical gear mold cavity directly, and the whole spur gear zone and almost the entire transition gear zone were located at the clamping section. Moreover, the difference between the calculated value (β) and the measured value (β_m) for the helix angle was less than 3% for the conventional range of 8~25°, which indicated that the accuracy of the helix angle for the target helical gear cavity, calculated by Equation (2), was sufficient. From Table 1, $\beta_{m\text{-}a}$, β_m, and β only increased by 3~4° as θ increased by 20°, which also indicated that it would be convenient and feasible to control the manufacturing accuracy of the helical tooth line of the cavity by the machine torsion angle θ.

Table 1. Measured and calculated crucial parameters for a spiral tooth line.

machine torsion angle θ (°)	20	40	60	80	100	120	140
axial travel of a tooth line L_p (mm)	9.732	9.747	9.614	9.572	9.563	9.452	9.632
circumferential travel of a tooth line S_p (mm)	0.762	1.421	2.100	2.745	3.482	4.152	4.908
torsion angle of a tooth line θ_p (°)	13.731	25.605	37.840	49.463	62.743	94.816	88.438
equivalent torsion angle θ_e (°)	21.163	39.405	59.040	77.512	98.415	118.730	137.726
helix angle on the addendum circle in the helical gear zone $\beta_{m\text{-}a}$ (°)	4.477	8.295	12.322	16.002	20.007	23.715	27.001
helix angle on the reference circle in the helical gear zone β_m (°)	4.496	7.780	11.568	15.042	18.840	22.374	25.525
helix angle on the reference circle in the torsion section β (°)	3.967	7.896	11.751	15.502	19.122	22.590	25.891

The results also suggested that the length (L) and diameter (D) of the torsion section had an important influence on the plastic torsional process. For a targeted helical gear cavity, increasing the D value could decrease the limit torsion angle (θ_{limit}) of the torsion sample. When the machine torsion angle (θ) was higher than θ_{limit}, cracks and other defects were more likely to occur on the surface of the torsion sample. As the L value was constant, a smaller θ was needed for a larger addendum circle diameter (d_a) and reference circle diameter (d) of the end face in Equation (2). Therefore, the parameters D and d_a should be adjusted carefully and a larger D was more favorable to increase θ_{limit}. Based on the good plasticity of S136 steel, its torsion-forming process could be completely carried out at room temperature. For $d_a < 10$ mm and the helix angles within 8~25°, the helical cylindrical gear cavity could be obtained by one-time twisting using the S136 steel material. For helix angles over 20°, intermediate annealing treatment was probably needed.

To further verify the novel gear mold fabricating method proposed in this paper, a series of injection molding experiments were carried out based on a complete plastic helical cylindrical gear injection mold (Figure 4). From Equation (2), θ of 104.980° was chosen to achieve the helix angle of 20°. The left-hand and right-hand gear cavities were twisted, respectively. Then the obtained two cavities were assembled into the plastic injection system. The result showed that the quality of the left and right helical cylindrical pinions was sufficient, which suggested that this novel method would be quite promising in solving the bottleneck problem in the helical gear fabricating process for the related industry.

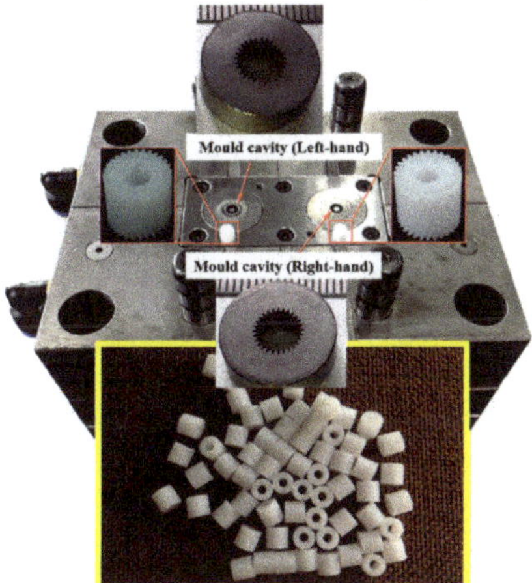

Figure 4. Injection mold system of a plastic helical cylindrical gear.

4. Conclusions

In this paper, a new method for manufacturing a helical gear cavity was proposed. LS-WEDM was used to form a spur gear cavity and then a helical gear cavity could be obtained by twisting the spur gear cavity. Several equations were derived to predict and evaluate the precision of the helical gear geometry. The results showed that this novel method would probably be promising for helical gear manufacturing in the future.

(1) A new method for the efficient and controllable manufacturing process of the helical cylindrical gear cavity was proposed. The experimental results showed that almost the whole torsion section was located in the helical gear zone with uniform deformation, which could be used as the cavity of the targeted helical gear mold directly, and the whole spur gear zone and almost the entire transition gear zone were located at the clamping section.

(2) A mathematical model was established for the relationship between the machine torsion angle and the helix angle on the reference circle, and the result showed that the difference between the helix angle calculated by the mathematical model and its measured value was close, which indicated that it is feasible to precisely control the helix angle of the helical cylindrical gear cavity by adjusting the machine torsion angle. Moreover, the helix angle on the reference circle only increased by 3~4° when the machine torsion angle increased by 20°, which was favorable for the accuracy control of the helical gear profile.

(3) For $d_a < 10$ mm and helix angles within 8~25°, the helical cylindrical gear cavity could be obtained by one-time twisting using the S136 steel material. For the helix angles over 20°, intermediate annealing treatment was probably needed.

Author Contributions: Conceptualization, X.W.; Validation, Z.Z.; Formal analysis, C.G.; Data curation, T.C.; Writing—original draft, B.W.; Supervision, L.Z.; Project administration, X.W.; Funding acquisition, X.W. All authors have read and agreed to the published version of the manuscript.

Funding: This work was supported by the National Natural Science Foundation of China under Grant [52205399 and 51975385]; Shenzhen Stable support B Plan under Grant [2020081405908002];

Basic and Applied Basic Research Fund of Guangdong Province under Grant [2019A1515111115]; Start-up funds for high-end talent research in Shenzhen [827-000737].

Data Availability Statement: The data presented in this study are available on request from the corresponding author.

Conflicts of Interest: The authors declare that they have no known competing financial interests or personal relationships that could have appeared to influence the work reported in this paper.

References

1. Kim, C.H. Durability improvement method for plastic spur gears. *Tribol. Int.* **2006**, *39*, 1454–1461. [CrossRef]
2. Landi, L.; Stecconi, A.; Morettini, G.; Cianetti, F. Analytical procedure for the optimization of plastic gear tooth root. *Mech. Mach. Theory* **2012**, *166*, 104496–104514. [CrossRef]
3. Apichartpattanasiri, S.; Hay, J.N.; Kukureka, S.N. A study of the tribological behaviour of polyamide 66 with varying injection-moulding parameters. *Wear* **2001**, *251*, 1557–1566. [CrossRef]
4. JYoon, D.; Cha, S.W.; Chong, T.H.; Ha, Y.W. Study on the Accuracy of Injection Molded Plastic Gear with the Assistance of Supercritical Fluid and a Pressurized Mold. *Polym. Plast. Technol. Eng.* **2007**, *46*, 815–820.
5. Ni, X.L.; Yin, H.Q.; Liu, L.; Yi, S.J.; Qu, X.H. Injection molding and debinding of micro gears fabricated by micro powder injection molding. *Int. J. Miner. Metall. Mater.* **2013**, *20*, 82–87. [CrossRef]
6. Wu, W.Q.; He, X.S.; Li, B.B.; Shan, Z.Y. An Effective Shrinkage Control Method for Tooth Profile Accuracy Improvement of Micro-Injection-Molded Small-Module Plastic Gears. *Polymers* **2022**, *14*, 3114. [CrossRef] [PubMed]
7. Tunalioglu, M.T.; Agca, B.V. Wear and Service Life of 3-D Printed Polymeric Gears. *Polymers* **2022**, *14*, 2064. [CrossRef] [PubMed]
8. Singh, P.K.; Siddhartha; Singh, A.K. An Investigation on the Thermal and Wear Behavior of Polymer Based Spur Gears. *Tribol. Int.* **2018**, *118*, 264–272. [CrossRef]
9. Düzcükoglu, H. PA 66 spur gear durability improvement with tooth width modification. *Mater. Des.* **2009**, *30*, 1060–1067. [CrossRef]
10. Imrek, H. Performance improvement method for Nylon 6 spur gears. *Tribol. Int.* **2009**, *42*, 503–510. [CrossRef]
11. Kalin, M.; Kupec, A. The dominant effect of temperature on the fatigue behaviour of polymer gears. *Wear* **2017**, *376*, 1339–1346. [CrossRef]
12. Chaubey, S.K.; Jain, N.K. State-of-art review of past research on manufacturing of meso and micro cylindrical gears. *Precis. Eng.* **2018**, *51*, 702–728. [CrossRef]
13. Feng, Z.P.; Shi, Z.Y.; Tong, A.G.; Lin, S.J.; Wang, P. A pitch deviation compensation technology for precision manufacturing of small modulus copper electrode gears. *Int. J. Adv. Manuf. Technol.* **2021**, *114*, 1031–1048. [CrossRef]
14. Wang, W.; Zhao, J.; Zhai, R.X. A forming technology of spur gear by warm extrusion and the defects control. *J. Manuf. Process.* **2016**, *21*, 30–38. [CrossRef]
15. Qi, Z.C.; Wang, X.X.; Chen, W.L. A new forming method of straight bevel gear using a specific die with a flash. *Int. J. Adv. Manuf. Technol.* **2019**, *100*, 3167–3183. [CrossRef]
16. Xia, F.; Li, H.; Liu, H.G.; Zhao, B.B.; Zhang, Z.Q.; Lu, D.H.; Chen, J.Q. Forward hot extrusion forming process of 4-lobe aluminum alloy helical surface rotor. *J. Cent. South Univ.* **2019**, *26*, 2307–2317. [CrossRef]
17. Li, J.; Wang, G.C.; Wu, T. Numerical-experimental investigation on the rabbit ear formation mechanism in gear rolling. *Int. J. Adv. Manuf. Technol.* **2017**, *91*, 3551–3559. [CrossRef]
18. Ma, Z.Y.; Luo, Y.X.; Wang, Y.Q. On the pitch error in the initial stage of gear roll-forming with axial-infeed. *J. Mater. Process. Technol.* **2018**, *252*, 659–672. [CrossRef]
19. Ma, Z.Y.; Luo, Y.X.; Wang, Y.Q.; Mao, J. Geometric design of the rolling tool for gear roll-forming process with axial infeed. *J. Mater. Process. Technol.* **2018**, *258*, 67–79. [CrossRef]

Article

Ultrasonic Plasticizing and Pressing of High-Aspect Ratio Micropillar Arrays with Superhydrophobic and Superoleophilic Properties

Shiyun Wu [1], Jianjun Du [1,*], Shuqing Xu [2], Jianguo Lei [2], Jiang Ma [2] and Likuan Zhu [2]

[1] School of Mechanical Engineering and Automation, Harbin Institute of Technology (Shenzhen), Shenzhen 518055, China; wusy626@163.com
[2] College of Mechatronics and Control Engineering, Shenzhen University, Shenzhen 518061, China
* Correspondence: jjdu@hit.edu.cn

Abstract: An ultrasonic plasticizing and pressing method (UPP) that fully utilizes ultrasonic vibration is proposed for fabricating thermoplastic polymer surface microstructures with high aspect ratios (ARs). The characteristics of UPP are elucidated based on the plasticization of the raw material, the melt flow, and the stress on the template microstructure during the forming process. Initially, the micronscale single-stage micropillar arrays (the highest AR of 4.1) were fabricated by using 304 stainless steel thin sheets with micronscale pore (through-hole) arrays as primary templates. Subsequently, anodic aluminum oxides (AAOs) with ordered nanoscale pore arrays were added as secondary templates, and the micro/nanoscale hierarchical micropillar arrays (the highest AR up to 24.1) were successfully fabricated, which verifies the feasibility and forming capability of UPP. The superiority and achievements of UPP are illustrated by comparing the prepared hierarchical micropillar arrays with those prepared in the previous work in four indexes: microstructure scale, aspect ratio, forming time, and preheating temperature of the raw material. Finally, the water contact angle (WCA) and oil droplet complete immersion time of the surface microstructures were measured by a droplet shape analyzer, and the results indicate that the prepared micropillar arrays are superhydrophobic and superoleophilic.

Keywords: ultrasonic vibration; surface microstructures; aspect ratio; hierarchical micropillar array; superhydrophobic

Citation: Wu, S.; Du, J.; Xu, S.; Lei, J.; Ma, J.; Zhu, L. Ultrasonic Plasticizing and Pressing of High-Aspect Ratio Micropillar Arrays with Superhydrophobic and Superoleophilic Properties. *Processes* **2024**, *12*, 856. https://doi.org/10.3390/pr12050856

Academic Editor: Sara Liparoti

Received: 28 March 2024
Revised: 21 April 2024
Accepted: 22 April 2024
Published: 24 April 2024

1. Introduction

Surface microstructures with the designed distribution of different micro- and nano-structural features present tunable functional properties, such as excellent optical [1–3], biological [4,5], and electrochemical [6–8] properties, and are widely used in the fields of biomedicine, energy, and environmental protection. However, it remains a great challenge for conventional processing techniques to fabricate tailorable micro- and nano-structures on a large scale. Compared with metals and inorganic non-metallic materials, polymer materials are inexpensive, diverse, and provide better corrosion resistance, electrical insulation, and biocompatibility. In addition, with the outstanding advantages of good processability and high replication accuracy, polymer materials have become the preferred raw material for large-scale, low-cost production of functional surfaces with micro- and nanostructures [9,10]. The template method is the mainstream method for large-scale production of polymer surface microstructures, and the main methods currently used in practical applications are the sol–gel method [11,12], UV-curable imprinting [13,14], hot embossing [15,16], and injection molding [17,18]. The sol–gel method is easy to carry out the chemical reaction, and the required temperature is low, but it is prone to defects such as micropores and bubbles, and the curing process is prone to shrinkage, making it difficult to control the precision of the product. UV-curable imprinting can obtain nanoscale structures without heating, but the mold or substrate used must be transparent and covered with an anti-adhesion layer, which severely limits its scope of application. Hot embossing allows for smaller-scale structures with high aspect ratios, but requires high temperature and pressure, and the polymer needs to be heated to the glass transition temperature, resulting in a long forming time. Injection molding has the advantages of short-cycle, large-scale, and low-cost production, but it has poor replication ability for nanoscale structures with high aspect ratios. In addition, it requires high temperature and pressure conditions and is difficult to achieve damage-free demolding. Therefore, there is still a demand to improve the existing or develop new methods for large-scale, low-cost, and high-quality production of surface microstructures.

The application of ultrasonic vibration as the primary or auxiliary energy source for the fabrication of polymer surface microstructures to enhance melt flow and filling capacity, improve microfabricated part performances, and reduce production costs has been widely noticed and studied. Mekaru et al. [19] introduced ultrasonic vibration into hot embossing, resulting in a substantial reduction or even disappearance of air bubbles during the embossing process, which significantly improved the microstructure replication accuracy. Sato et al. [20] applied ultrasonic vibration to the mold in injection molding, while Qiu et al. [21] applied ultrasonic vibration directly to the polymer melt inside microcavities, both of which significantly improved the forming quality of the microstructures, but the latter was more effective. Lee et al. [22,23] rapidly replicated nanostructures and micronanoscale hierarchical structures by room temperature ultrasonic embossing. Li et al. [24] rapidly prepared a nanowire array on polyvinylidene fluoride polymer surfaces at room temperature by ultrasonic loading. Liang et al. [25] prepared a microgroove array with hydrophobicity by ultrasonic powder molding. Pan et al. [26] fabricated a high-depth (~240 μm) polypropylene micro-square pore array using ultrasonic plasticization micro-injection molding. Room-temperature ultrasonic embossing has high replication accuracy for micro- and nano-structures, but it is difficult to emboss microstructures with high aspect ratios, and the templates are mostly limited to high-strength and high-hardness materials. Ultrasonic powder molding is a simple and low-cost process, but for microstructures with high-convex features, the ends are prone to incompletely melted powder particles [27]. For ultrasonic plasticization micro-injection molding, although the flowability of the polymer melt is improved [28], it is still difficult to form nanoscale surface microstructures with high aspect ratios.

Here, an ultrasonic plasticizing and pressing method (UPP) is proposed for large-scale fabrication of micron- and nano-scale surface microstructures with high aspect ratios. The proposed method fully exerts the effect of ultrasonic vibration, which is conducive to improving the filling capability of the polymer melt for micro- and nano-scale pores. In addition, it is conducive to the exclusion of residual air in the microcavities, reducing the formation of air bubbles and thus improving the forming quality of the surface microstructures.

2. Materials and Methods

2.1. The UPP Process

The process of UPP consists of the three following stages:

(1) Mold assembly and raw material addition. As shown in Figure 1a, the bottom mold, template, and top mold are mounted and fixed sequentially on the worktable of the ultrasonic loading system, and the punching port of the top mold is aligned with the ultrasonic horn. Then, the thermoplastic polymer pellets are added to the inner step surface of the top mold.

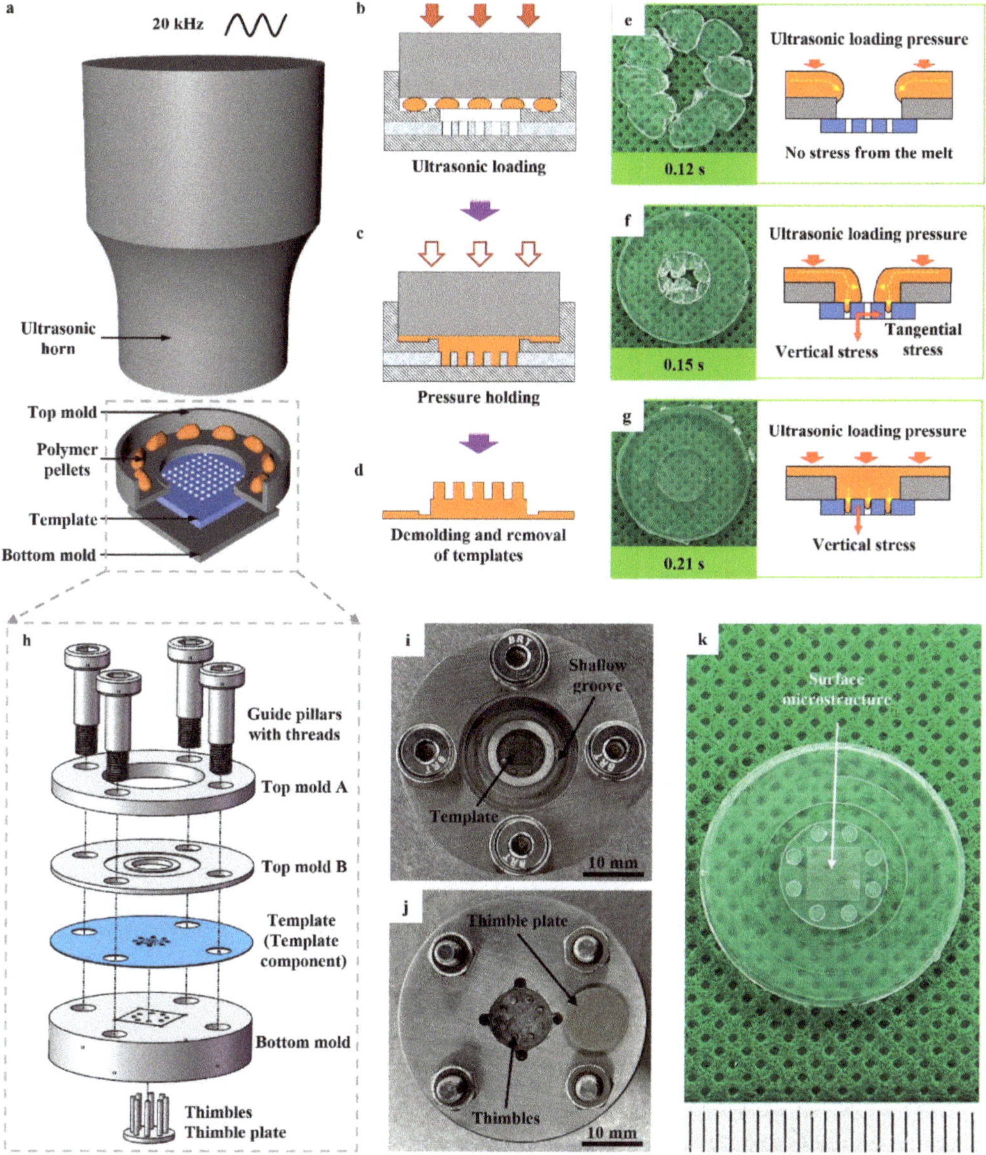

Figure 1. UPP process and the mold. (**a–d**) The schematic illustration of the UPP process. (**e–g**) The melt flow and stress diagrams. (**h**) The exploded diagram of the mold. (**i,j**) The top and bottom of the mold after installation, respectively. (**k**) The sample with a surface microstructure.

(2) Setting process parameters and ultrasonic loading. The process parameters, such as resonant frequency, ultrasonic amplitude, ultrasonic trigger pressure and loading pressure, ultrasonic duration time, and pressure holding time, are set. By pressing the start button, the ultrasonic horn rapidly moves downward and presses against the polymer pellets (Figure 1b), and ultrasonic loading is automatically initiated when the pressure reaches the trigger pressure. Under high-frequency vibration, the thermoplastic polymer pellets are rapidly plasticized and melted due to frictional and viscoelastic heat [29–31], and then

fill the micropores of the template. After ultrasonic loading, the melt in the micropores is rapidly cooled and solidified during pressure holding (Figure 1c).

(3) Demolding and removal of templates. After pressure holding, demolding, and removal of templates to obtain the surface microstructures (Figure 1d).

2.2. Materials, Templates, and Equipment

The raw material was polypropylene pellets with a size of about $3 \times 2.5 \times 5$ mm^3, which were supplied by Korea Chungnam Lotte Chemical Co., Seoul, Republic of Korea. The primary template with a micropore (through-hole) array was obtained by laser cutting a 304 stainless steel thin sheet with a thickness of 0.2 mm. Two primary templates were designed, one with a pore diameter of 100 μm and a pitch of 150 μm, and the other with a pore diameter of 50 μm and a pitch of 100 μm. In order to facilitate the description and differentiation, the above two templates are named D100-array-LC and D50-array-LC after laser cutting, respectively, and the latter is renamed D50-array-LCM after re-cutting by adjusting the process parameter. The secondary templates were made of anodic aluminum oxide (AAO) with ordered pore arrays, including two specifications, one with a pore size of 390 nm and a pitch of 450 nm, and the other with a pore size of 250 nm and a pitch of 450 nm, and the thickness of both specifications was 50 μm. These two AAO templates are named 390 nm-AAO and 250 nm-AAO, respectively.

The equipment used in this paper is an ultrasonic loading system, which mainly consists of a pneumatic piston, a transducer, a horn, a worktable, an air source, and a power supply. The pneumatic piston converts the air pressure into the mechanical pressure of the horn. The piezoelectric ceramic of the transducer converts the alternating current into high-frequency mechanical vibration, which can realize the ultrasonic loading on the sample through the horn. The resonant frequency and maximum amplitude of ultrasonic vibration are 20 kHz and 60 μm, respectively.

Figure 1h shows the mold designed for UPP, which mainly consists of the bottom mold, thimbles and thimble plate, template (template component), top mold B, top mold A, and the guide pillars with threads. A shallow groove was designed on the top of mold B to hold polymer pellets and to prevent the polymer pellets from being vibrated down onto the template microstructure at the moment of initiating ultrasonic vibration. The thimbles and thimble plate were mounted in the deep groove on the lower surface of the bottom mold for quick ejection of the sample. After machining and assembly, the top and bottom of the mold are shown in Figures 1i and 1j, respectively. Figure 1k shows the sample after demolding, and its central region is the formed surface microstructure.

2.3. Fabrication of Micropiller Arrays

The D100-array-LC, D50-array-LC, and D50-array-LCM templates were used in UPP, respectively, for fabricating the microscale single-stage micropillar arrays. The number of polypropylene pellets was fixed at 8, and the total weight was about 0.18 g. Setting process parameters on the ultrasonic loading system: the ultrasonic trigger pressure, loading pressure, amplitude, and pressure holding time were set to 500 N, 300 kPa, 60 μm, and 5 s, respectively. The ultrasonic duration time for the D100-array-LC template was 0.28 s, while for both the D50-array-LC and D50-array-LCM templates, it was 0.3 s. The micro/nanoscale hierarchical micropillar arrays were prepared by combining D100-array-LC (primary template) with 390 nm-AAO (secondary template), D100-array-LC with 250 nm-AAO, and D50-array-LCM with 250 nm-AAO in UPP. The same process parameters were employed as for preparing the corresponding single-stage micropillar array (using the same primary template). The residual AAO templates embedded in the sample surface were dissolved in the 15% wt NaOH solution within 10 min. The above process parameters were set based on our extensive preliminary experimental results.

2.4. Characterization

The surface microstructures prepared by UPP were examined under scanning electron microscopy (SEM, Quanta FEG450, FEI, Hillsboro, OR, USA) after spray gold treatment. The hydrophobicity of the surface microstructures was characterized by water contact angle (WCA), which was measured by a droplet shape analyzer (DSA100S, Krüss, Hamburg, Germany) with a water droplet volume of 2.5 μL. The lipophilicity was characterized by measuring the time for an oil droplet (micro-molecule lube oil supplied by Cylion Technology Int'L Co. Ltd., Shenzhen, China) with a volume of 1 μL to completely immerse into the surface microstructures; thus, the entire process of oil droplet immersion into the surface microstructures was recorded.

3. Results and Discussion

3.1. The Characteristics and Advantages of UPP

The left sides of Figure 1e–g show the results of polypropylene pellets being plasticized and filling the mold cavities when the ultrasonic duration times were 0.12 s, 0.15, and 0.21 s, respectively. Based on the results, the flow of the melt during ultrasonic loading can be categorized into three stages, as shown on the right sides of Figures 1e, 1f and 1g, respectively. In the first stage, the polymer pellets are plasticized. Under high-frequency vibration, the surface of polymer pellets rubs violently against the horn and mold, generating frictional heat, while viscoelastic heat is generated within the material due to high-frequency alternating stress loads. The frictional and viscoelastic heat leads to a rapid increase in the temperature of polymer pellets with low thermal conductivity, which are plasticized and melted. In the second stage, the melt flows from the top mold to the upper surface of the template and converges to the template center while gradually diverging to the micro-cavity holes. At this stage, the template microstructure is not only subjected to the vertical stress from the melt but also to the tangential stress generated by the converging flow of the melt. Since the ultrasonic loading pressure is mainly applied to the melt on the upper surface of the top mold and is transferred axially to the worktable, whether vertical stress or tangential stress is much smaller than the ultrasonic loading pressure. In the third stage, after the melt has completely covered the upper surface of the template, it continues to fill the template micro-cavity holes. The template microstructure is mainly subjected to vertical stress from the melt, which is comparable in magnitude to ultrasonic loading pressure.

From the melt flow diagrams, it can be seen that during ultrasonic loading, the melt and the micro-cavity holes are always directly under the horn, and are subject to the continuous action of ultrasonic longitudinal vibration (main vibration). This can fully utilize the role of ultrasonic vibration. On the one hand, it effectively reduces the melt viscosity [32,33], thus improving the filling ability of the melt to the micro-cavity holes [34–36]. On the other hand, it is conducive to the elimination of residual air inside the micro-cavity holes [19,37], which reduces the formation of air bubbles and thus improves the forming quality of surface microstructures. From the stress diagrams, it can be seen that for the templates used in this work, only the tangential stress may cause deformation or fracture of the template microstructure, but this stress is much smaller than the ultrasonic loading pressure. In addition, UPP avoids violent friction between the polymer pellets or substrate and the template microstructure. Therefore, combined with the results of Figure S1, it can be demonstrated that UPP is not prone to deformation or even fracture of the template microstructure while ensuring sufficiently large melt filling pressure, which is favorable for the reuse of the template.

3.2. Micronscale Single-Stage Arrays

Figure 2a shows the D100-array-LC template, where the upper surface (laser-cut surface) of the template is outside the red box and the lower surface is inside the red box. The diameters of the circular holes on the upper and lower surfaces are about 115.9 μm and 106.0 μm, respectively, thus the template has a demolding angle of 1.42°. The top and side views of the single-stage column array prepared by using the D100-array-LC template are shown in Figures 2d and 2g, respectively. In Figure 2g, the solid box is a close-up view of the dashed box. It can be found that the diameter, height, and pitch of the prepared column array are consistent with the corresponding structural parameters of the template, respectively. The column array is structurally intact and aesthetically pleasing. The results indicate that the polypropylene melt filled the micropores adequately under the continuous action of ultrasound vibration. In Figure 2j, the micropore array of the D100-array-LC template remains intact and unobstructed after multiple uses, indicating that the template can be recycled.

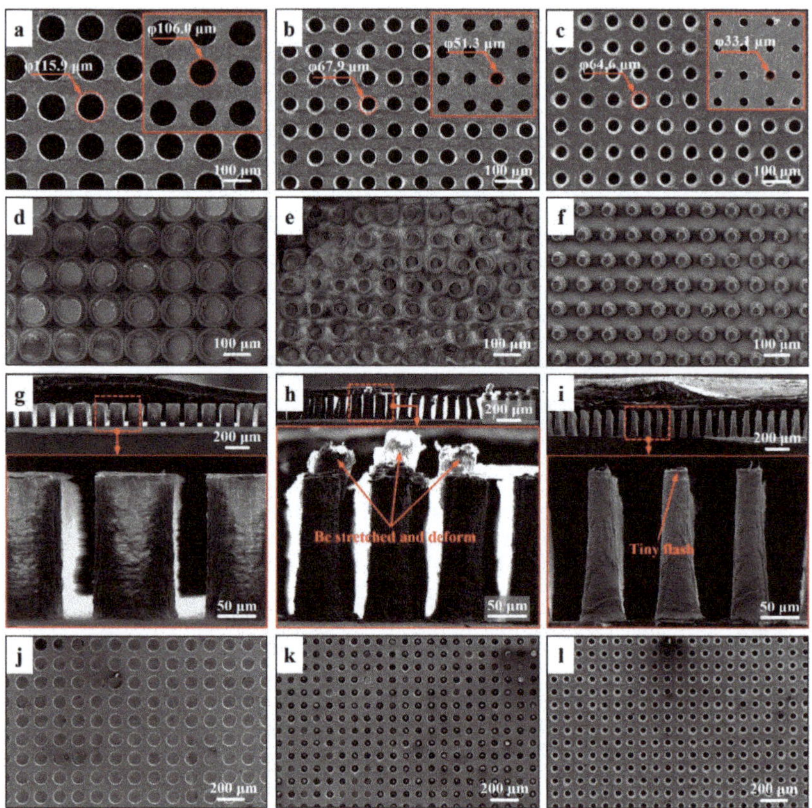

Figure 2. The preparation of single-stage micropillar arrays. (**a–c**) The surface morphologies of D100-array-LC, D50-array-LC, and D50-array-LCM templates, respectively. (**d–f**) Top views of φ110 μm column, φ67 μm column, and φ64-33 μm cone-column arrays, respectively. (**g–i**) Side views of the micropillar arrays, respectively. (**j–l**) The surface morphologies of the templates after multiple uses, respectively.

As shown in Figure 2b, the diameters of the circular holes on the upper and lower surfaces of the D50-array-LC template are about 67.9 μm and 51.3 μm, respectively. The demolding angle of the D50-array-LC template is 2.38°. Figure 2e,h show the top and side views of the column array obtained by using this template, respectively. It can be observed that the polypropylene melt was able to completely fill the micropores, and the array structure was complete after solidification. However, some of the columns were stretched and deformed due to excessive demolding resistance during ejection. Figure 2k demonstrates that after multiple uses, the micropores of the D50-array-LC template were gradually blocked by the polypropylene pillars that were pulled off due to excessive demolding resistance. Therefore, it is necessary to reduce the roughness of the inner surface of the micropores or increase the demolding angle of the micropores by adjusting the laser cutting parameters of the template. In Figure 2c, the diameters of the circular holes on the upper and lower surfaces of the D50-array-LCM template are approximately 64.6 μm and 33.1 μm, respectively. Consequently, the demolding angle is increased to 4.5°. As shown in Figure 2f,i, the cone-column array prepared by using the D50-array-LCM template is neat and aesthetic, with no cone columns being stretched or deformed. The overall forming quality is very good, although there is a tiny flash on top of the cone column. Figure 2l shows the micropore array of the D50-array-LCM template after multiple uses; it remains intact and unobstructed. The results indicate that the template with a proper demolding angle can be recycled in UPP and achieve damage-free demolding of the micropillar array.

In this section, a single-stage column array with a diameter of about 110 μm, a height of 200 μm, and a pitch of 150 μm, as well as a single-stage cone-column array with a top diameter of about 33 μm, a root diameter of about 64 μm, a height of 200 μm, and a pitch of 100 μm, were successfully fabricated by UPP.

3.3. Micro/Nanoscale Hierarchical Arrays

The successful fabrication of a micronscale micropillar array with a high aspect ratio (~4.1) indicates that the polymer melt has a very strong ability to fill micropores during UPP. Therefore, multilevel templates can be set up for the fabrication of hierarchical array structures. In Figure 3a, a secondary template with nanoscale pores is added below the primary template with micron-scale pores. When ultrasonic loading is initiated, the polymer pellets are plasticized and melted, rapidly filling the micropores of the two-stage templates under ultrasonic vibration. After pressure holding, the sample is demolded, and the residual templates are removed to obtain the micro/nanoscale hierarchical arrays.

Figure 3c,d show the hierarchical micropillar arrays prepared by combining the D100-array-LC template (primary template) with the 390 nm-AAO template (secondary template, Figure 3b). It can be found that the top of the column with an intact array structure is covered with a nanowire array. The results indicate that under the continuous action of the horn and ultrasonic vibration, the polypropylene melt can continue to fill the ordered nanopore array of the secondary template after completely filling the micropores of the primary template, which adequately demonstrates that UPP has superb micro- and nano-pore filling ability. Obviously, the formed nanowires are severely agglomerated. The residual bending stresses of the polypropylene nanowires after solidification at room temperature were released upon dissolution of the AAO template [24], and the pore margin of the 390 nm-AAO template was much smaller relative to the pore diameter, thus leading to the bending of the high aspect ratio nanowires and the formation of severe agglomerates in various localized regions. Figure 3f,g show the hierarchical micropillar arrays prepared by combining the D100-array-LC template with the 250 nm-AAO template (Figure 3e). Although nanowire agglomeration still exists, it is slight as the pore margin of the 250 nm-AAO template is comparable to the pore diameter.

Figure 3h–j show the hierarchical micropillar arrays prepared by combining the D50-array-LCM template with the 250 nm-AAO template. The nanowire array (diameter 250 nm, height about 5 μm, and aspect ratio up to 20) covers the top of the cone-column. In the center region, the nanowires are well defined with slight agglomeration, while in the edge region, dumping and severe agglomeration of nanowires are present. Since the micropores of the D50-array-LCM template were small and dense, and the microporous cutting depth of 200 μm was large relative to the pore diameter, tiny localized warping deformations inevitably occurred in the micropore array after laser cutting. This warping deformation resulted in gaps between the lower surface of the micropore array and the upper surface of the bottom mold or secondary template. During single-stage cone-column array preparation, the polypropylene melt filled the gaps under ultrasonic vibration; thus, tiny flashes were formed at the edge of the cone-column top, which can be clearly observed in Figure 2i. While hierarchical micropillar arrays were being prepared, the polypropylene melt not only filled the gaps but also filled the ordered nanopore array underneath the gaps. The solidified flashes were pulled and compressed to shape under the constraints of the micropores of the primary template during demolding. As a result, the nanowires formed on the flash surface deformed along with the flash, leading to dumping and severe agglomeration of the nanowires in the edge region. The formation of nanowires on the flash surface caused by machining defects of the template exactly demonstrates the superb micro- and nano-pore filling ability of the polymer melt in UPP.

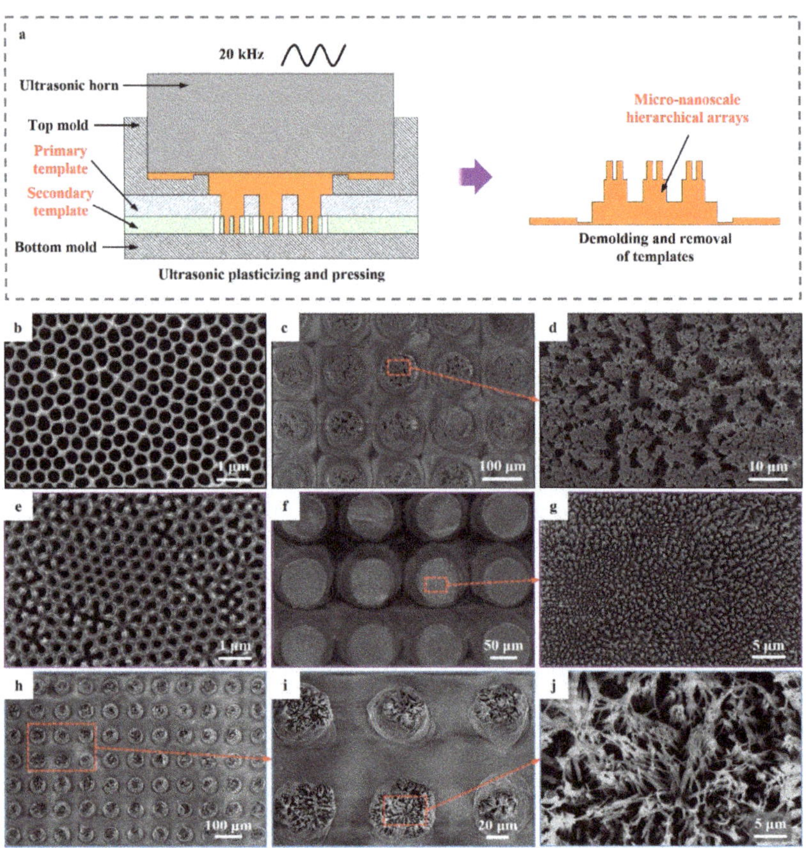

Figure 3. The preparation of micro-nanoscale hierarchical micropillar arrays. (**a**) The schematic illustration of preparing hierarchical arrays. (**b**) The surface morphology of the 390 nm-AAO template. (**c**) The φ110 μm and φ390 nm hierarchical arrays. (**d**) A close-up view of the selected area in (**c**). (**e**) The surface morphology of the 250 nm-AAO template. (**f**) The φ110 μm and φ250 nm hierarchical arrays. (**g**) A close-up view of the selected area in (**f**). (**h**) The φ64-33 μm and φ250 nm hierarchical arrays. (**i,j**) are close-up views of the selected area in (**h,i**), respectively.

In this section, φ110 μm and φ390 nm, φ110 μm and φ250 nm, and φ64-33 μm and φ250 nm hierarchical micropillar arrays were successfully fabricated, which verifies the capability of UPP to fabricate micro/nanoscale hierarchical surface microstructures with high aspect ratios.

3.4. Comparison with Other Methods for Fabricating Micropillar Arrays

To demonstrate the superiority of UPP, the previous work on the fabrication of hierarchical micropillar arrays by the template method and the results thereof have been summarized in Table S1. Microstructure scale, aspect ratio, forming time, and preheating temperature of raw material are the key indexes for evaluating the merits and drawbacks of a forming method. The microstructure scale and aspect ratio reflect the forming capability of the method, while the forming time and preheating temperature of the raw material directly affect the production cost. Therefore, the above indexes are extracted from Table S1 for comparison, as shown in Figure 4. Among the previous results in Table S1, the aspect ratio of primary structure (micron-scale) of the hierarchical micropillar arrays prepared in this work is within the top 25% (Figure 4a), while the aspect ratio of secondary structure (nanoscale) is the highest, up to 20 (Figure 4b). The sum of the aspect ratios of primary and secondary structures in this work is 24.1, which is only 2.8 smaller than that of the highest (nanoimprinting, No.5). However, UPP has the shortest forming time (only 5.3 s; Figure 4c) and does not need to heat the raw material (at room temperature; Figure 4d), whereas nanoimprinting (No.5) requires a forming time of 20 min and heating the raw material up to 175 °C. In other aspects, some methods (No.1, No.3, No.4, No.7, No.8, and No.10) need two-step preparation of the primary and secondary structures, some methods (No.3, No.4, and No.6–10) are limited to the light-cured materials, and some methods (No.4, No.5, No.8, No.11, No.14, No.15, and No.17–19) require a vacuum environment or preheating of the mold, while UPP can form multilayer structures in one step at room temperature without preheating the mold. Therefore, UPP has obvious superiority in mass production of polymer surface microstructures due to its superb forming capability of microstructures and nanostructures, simple process, short production cycle, and low cost.

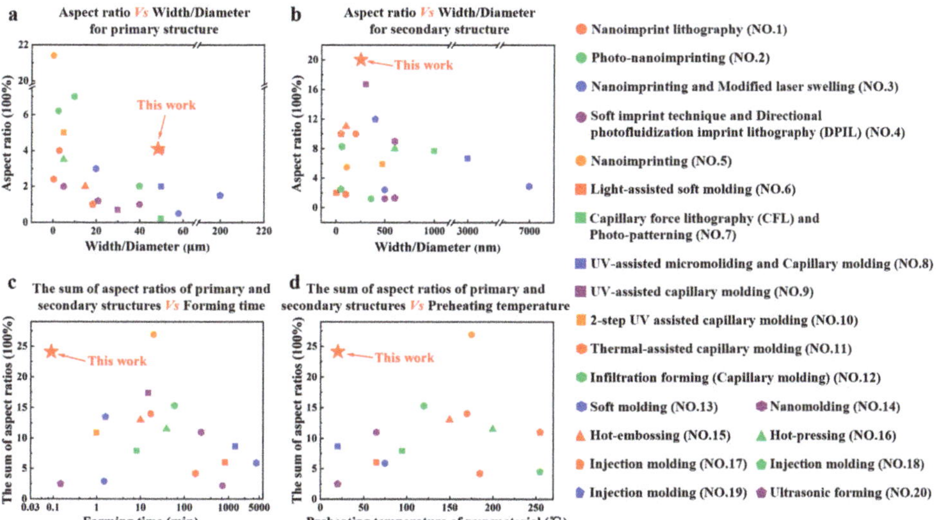

Figure 4. Comparisons of the hierarchical micropillar arrays prepared in this work with those prepared in the previous work. (**a**) Aspect ratio vs. width or diameter for primary structure. (**b**) Aspect ratio vs. width/diameter for secondary structure. (**c**) The sum of the aspect ratios of primary and secondary structures vs. forming time. (**d**) The sum of the aspect ratios vs. preheating temperature of the raw material.

Ultrasonic molding technology, which utilizes ultrasonic vibration as the main energy source to plasticize thermoplastic polymers, has become a promising microstructure replication technology with the advantages of short forming timed, simple equipment, and low cost. Table 1 summarizes the results of the previous work on the preparation of single-stage or hierarchical micropillar arrays by ultrasonic molding technology. Among the previous results, the aspect ratios of the micropillar arrays prepared in this work are the highest and far ahead, regardless of the micronscale or nanoscale. In addition, UPP has the shortest ultrasonic duration. As shown in Table 1, only two previous works have successfully fabricated nanoscale micropillar arrays, and even less, only one previous work has successfully fabricated micro/nanoscale hierarchical micropillar arrays, but the aspect ratios of the primary and secondary structures are only 1.2 and 1.3, respectively. Therefore, ultrasonic molding technology is almost blank in the preparation of single-stage or hierarchical micropillar arrays with high aspect ratios, whereas UPP achieves a significant breakthrough in this aspect, fast-tracking the development and application of ultrasonic molding technology.

Table 1. The results of the previous work in the preparation of micropillar arrays by ultrasonic molding technology.

No.	Technique	Structural Levels	Primary Structure/Micron-Scale		Secondary Structure /Nanoscale		Ultrasonic Duration Time (s)	Mold Temperature (°C)	Raw Material	Mold/Template	Reference
			Width/Diameter (μm)	Aspect Ratio	Width/Diameter (nm)	Aspect Ratio					
1	Ultrasonic-assisted hot embossing	Single-stage	250	0.4	/	/	2	Room temperature	Polymethyl methacrylate (PMMA)	304 stainless steel mold	[38]
2	Ultrasonic embossing	Single-stage	100	2.7	/	/	1	Room temperature	Polyethylene terephthalate (PET)	6061 aluminum mold	[39]
3	Ultrasonic hot embossing	Single-stage	280	0.5	/	/	Not reported	Room temperature	High-density polyethylene (HDPE)	Aluminum mold	[40]
4	Ultrasonic micromolding	Single-stage	68.5	0.8	/	/	2.5	Room temperature	Polymethyl methacrylate (PMMA)	Nickel micro-mold	[22]
5	Micro ultrasonic powder molding (micro-UPM)	Single-stage	108.4	1.7	/	/	4.5	Room temperature	Ultra-high-molecule weight polyethylene (UHMWPE)	Printed circuit board (PCB)	[41]
6	Ultrasonic micro-moulding	Single-stage	409	2.1	/	/	4	90	Polypropylene (PP)	Metal mold	[42]
7	Ultrasonic plasticization microinjection molding (UPMIM)	Single-stage	174.5	2.7	/	/	2	90	Polypropylene (PP)	316 stainless steel mold	[43]
8	Ultrasonic loading	Single-stage	/	/	200	5	0.7	Room temperature	Polyvinylidene fluoride (PVDF)	AAO template	[24]
9	Ultrasonic forming	Hierarchical	20.9	1.2	600	1.3	3	Room temperature	Polyethylene (PE)	Nickel nano–micro mold	[23]
10	Ultrasonic plasticizing and pressing (UPP)	Hierarchical	48.5	4.1	250	20	0.3	Room temperature	Polypropylene (PP)	(i) 304 stainless steel template (primary structure). (ii) AAO template (secondary structure).	This work

3.5. Wettability

As shown in Figure 5a–f, the WCA of the polypropylene original surface is 85.9°, while the WCAs of the φ110 μm column array and φ64-33 μm cone-column array are 147.6° and 153.0°, which are 71.83% and 78.11% higher than the WCA of the original surface, respectively. Compared with the corresponding single-stage micropillar arrays, the WCAs of φ110 μm and φ390 nm, φ110 μm and φ250 nm, and φ64-33 μm and φ250 nm hierarchical micropillar arrays are further improved due to the presence of nanowire arrays, which are 154.5°, 152.1°, and 157.1°, respectively. According to the Cassie–Baxter model [44], when air in the gaps between the microstructures prevents the droplets from filling the surface microstructure, the WCA is expressed by the following formula:

$$\cos \theta_{CB} = f(1 + \cos \theta_Y) - 1 \tag{1}$$

where θ_{CB} is the WCA, θ_Y is the intrinsic WCA on a structure-free surface, and f is the fraction of the solid surface in contact with the liquid. The dense micropillars and nanowires effectively increased the water–air contact area between the droplets and the solid surface and changed the contact state, resulting in a significant decrease in the f value. Therefore, the WCAs of the micropillar arrays were improved to different degrees.

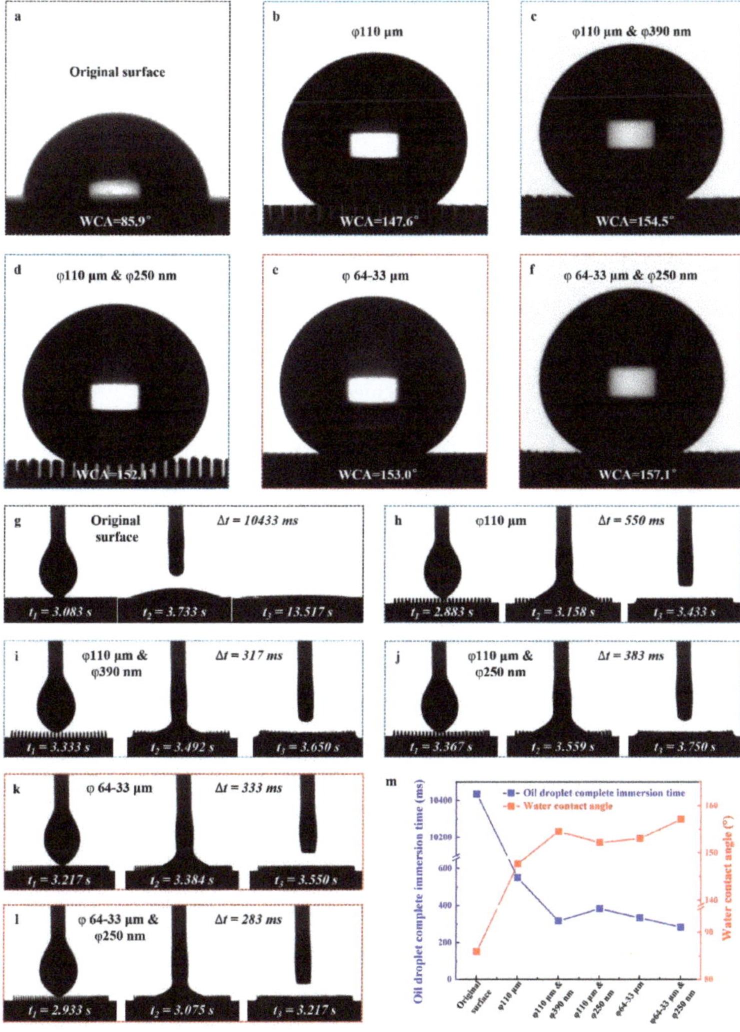

Figure 5. Wettability of the micropillar arrays. (**a–f**) The WCAs of the polypropylene original surface and five prepared micropillar arrays, respectively. (**g–l**) The states and corresponding moments of the oil droplet immersing into the original surface and five prepared micropillar arrays, respectively. (**m**) The trends of the WCA and oil droplet complete immersion time for different micropillar arrays.

Figure 5g–l show three states and the corresponding moments during the immersion of oil droplets into the original surface and five micropillar arrays prepared in this work, respectively, including the states of beginning contact, a certain moment during the period, and complete immersion. It can be noticed that it was difficult for the oil droplet to completely immerse into the polypropylene original surface, and the time required for the state at the moment t_3 was 10,433 ms (Figure 5g). However, for the prepared micropillar arrays, it was easy to achieve complete immersion of the oil droplet, and the time required for $\varphi110$ μm and $\varphi250$ hierarchical micropillar arrays was the shortest, only 283 ms (Figure 5l). As shown in Figure 5m, the trends of the WCA and the oil droplet complete immersion time are exactly opposite. The results indicate that the presence of prepared micropillar arrays ensures complete immersion of the oil, and the larger the contact angle, the faster the immersion rate. In conclusion, the surface microstructures prepared in this work have superhydrophobic and superoleophilic properties.

4. Conclusions

(1) UPP makes full use of ultrasonic vibration, avoids violent friction between the raw material and the template microstructure, and is therefore particularly suitable for reproducing surface microstructures with high aspect ratios from templates.

(2) Micron-scale single-stage micropillar arrays with an aspect ratio of 4.1 and micro-nanoscale hierarchical micropillar arrays with an aspect ratio of 24.1 were successfully prepared, verifying the feasibility and forming capability of UPP.

(3) Compared with other template methods for fabricating polymer surface microstructures, UPP possesses the advantages of superb forming capability of micro- and nanostructures, simple process, short production cycle and high cost-effectiveness. Therefore, it has potential uses in both research and application.

(4) UPP has achieved a significant breakthrough in ultrasonic molding technology in the fabrication of micropillar arrays with high aspect ratios.

(5) The prepared PP micropillar arrays have superhydrophobic and superoleophilic properties.

Supplementary Materials: The following supporting information can be downloaded at: https://www.mdpi.com/article/10.3390/pr12050856/s1, References [45–64] are cited in Supplementary Materials.

Author Contributions: Methodology, S.W.; Validation, S.W.; Formal analysis, S.X.; Investigation, S.W. and S.X.; Resources, J.D.; Writing—original draft, S.W.; Writing—review & editing, S.W., J.D. and J.M.; Visualization, J.L. and L.Z.; Supervision, J.D.; Funding acquisition, J.D., J.L., J.M. and L.Z. All authors have read and agreed to the published version of the manuscript.

Funding: The current work was supported, in part, by the Guangdong HUST Industrial Technology Research Institute, the Guangdong Provincial Key Laboratory of Manufacturing Equipment Digitization (2023B1212060012), and Shenzhen Stable Support B Plan (Grant No. 20231120184926001).

Data Availability Statement: Data are contained within the article and Supplementary Materials.

Conflicts of Interest: The authors declare that they have no known competing financial interests or personal relationships that could have appeared to influence the work reported in this paper.

References

1. Peng, Y.J.; Huang, H.X.; Xie, H. Rapid fabrication of antireflective pyramid structure on polystyrene film used as protective layer of solar cell. *Sol. Energy Mater. Sol. Cells* **2017**, *171*, 98–105. [CrossRef]
2. Fu, J.; Li, Z.; Li, X.; Sun, F.; Li, L.; Li, H.; Zhao, J.; Ma, J. Hierarchical porous metallic glass with strong broadband absorption and photothermal conversion performance for solar steam generation. *Nano Energy* **2023**, *106*, 108019. [CrossRef]
3. Tadepalli, S.; Slocik, J.M.; Gupta, M.K.; Naik, R.R.; Singamaneni, S. Bio-optics and bio-inspired optical materials. *Chem. Rev.* **2017**, *117*, 12705–12763. [CrossRef] [PubMed]
4. Bandara, C.D.; Singh, S.; Afara, I.O.; Wolff, A.; Tesfamichael, T.; Ostrikov, K.; Oloyede, A. Bactericidal effects of natural nanotopography of dragonfly wing on escherichia coli. *ACS Appl. Mater. Interfaces* **2017**, *9*, 6746–6760. [CrossRef] [PubMed]

5. Tsui, K.H.; Li, X.; Tsoi, J.K.H.; Leung, S.F.; Tang, L.; Chak, W.Y.; Zhang, C.; Chen, J.; Cheung, G.S.P.; Fan, Z. Low-cost, flexible, disinfectant-free and regular-array three-dimensional nanopyramid antibacterial films for clinical applications. *Nanoscale* **2018**, *10*, 10436–10442. [CrossRef] [PubMed]
6. Hu, D.; Deng, Y.; Jia, F.; Jin, Q.; Ji, J. Surface charge switchable supramolecular nanocarriers for nitric oxide synergistic photodynamic eradication of biofilms. *ACS Nano* **2020**, *14*, 347–359. [CrossRef] [PubMed]
7. Ye, S.; Cheng, C.; Chen, X.; Chen, X.; Shao, J.; Zhang, J.; Hu, H.; Tian, H.; Li, X.; Ma, L.; et al. High-performance piezoelectric nanogenerator based on microstructured P (VDF-TrFE)/BNNTs composite for energy harvesting and radiation protection in space. *Nano Energy* **2019**, *60*, 701–714. [CrossRef]
8. Zhang, Y.; Han, F.; Hu, Y.; Xiong, Y.; Gu, H.; Zhang, G.; Zhu, P.; Sun, R.; Wong, C.P. Flexible and highly sensitive pressure sensors with surface discrete microdomes made from self-assembled polymer microspheres array. *Macromol. Chem. Phys.* **2020**, *221*, 2000073. [CrossRef]
9. Attia, U.M.; Marson, S.; Alcock, J.R. Micro-injection moulding of polymer microfluidic devices. *Microfluid. Nanofluid.* **2009**, *7*, 1–28. [CrossRef]
10. Rytka, C.; Kristiansen, P.M.; Neyer, A. Iso- and variothermal injection compression moulding of polymer micro-and nanostructures for optical and medical applications. *J. Micromech. Microeng.* **2015**, *25*, 065008. [CrossRef]
11. Hench, L.L.; West, J.K. The sol-gel process. *Chem. Rev.* **1990**, *90*, 33–72. [CrossRef]
12. Dudem, B.; Heo, J.H.; Leem, J.W.; Yu, J.S.; Im, S.H. $CH_3NH_3PbI_3$ planar perovskite solar cells with antireflection and self-cleaning function layers. *J. Mater. Chem. A* **2016**, *4*, 7573–7579. [CrossRef]
13. Zhang, Y.; Lin, C.T.; Yang, S. Fabrication of hierarchical pillar arrays from thermoplastic and photosensitive SU-8. *Small* **2010**, *6*, 768–775. [CrossRef] [PubMed]
14. Jeong, H.E.; Suh, K.Y. On the role of oxygen in fabricating microfluidic channels with ultraviolet curable materials. *Lab Chip* **2008**, *8*, 1787–1792. [CrossRef]
15. Deshmukh, S.S.; Goswami, A. Hot embossing of polymers—A review. *Mater. Today Proc.* **2020**, *26*, 405–414. [CrossRef]
16. Li, J.M.; Liu, C.; Peng, J. Effect of hot embossing process parameters on polymer flow and microchannel accuracy produced without vacuum. *J. Mater. Process. Technol.* **2008**, *207*, 163–171. [CrossRef]
17. Giboz, J.; Copponnex, T.; Mélé, P. Microinjection molding of thermoplastic polymers: A review. *J. Micromech. Microeng.* **2007**, *17*, R96. [CrossRef]
18. Sha, B.; Dimov, S.; Griffiths, C.; Packianather, M.S. Investigation of micro-injection moulding: Factors affecting the replication quality. *J. Mater. Process. Technol.* **2007**, *183*, 284–296. [CrossRef]
19. Mekaru, H.; Nakamura, O.; Maruyama, O.; Maeda, R.; Hattori, T. Development of precision transfer technology of atmospheric hot embossing by ultrasonic vibration. *Microsyst. Technol.* **2007**, *13*, 385–391. [CrossRef]
20. Sato, A.; Ito, H.; Koyama, K. Study of application of ultrasonic wave to injection molding. *Polym. Eng. Sci.* **2009**, *49*, 768–773. [CrossRef]
21. Qiu, Z.; Yang, X.; Zheng, H.; Gao, S.; Fang, F. Investigation of micro-injection molding based on longitudinal ultrasonic vibration core. *Appl. Opt.* **2015**, *54*, 8399–8405. [CrossRef] [PubMed]
22. Yu, H.W.; Lee, C.H.; Jung, P.G.; Shin, B.S.; Kin, J.H.; Hwang, K.Y.; Ko, J.S. Polymer microreplication using ultrasonic vibration energy. *J. Micro/Nanolithogr. MEMS MOEMS* **2009**, *8*, 021113. [CrossRef]
23. Lee, C.H.; Jung, P.G.; Lee, S.M.; Park, S.H.; Shin, B.S.; Kim, J.H.; Hwang, K.Y.; Ko, J.S. Replication of polyethylene nano-micro hierarchical structures using ultrasonic forming. *J. Micromech. Microeng.* **2010**, *20*, 035018. [CrossRef]
24. Li, Z.; Fu, J.; Zhang, L.; Ma, J.; Shen, J. Rapid Forming of Nanowire Array on Polyvinylidene Fluoride Polymer Surfaces at Room Temperature by Ultrasonic Loading. *J. Adv. Eng. Mater.* **2023**, *25*, 2200700. [CrossRef]
25. Liang, X.; Liu, Y.; Chen, S.; Ma, J.; Wu, X.; Shi, H.; Fu, L.; Xu, B. Fabrication of microplastic parts with a hydrophobic surface by micro ultrasonic powder moulding. *J. Manuf. Process.* **2020**, *56*, 180–188. [CrossRef]
26. Pan, L.; Wu, W.; Jia, C.; Liu, J. Ultrasonic plasticization micro-injection molding technology: Characterization of replication fidelity and wettability of high-depth microstructure. *J. Polym. Res.* **2023**, *30*, 349. [CrossRef]
27. Liang, X.; Wu, X.; Xu, B.; Ma, J.; Liu, Z.; Peng, T.; Fu, L. Phase structure development as preheating UHMWPE powder temperature changes in the micro-UPM process. *J. Micromech. Microeng.* **2016**, *26*, 015014. [CrossRef]
28. Jiang, B.; Zou, Y.; Liu, T.; Wu, W. Characterization of the fluidity of the ultrasonic plasticized polymer melt by spiral flow testing under micro-scale. *Polymers* **2019**, *11*, 357. [CrossRef]
29. Jiang, B.; Peng, H.; Wu, W.; Jia, Y.; Zhang, Y. Numerical simulation and experimental investigation of the viscoelastic heating mechanism in ultrasonic plasticizing of amorphous polymers for micro injection molding. *Polymers* **2016**, *8*, 199. [CrossRef]
30. Janer, M.; Planta, X.; Riera, D. Ultrasonic moulding: Current state of the technology. *Ultrasonics* **2020**, *102*, 106038. [CrossRef]
31. Wu, W.; Peng, H.; Jia, Y.; Jiang, B. Characteristics and mechanisms of polymer interfacial friction heating in ultrasonic plasticization for micro injection molding. *Microsyst. Technol.* **2017**, *23*, 1385–1392. [CrossRef]
32. Chen, J.; Chen, Y.; Li, H.; Lai, S.Y.; Jow, J. Physical and chemical effects of ultrasound vibration on polymer melt in extrusion. *Ultrason. Sonochem.* **2010**, *17*, 66–71. [CrossRef]
33. Peshkovskii, S.L.; Friedman, M.L.; Tukachinskii, A.I.; Vinogradov, G.V.; Enikolopian, N.S. Acoustic cavitation and its effect on flow in polymers and filled systems. *Polym. Compos.* **1983**, *4*, 126–134. [CrossRef]

34. Yang, Y.J.; Huang, C.C.; Lin, S.K.; Tao, J. Characteristics analysis and mold design for ultrasonic assisted injection molding. *J. Polym. Eng.* **2014**, *34*, 673–681. [CrossRef]
35. Yang, Y.J.; Huang, C.C.; Tao, J. Application of ultrasonic-assisted injection molding for improving melt flowing and floating fibers. *J. Polym. Eng.* **2016**, *36*, 119–128. [CrossRef]
36. Yang, Y.J.; Huang, C.C. Effects of ultrasonic injection molding conditions on the plate processing characteristics of PMMA. *J. Polym. Eng.* **2018**, *38*, 905–914. [CrossRef]
37. Liu, S.J.; Dung, Y.T. Hot embossing precise structure onto plastic plates by ultrasonic vibration. *Polym. Eng. Sci.* **2005**, *45*, 915–925. [CrossRef]
38. Chang, C.Y.; Yu, C.H. A basic experimental study of ultrasonic assisted hot embossing process for rapid fabrication of microlens arrays. *J. Micromech. Microeng.* **2015**, *25*, 025010. [CrossRef]
39. Zhu, J.; Tian, Y.; Yang, C.; Cai, L.; Wang, F.; Zhang, D.; Liu, X. Low-cost and fast fabrication of the ultrasonic embossing on polyethylene terephthalate (PET) films using laser processed molds. *Microsyst. Technol.* **2017**, *23*, 5653–5668. [CrossRef]
40. Liao, S.; Gerhardy, C.; Sackmann, J.; Schomburg, W.K. Tools for ultrasonic hot embossing. *Microsyst. Technol.* **2015**, *21*, 1533–1541. [CrossRef]
41. Liang, X.; Li, B.; Wu, X.; Shi, H.; Zeng, K.; Wang, Y. Micro UHMW-PE column array molded by the utilization of PCB as mold insert. *Circuit World* **2013**, *39*, 95–101. [CrossRef]
42. Gülür, M.; Brown, E.; Gough, T.; Romano, J.M.; Penchev, P.; Dimov, S.; Whiteside, B. Ultrasonic micromoulding: Process characterisation using extensive in-line monitoring for micro-scaled products. *J. Manuf. Process.* **2020**, *58*, 289–301. [CrossRef]
43. Pan, L.; Wu, W.; Liu, J.; Li, X. Ultrasonic plasticization microinjection precision molding of polypropylene microneedle arrays. *ACS Appl. Polym. Mater.* **2023**, *5*, 9354–9363. [CrossRef]
44. Cassie, A.; Baxter, S. Wettability of Porous Surfaces. *Trans. Faraday Soc.* **1944**, *40*, 546–551. [CrossRef]
45. Raut, H.K.; Baji, A.; Hariri, H.H.; Parveen, H.; Soh, G.S.; Low, H.Y.; Wood, K.L. Gecko-inspired dry adhesive based on micro-nanoscale hierarchical arrays for application in climbing devices. *ACS Appl. Mater. Interfaces* **2018**, *10*, 1288–1296. [CrossRef] [PubMed]
46. Yanagishita, T.; Sou, T.; Masuda, H. Micro-nano hierarchical pillar array structures prepared on curved surfaces by nanoimprinting using flexible molds from anodic porous alumina and their application to superhydrophobic surfaces. *RSC Adv.* **2022**, *12*, 20340–20347. [CrossRef] [PubMed]
47. Shao, J.; Ding, Y.; Wang, W.; Mei, X.; Zhai, H.; Tian, H.; Li, X.; Liu, B. Generation of fully-covering hierarchical micro-/nano-structures by nanoimprinting and modified laser swelling. *Small* **2014**, *10*, 2595–2601. [CrossRef] [PubMed]
48. Choi, J.; Cho, W.; Jung, Y.S.; Kang, H.S.; Kin, H.T. Direct fabrication of micro/nano-patterned surfaces by vertical-directional photofluidization of azobenzene materials. *ACS Nano* **2017**, *11*, 1320–1327. [CrossRef] [PubMed]
49. Ho, A.Y.Y.; Yeo, L.P.; Lam, Y.C.; Rodriguez, I. Fabrication and analysis of gecko-inspired hierarchical polymer nanosetae. *ACS Nano* **2011**, *5*, 1897–1906. [CrossRef] [PubMed]
50. Greiner, C.; Arzt, E.; Campo, A.D. Hierarchical gecko-like adhesives. *Adv. Mater.* **2009**, *21*, 479–482. [CrossRef]
51. Aksak, B.; Murphy, M.P.; Sitti, M. Adhesion of biologically inspired vertical and angled polymer microfiber arrays. *Langmuir* **2007**, *23*, 3322–3332. [CrossRef] [PubMed]
52. Murphy, M.P.; Aksak, B.; Sitti, M. Adhesion and anisotropic friction enhancements of angled heterogeneous micro-fiber arrays with spherical and spatula tips. *J. Adhes. Sci. Technol.* **2007**, *21*, 1281–1296. [CrossRef]
53. Murphy, M.P.; Kim, S.; Sitti, M. Enhanced adhesion by gecko-inspired hierarchical fibrillar adhesives. *ACS Appl. Mater. Interfaces* **2009**, *1*, 849–855. [CrossRef] [PubMed]
54. Ma, Z.; Jiang, C.; Li, X.; Ye, F.; Yuan, W. Controllable fabrication of periodic arrays of high-aspect-ratio micro-nano hierarchical structures and their superhydrophobicity. *J. Micromech. Microeng.* **2013**, *23*, 095027. [CrossRef]
55. Jeong, H.E.; Lee, J.K.; Kim, H.N.; Moon, S.H.; Suh, K.Y. A nontransferring dry adhesive with hierarchical polymer nanohairs. *Proc. Natl. Acad. Sci. USA* **2009**, *106*, 5639–5644. [CrossRef] [PubMed]
56. Lee, D.Y.; Lee, D.H.; Lee, S.G.; Cho, K. Hierarchical gecko-inspired nanohairs with a high aspect ratio induced by nanoyielding. *Soft Matter* **2012**, *8*, 4905–4910. [CrossRef]
57. Kustandi, T.S.; Samper, V.D.; Ng, W.S.; Chong, A.S.; Gao, H. Fabrication of a gecko-like hierarchical fibril array using a bonded porous alumina template. *J. Micromech. Microeng.* **2007**, *17*, N75. [CrossRef]
58. Jin, K.J.; Tian, Y.; Erickson, J.S.; Puthoff, J.; Autumn, K.; Pesika, N.S. Design and Fabrication of Gecko-Inspired Adhesives. *Langmuir* **2012**, *28*, 5737–5742. [CrossRef] [PubMed]
59. Sitti, M. High aspect ratio polymer micro/nano-structure manufacturing using nanoembossing, nanomolding and directed self-assembly. In Proceedings of the 2003 IEEE/ASME International Conference on Advanced Intelligent Mechatronics, Kobe, Japan, 20–24 July 2003.
60. Li, K.; Hernandez-Castro, J.A.; Turcotte, K.; Morton, K.; Veres, T. Superhydrophobic thermoplastic surfaces with hierarchical micro-nanostructures fabricated by hot embossing. In Proceedings of the 20th IEEE International Conference on Nanotechnology, Montreal, QC, Canada, 29–31 July 2020.
61. Bhushan, B.; Lee, H. Fabrication and characterization of multi-level hierarchical surfaces. *Faraday Discuss.* **2012**, *156*, 235–241. [CrossRef]

62. Puukilainen, E.; Rasilainen, T.; Suvanto, M.; Pakkanen, T.A. Superhydrophobic polyolefin surfaces: Controlled micro-and nanostructures. *Langmuir* **2007**, *23*, 7263–7268. [CrossRef]
63. Huovinen, E.; Hirvi, J.; Suvanto, M.; Pakkanen, T.A. Micro-micro hierarchy replacing micro–nano hierarchy: A precisely controlled way to produce wear-resistant superhydrophobic polymer surfaces. *Langmuir* **2012**, *28*, 14747–14755. [CrossRef] [PubMed]
64. Zhou, M.; Xiong, X.; Jiang, B.; Weng, C. Fabrication of high aspect ratio nanopillars and micro/nano combined structures with hydrophobic surface characteristics by injection molding. *Appl. Surf. Sci.* **2018**, *427*, 854–860. [CrossRef]

 *processes*

MDPI

Article

Validation of Fluid Flow Speed Behavior in Capillary Microchannels Using Additive Manufacturing (SLA Technology)

Victor H. Cabrera-Moreta [1,2,*], Jasmina Casals-Terré [1] and Erick Salguero [2]

[1] Laboratory of Microsystems and Nanotechnology, Mechanical Engineering Department, Polytechnic University of Catalonia (UPC), Colom Street 11, 08222 Terrassa, Spain; jasmina.casals@upc.edu

[2] Mechanical Engineering Department, Universidad Politécnica Salesiana, Quito 170517, Ecuador; esalgueror@est.ups.edu.ec

* Correspondence: victor.cabrera.moreta@upc.edu or vcabrera@ups.edu.ec

Abstract: This research explores fluid flow speed behavior in capillary channels using additive manufacturing, focusing on stereolithography (SLA). It aims to validate microchannels fabricated through SLA for desired fluid flow characteristics, particularly capillary-driven flow. The methodology involves designing, fabricating, and characterizing microchannels via SLA, with improvements such as an air-cleaning step facilitating the production of microchannels ranging from 300 to 1000 µm. Experimental validation assesses fluid flow speed behavior across channels of varying dimensions, evaluating the impact of channel geometry, surface roughness, and manufacturing parameters. The findings affirm the feasibility and efficacy of SLA in producing microchannels with consistent and predictable fluid flow behavior between 300 to 800 µm. This study contributes insights into microfluidic device fabrication techniques and enhances the understanding of fluid dynamics in capillary-driven systems. Overall, it underscores the potential of additive manufacturing, specifically SLA, in offering cost-effective and scalable solutions for microfluidic applications. The validated fluid flow speed behavior in capillary channels suggests new avenues for developing innovative microfluidic devices with improved performance and functionality, marking a significant advancement in the field.

Keywords: additive manufacturing; stereolithography (SLA); microchannels; capillary-driven

Citation: Cabrera-Moreta, V.H.; Casals-Terré, J.; Salguero, E. Validation of Fluid Flow Speed Behavior in Capillary Microchannels Using Additive Manufacturing (SLA Technology). *Processes* **2024**, *12*, 1066. https://doi.org/10.3390/pr12061066

Academic Editors: Guoqing Zhang, Zejia Zhao and Wai Sze Yip

Received: 3 April 2024
Revised: 30 April 2024
Accepted: 7 May 2024
Published: 23 May 2024

1. Introduction

Figure 1 demonstrates the exponential growth of additive manufacturing and 3D printing in microfluidics over the past two decades [1]. This trend highlights the increasing recognition of 3D printing as a powerful tool in microfluidic research, enabling the fabrication of intricate devices with tailored designs and functionalities. This evolution promises to revolutionize microfluidic experimentation and applications, ushering in a new era of innovation and discovery in the field [2].

Additive manufacturing has emerged as an appealing option for producing microfluidic chips across various applications. Compared to traditional methods like soft lithography, additive manufacturing offers a straightforward, rapid, and cost-effective approach to creating microstructures. One of its key advantages lies in the accessibility and affordability of the necessary equipment, making it feasible for both industry and research laboratories. Over recent years, a range of rapid prototyping methods, including 3D printing, have surfaced as invaluable tools for developing microfluidic devices [1]. Figure 2 provides an overview of the additive manufacturing technologies currently accessible in the market.

Despite the array of additive manufacturing technologies available, research will primarily concentrate on SLA, Polyjet (MultiJet), or FDM due to their accessibility and cost-effectiveness. These technologies will be compared based on their respective advantages and disadvantages within the field. Table 1 presents a condensed summary of the most pertinent information concerning these listed technologies [2–6].

Table 1 highlights crucial characteristics for selecting the optimal technology for future micromixers. The initial cost is a major consideration, with SLA offering the lowest startup expense. Resolution varies among technologies but is generally acceptable for microdevices. Polyjet and FDM have an advantage over SLA as they do not require post-processing [2].

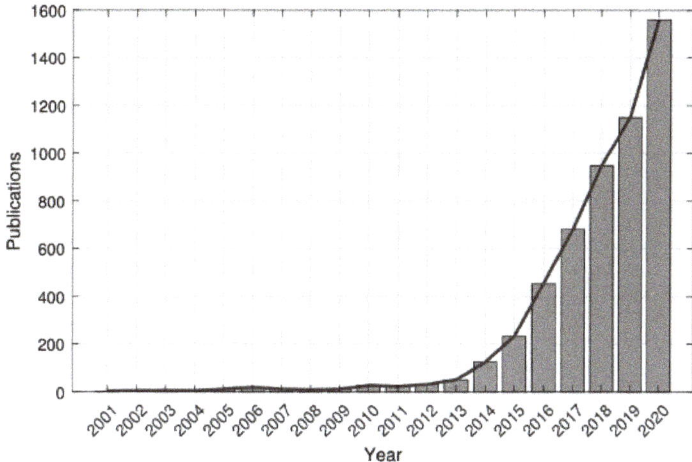

Figure 1. Trends of 3D printing, microfluidic, and additive manufacturing published articles [2].

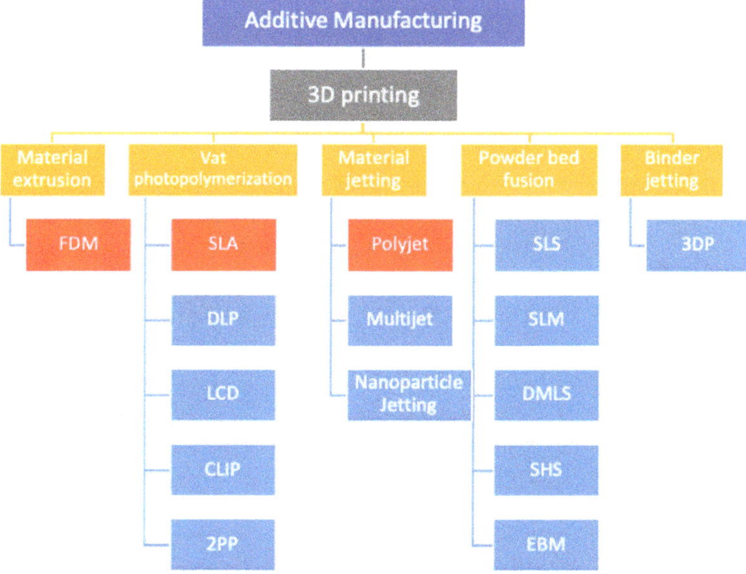

Figure 2. 3D printing categories and technologies [2,3].

All technologies exhibit an acceptable building rate, significantly reducing manufacturing time to a few hours for final prototypes compared to traditional methods. Ongoing research employs additive manufacturing for microfluidic devices, addressing manufacturing restrictions to broaden application possibilities. Challenges include improving surface quality, post-processing, and reducing device costs [7,8].

SLA printing technology, known for its cost-effectiveness and reasonable resolution (ranging from 0.025 to 0.1 mm), stands out as one of the most utilized methods in microfabrication. Post-processing typically involves washing with isopropanol followed by UV post-curing [9], though this step may be viewed as a limitation of the process. The specific time and configuration parameters for post-processing vary depending on the application, representing an important yet uncertain parameter to define for the current study. Given the significance of SLA technology in microdevice manufacturing, Table 2 provides a summarized literature review of research utilizing SLA as a manufacturing process, including miniaturized SLA arrays [1,10,11].

Table 1. 3D printed mixers [2–6].

Additive Manufacturing Technology	SLA (Stereolithography)	Polyjet	FDM (Fused Deposition Modeling)
Cost of main equipment	$2500–$3400	$60,000–$300,000	$1000
Post-processing equipment cost	Wash: $499 Cure: $699	None	None
Material required	Resin Consumable resin tank Isopropyl Alcohol	Liquid photopolymer Caustic soda solution	PLA
Layer print height	0.025 mm 0.05 mm 0.100 mm	High Quality High Mix High Speed	0.15 mm 0.20 mm 0.30 mm
Build rate	0.18–1.16	0.10–0.19	0.06–0.33

Table 2. 3D-printed SLA devices.

Active method	Channel range	Application	Material	Ref.	Year
Syringe pump	100–500 µm	Microfluidic droplet	Grey Resin	[12]	2021
Syringe pump	280–310 µm	SLA 3D printed droplet generators	Clear Resin	[1]	2021
Mini centrifuge	1–2 mm	Platform for multiplexed molecular detection of SARS-CoV-2.	Clear Resin	[13]	2021
Syringe pump	1000 µm	Laboratory experiments	Clear Resin	[14]	2020
Syringe pump	1.58 mm	Controlled Microdroplets	Clear Resin	[15]	2019
Syringe pump	100–500 µm	Modular microfluidic for emulsion droplets	Curable polymer	[16]	2019

This research endeavors to investigate the feasibility of utilizing additive manufacturing, specifically SLA, to fabricate microchannels for determining fluid velocity. By leveraging insights gained from previous micromixer devices, the study aims to delineate its scope effectively.

Early studies primarily examined flows within straight pipes of varying geometries and used active methods to operate them, such as syringe pumps. Interest then shifted to investigating flow behavior in capillary tubes under microgravity conditions. Recent research has focused on hydrodynamics and mass transfer in microchannels with diameters <1 mm, crucial for bio- and micro-fluidic applications. Experimental parameters include inlet geometry, cross-sections, flow rates, and fluid properties. Studies typically concentrate on low flow velocities dominated by surface forces, resulting in intermittent flow patterns [17]. Researchers have explored various miniature geometries, including straight channels with circular or rectangular cross-sections. These studies have shown that different geometries exhibit distinct flow characteristics and performance. For instance, straight channels with rectangular cross-sections may offer advantages over those with circular cross-sections in terms of fluid dynamics and mass transfer efficiency [18]. An application could involve utilizing capillary-driven flow in microchannels for precise fluid control in lab-on-a-chip devices, benefiting medical diagnostics and chemical analysis [17].

Surface roughness is a factor that can alter fluid behavior within microchannels. The impact of this factor could change the characteristics of a material from hydrophobic to hydrophilic [19,20]. However, for this study, the surface of the resin used was not affected. The material exhibited hydrophilic behavior according to the results obtained.

The ultimate objective is to develop a passive microfluidic device suitable for deployment as a mobile lab-on-chip, obviating the need for an external power source. This endeavor requires an alternative manufacturing approach capable of producing intricate geometries to exploit the laminar flow behavior inherent in microdevices. SLA technology emerges as a promising method due to its precision and capability to achieve the desired outcomes. Ensuring compact dimensions is paramount to facilitate portability and ease of handling for the device.

2. Materials and Methods

2.1. Device Printing Process

The traditional and recommended SLA 3D printing process includes four main steps: digital design, printing, washing, and curing [15]. However, microchannels under 400 μm the channels showed a material stagnation. As a result, an improvement in the process was added.

The recommended printing workflow is shown in Figure 3. The process includes an air-cleaning step. The effectiveness of the process variation is evident in the results, as clear channels up to 300 μm and beyond are obtained.

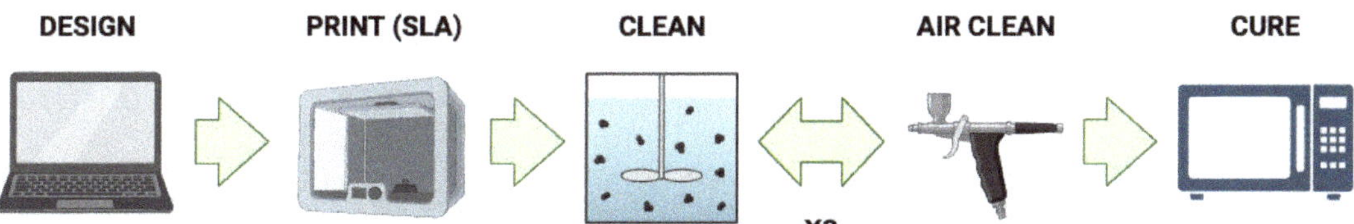

Figure 3. Optimized 3D printing process.

Design: The initial step involves creating a 3D model of the mixer using parametric software, specifically SOLIDWORKS 2024. Once the design is complete, it is exported as a stereolithography file (STL) for further processing, with a straight channel of 25 mm in length and square section of 300 to 1000 μm. Figure 4 shows an example of the designed device. Printing: The FORM 3+ printer, manufactured by Formlabs, is used for this stage. SLA technology is employed to manufacture the models, with a printing resolution set as an adaptative resolution. The adaptive printing method adapts each layer into the proper resolution from 25 to 100 μm). The printer has a maximum build dimension of 145 × 145 mm flat and 185 mm height. Devices are printed using CLEAR RESIN (by Formlabs Inc., Somerville, MA, USA) chosen for its ability to provide clear images of the process.

Cleaning: Post-printing, a crucial cleaning process using isopropyl alcohol, is undertaken to remove any uncured resin that could potentially clog the channels. An automated washer, the FORMWASH, is employed for this task.

Air Cleaning: An air-cleaning process is incorporated to ensure thorough cleaning. The washer conducts a 5-minute wash cycle followed by air cleaning at 4 psi pressure, repeated three times. After this, the device undergoes final curing. This stage allows for the obtaining of a clean and more accurate channel size before UV exposure. The cleaning regimen consisted of intervals set every 5 min for a total of 3 iterations. Despite conducting additional cycles beyond the prescribed 3, discernible enhancements in the outcomes were not observed. It was elucidated that the third iteration of air cleaning sufficed in eliminating all residual uncured resin.

Curing: The curing process takes place in a FORMCURE curing chamber, where the device is exposed to ultraviolet rays. Curing occurs at 60° for a duration of 15 min to ensure optimal strength and stability of the model. To prevent device deformation due to heat and UV rays, tempered glass sheets were used to cover the devices.

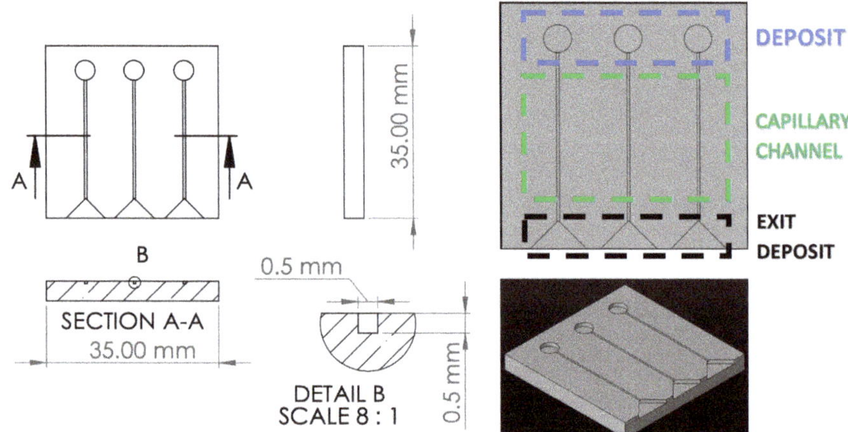

Figure 4. Reference images of test devices.

2.2. Device Setup and Data Collect

2.2.1. Device Preparation

The device is designed with a closed channel to achieve optimal capillary flow. However, additive SLA technology does not allow for the fabrication of closed channels. Therefore, open channels were designed, as shown in Figure 4, where side walls and the base were fabricated. The channel was closed by adding the top wall using a specialized hydrophilic adhesive to enhance capillary flow. This adhesive is uniformly applied to the device surfaces, promoting optimal wetting of the channels and facilitating smooth fluid flow. Specification: 3M 9984 Diagnostic Microfluidic Surfactant Free Hydrophilic Film.

The prepared device undergoes thorough cleaning with high-purity distilled water (Grade 0) to eliminate any lingering contaminants or particles. Following this, it is dried using a controlled stream of compressed air at 5 psi. This meticulous cleaning and drying process is conducted before each new experiment to ensure consistent and reliable outcomes.

2.2.2. Experimental Setup

Before commencing each experiment, the microdevice is carefully placed on a precision-leveled platform to mitigate any potential disruptions resulting from tilting or uneven surfaces. The initial experiments, where the stabilization process on the platform was not considered, lacked stability and consistency in the values obtained. Upon leveling the platform using leveling tools, the results began to exhibit greater coherence and repeatability. These results were not included in the research as they were considered a minor factor.

A precisely measured 20 µL volume of high-purity distilled water (Grade 0) is dispensed into the reservoir of each channel using a calibrated pipette, as illustrated in Figure 4. This distilled water is pre-prepared with a biocompatible vegetable dye to facilitate the visualization of fluid flow. Upon dispensing into the reservoir, the water is driven through the channel by capillary force.

2.3. Video and Image Processing

Image and Video Capture: Utilizing a USB digital microscope (Dino-Lite Digital Microscope AF4515ZTL, AnMo Electronics Corporation, Taipei, Taiwan), high-resolution images and videos were captured. To optimize illumination for image and video capture, a smartphone with a white light screen was strategically positioned beneath the device (Figure 5b).

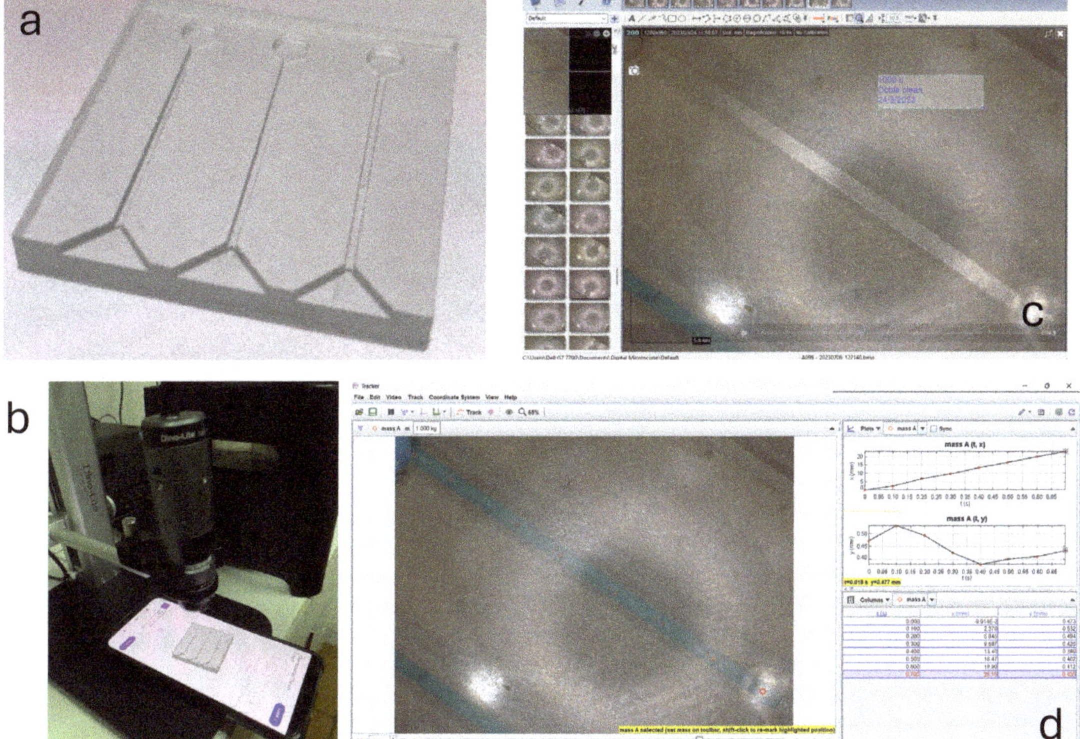

Figure 5. Data analysis process (**a**) prepare device (**b**) setup of experiment (**c**) capture images with microscope (**d**) obtain data.

Software Utilization: The DINO CAPTURE 2.0 software was employed to save image and video files, ensuring standardized and organized data storage. Additionally, it facilitated the real-time observation of the microfluidic processes.

Channel Size: To ascertain the actual dimensions of the channels, a profilometer (Bruker brand, model Dektak XT) was utilized. Validation was conducted through image correlation with the Dino-Lite Digital Microscope (AnMo Electronics Corporation, Taipei, Taiwan).

Angle Measurements of the Drop Over Surfaces (Resin and Adhesive): To precisely measure contact angles, we employed a USB microscope for image capture, then conducted a thorough analysis using the freely available image processing software, IMAGE J (Version 1.54i). The revealed contact angles averaged 49.2° with a standard deviation of 2.93°. Additionally, we determined that the static contact angle on the adhesive used in our study averaged 54.8° with a standard deviation of 3.84°. These measurements offer valuable insights into the wetting behavior of various materials within our microdevice, providing essential data for assessing fluid dynamics and surface interactions. The data were utilized for calculations in the article.

2.4. Data Analysis

Velocity Measurements: The velocity data were derived from the videos captured using TRACKER (version 6). Velocity analysis was conducted using image and video analysis techniques. The experiment was recorded using video and image processing. The data were analyzed using the software TRACKER. Image processing was carried out over a 0.1-s interval, during which the position of fluid was compared to its previous position. This allowed for data tabulation and the establishment of flow velocity at different points. This allowed for the computation of fluid velocities within the microchannels.

Graphical Representation: MATLAB R2023a was utilized for data processing and generating graphical representations. Graphs and figures were created to illustrate the experimental findings, facilitating data interpretation and comparison.

By following this refined methodology, we ensure precise device preparation, optimal surface treatment, controlled experimental setup, precise data acquisition, and analysis in our capillary-driven microdevice mixer study, all of which are essential for accurate and replicable results in capillary-driven microdevice mixing studies. The combination of advanced imaging techniques and powerful software tools enabled the accurate quantification of contact angles, fluid velocities, color changes, and mixing patterns, providing a thorough understanding of the device's performance and its comparison with simulation data.

2.5. Govern Equations

2.5.1. Capillary Pressure

$$\Delta P = \gamma \left(\frac{\cos\left(\theta_{left}\right) + \cos\left(\theta_{right}\right)}{w} + \frac{\cos(\theta_{top}) + \cos(\theta_{bottom})}{h} \right) \tag{1}$$

ΔP—Capillary Pressure.
$\cos\left(\theta_{top}\right)$—contact angle with adhesive = 54.8°.
$\cos\left(\theta_{left}\right)$, $\cos\left(\theta_{right}\right)$, $\cos(\theta_{bottom})$—contact angle with resin = 49.2°.
w—channel depth = 300 μm to 1000 μm .
h—channel width = 300 μm to 1000 μm.
γ—surface tension of liquid [N/m] = 0.07 N/m N/m = 70 mN/m.

2.5.2. Fluid Flow

$$Q = \Delta P \frac{h_0{}^3 w}{12\mu L} \gamma \left[1 - 0.63 \frac{h_0}{w} \right] \tag{2}$$

ΔP—Capillary Pressure.
h_0—channel depth = 300 μm to 1000 μm.
w—channel width = 300 μm to 1000 μm.
L—channel length = 25 mm.
γ—surface tension of liquid [N/m] = 0.07 N/m N/m = 70 mN/m.
μ—Kinematic viscosity of water = 8.9 × 10⁻⁴ Pa s.

2.5.3. Velocity

$$V = \frac{Q}{A} \tag{3}$$

Q—Fluid flow.
A—Channel section = width and depth size between 300 µm to 1000 µm.

3. Results

3.1. Printed Results

Following a series of iterative printing tests and refinements, we present the finalized design of the proposed device, as depicted in Figures 6 and 7. This culmination of multiple printing experiments represents our commitment to achieving the optimal configuration and functionality of the device, ensuring that it meets the highest standards of performance and precision.

The enhancements made to the printing process yielded significant results. Initially, as depicted in Figure 6, resin stagnation was noticeable within the channels, particularly those under 500 µm in width. The printing method suggested by the provider proved insufficient in achieving the desired channel specifications, resulting in resin stagnation that impeded fluid flow. Following the implementation of an air-cleaning process, notable improvements were observed (Figure 7). Channels ranging from 300 µm to 1000 µm in width were successfully produced. However, further investigation revealed that 400 µm channels represent the smallest size attainable for obstruction-free applications. Air cleaning proved instrumental in facilitating the creation of smaller channel sizes, ensuring smoother fluid flow. Its importance lies in its efficacy in eliminating stagnation residue introduced during the printing process.

The occurrence of channel clogging arises from the challenge of effectively removing residual resin from within the channels using current processing methods. To address this issue, we are exploring various post-processing techniques aimed at efficiently eliminating excess material. Methods such as sandblasting and utilizing alternative cleaning agents besides isopropyl alcohol are being considered to be potential solutions to mitigate this problem.

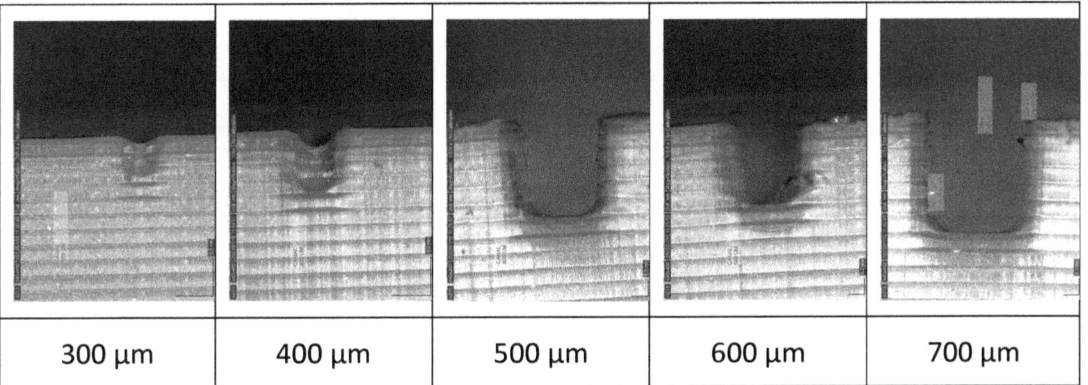

Figure 6. Printed Devices with traditional process.

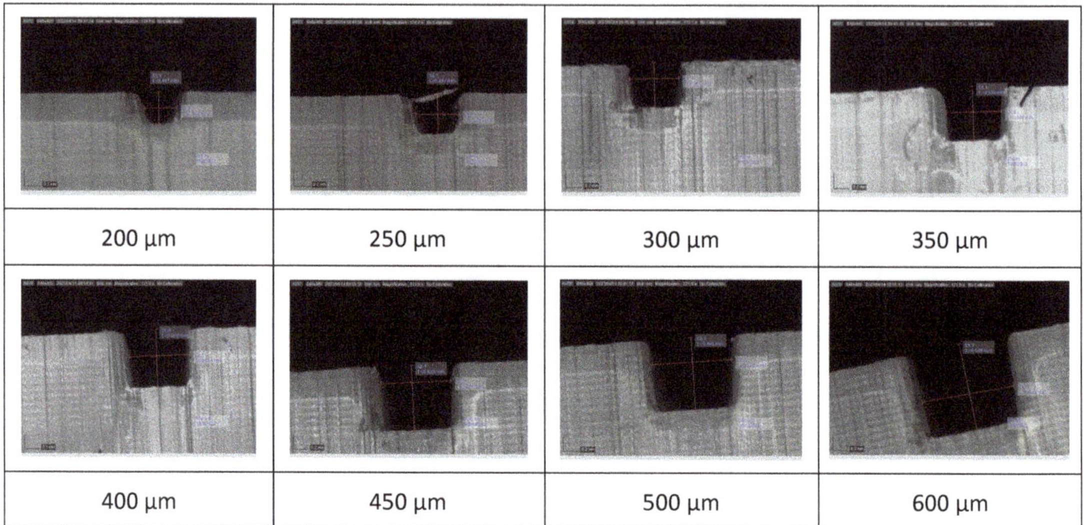

<table>
<tr><td>200 μm</td><td>250 μm</td><td>300 μm</td><td>350 μm</td></tr>
<tr><td>400 μm</td><td>450 μm</td><td>500 μm</td><td>600 μm</td></tr>
</table>

Figure 7. Printed Devices with proposed process.

The channel measurements were validated using both a USB microscope and a scanning electron microscope (SEM), each possessing the following characteristics.

Scan time: 400 s for each zone.

- Length scan: 20,000–50,000 μm.
- Scan resolution: 0.0166667–0.0462954 μm.
- Scan type: Standard.
- Needle force: 3 mg.
- Needle range scan: 1 mm.
- Needle type: radius 25 μm.
- Correction: quadratic and removal of curvature.

Figure 8 illustrates the measurement results of the obtained channels. It allows for comparison between the channel size designed using software and the size obtained after printing.

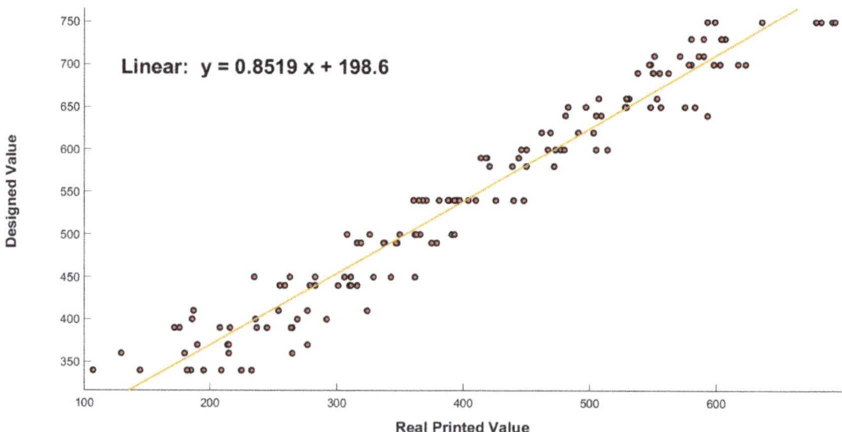

Figure 8. Printed results compared to Designed results.

In general, we can observe data dispersion across all measurements. Nevertheless, a linear relationship exists. Thus, we were able to establish the equation that enables the determination of the channel size to be designed in order to achieve a specific actual channel value.

$$y = 0.8519x + 198.6 \tag{4}$$

The correlation coefficient was established between the designed value variables compared to the actual width measurement obtained on the device after printing. The correlation coefficient was 0.9747, indicating a strong positive correlation between the two variables. Consequently, they tend to change in the same direction.

The proposed equation has been validated and is currently being utilized in devices to determine the size of channels. The variation in the proposed channel size compared to the actual size is attributed to several factors. Among the main factors is that the equipment used lacks recommendations for printing elements on a micro scale. Thus, the tolerance exhibited by the printing at channels of this size becomes evident and, therefore, warrants an adjustment.

Based on the printed devices, it was found that channels up to 150 µm wide can be printed. However, the channels are not uniform and exhibit stagnant resin. For this reason, it is recommended that channels be used between 400 and 800 µm wide to avoid resin blockage. This was verified by conducting tests with fluid through the channels. These experiments were not included in the presented results.

3.2. Velocity of the Fluid During the Experiment

Once the devices were printed and prepared for capillarity tests, velocities were measured in channels of 25 mm length and square sections ranging from 300 to 1000 µm. Throughout the channels, it was observed that the velocity maintained its stability consistently. Subsequently, velocity data for each of the channels were recorded and plotted in Figure 9.

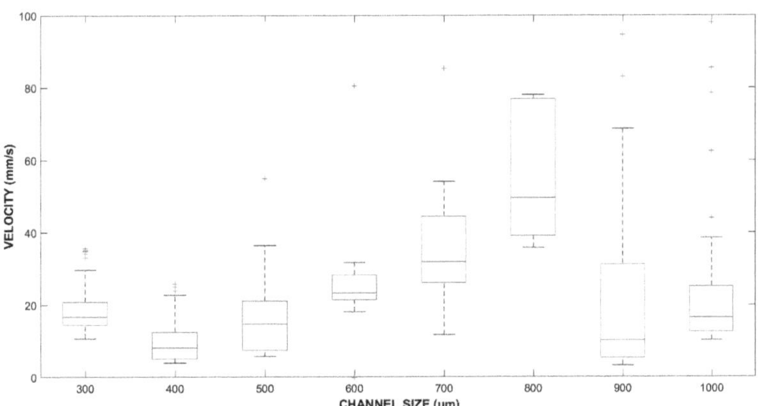

Figure 9. Boxplot of velocity (mm/s) in different channel size.

In Table 3, a summary of the data obtained from the various channels between 300 and 1000 µm is presented.

Table 3. Experimental data velocity for different channel size.

Channel Size	300	400	500	600	700	800	900	1000
Mean (mm/s)	18.62	9.24	17.77	26.89	37.75	61.16	25.53	26.32
Standard Deviation (mm/s)	6.54	4.78	14.09	18.64	22.63	33.37	34.01	24.12

Both the resin and the cover layer demonstrate hydrophilic properties, leading to enhanced fluid interaction. The larger surface area of the walls facilitates increased contact with the fluid, resulting in a greater thrust force to propel it forward. This relationship is illustrated in Figure 8, where velocity demonstrates a corresponding increase with surface area increment. However, for channels exceeding 800 µm or falling below 300 µm, additional factors influence velocity. Through our research, we have established that velocities between 400 µm and 800 µm are stable based on the data analysis.

As an experimental process, various factors, such as printing techniques and other variables, can influence the final device, leading to a degree of uncertainty in predicting the speed of future devices. While current technology may not eliminate this uncertainty, ensuring the reliability of experiments through meticulous data analysis and statistical validation enhances the confidence in the obtained results, as shown in Table 4.

Figure 10 illustrates the relationship between the calculated capillary pressure and the experimentally obtained velocity in the devices, compared with the channel size (300 µm to 1000 µm). It is evident that printing and surface roughness result in an exponential increase in velocity between 400 and 800 µm. According to previous studies, roughness smaller than the drop size generates greater surface hydrophilicity [20]. For our study, the measured roughness was 4.99 µm with a standard deviation of 0.87 µm. The roughness is much smaller than the drop size (3 mm radius). Variations in roughness could alter velocity results through the principle of capillarity [19].

Table 4. Data of capillary data for different channel sizes.

Width Height (µm)	P (Pa)	Q (mm³/s)	V (mm/s)	V Exp (mm/s)
300	0.63149	0.0071	0.0788	0.1862
400	0.47362	0.0168	0.105	0.0924
500	0.3789	0.0328	0.1313	0.1777
600	0.31575	0.0567	0.1575	0.2689
700	0.27064	0.09	0.1838	0.3775
800	0.23681	0.1344	0.21	0.6116
900	0.2105	0.1914	0.2363	0.2553
1000	0.18945	0.2625	0.2625	0.2632

In general, this research document showcases the feasibility of utilizing additive manufacturing equipment to produce microchannels (ranging from 300 to 1000 µm) for device applications. It was possible to establish the proposed design parameters for obtaining channels of the required dimensions. Additionally, it was confirmed that the obtained channels are suitable for capillary action applications. Moreover, the study demonstrates the potential of additive manufacturing techniques in microfluidic device fabrication, highlighting their adaptability and precision in producing microstructures with specific geometrical requirements.

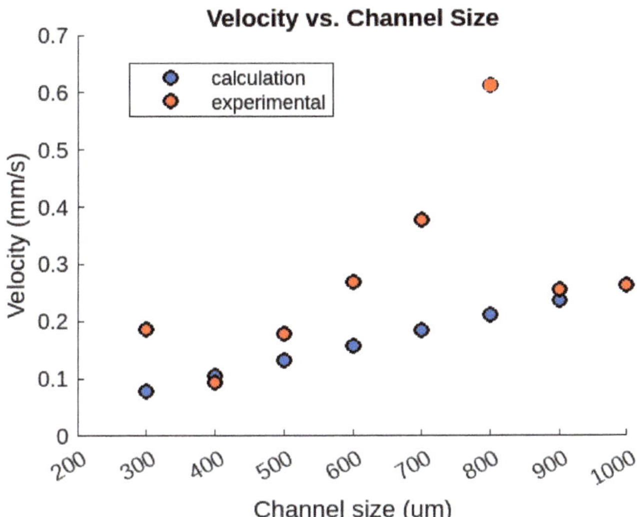

Figure 10. Comparative graph between calculation and experimental data.

4. Conclusions

It has been determined that to obtain microchannels of 300 µm or larger, a proper washing and curing process must be followed to prevent material stagnation. Adding a triple-compressed air-cleaning process at 5 bar between the washing and curing steps allows for the attainment of the expected channel dimensions.

This study successfully validated the use of stereolithography (SLA) 3D-printing technology for fabricating microchannel devices with consistent capillary-driven fluid flow behavior. The key findings showed that channels ranging from 300 to 1000 µm could be reliably produced using an optimized SLA printing process with an added air-cleaning step. The equation $y = 0.8519\,x + 198.6$ was derived to predict the actual printed channel size based on the designed dimensions, accounting for fabrication inaccuracies at the microscale.

The mean fluid velocities measured experimentally aligned well with theoretical calculations, validating the flow behavior in the printed channels. For instance, the 500 µm channel exhibited a mean velocity of 17.77 mm/s, closely matching the calculated value of 13.13 mm/s. This concordance between experimental data and theory substantiates the suitability of SLA printing for realizing precisely designed microfluidic components driven by capillary forces. Experimental data demonstrated that fluid velocities increased exponentially with larger channel sizes, reaching up to 61.16 mm/s for 800 µm channels.

According to the comparison between mean experimental velocity values and theoretical ones, there is a correlation up to the 800 µm channel. Channels ranging from 400 to 800 µm present a standard deviation between 4.78 and 33.37 mm/s. Hence, we can infer that for channels within these ranges, repeatability in results can be achieved. Channels larger than 800 µm exhibit a deviation greater than 33.37 mm/s, assuming a variation in the processes. This is because external variables such as gravity, material roughness, and channel size significantly affect the process, rendering it unpredictable.

In conclusion, this study delineates an appropriate printing and post-processing procedure that enables the production of channels from 300 µm. To obtain the desired channel values, it is important to use the determined formula to define the channel size to be designed. This could contribute to the development of future devices. Additionally, it is recommended to work with channels between 300 and 800 µm, which exhibit stable capillary-driven fluid velocity values. This could contribute to the development of future devices.

Author Contributions: V.H.C.-M. executed the conceptualization, investigation, data analysis, manuscript preparation and wrote the original draft; J.C.-T. supervised, reviewed, and edited the manuscript. E.S. validated the manuscript. All authors have read and agreed to the published version of the manuscript.

Funding: This work is funded by the Universidad Politécnica Salesiana through Research Project (035-01-2024-01-30). Additionally, it received support from the Ministerio de Ciencia e Innovación (PID2020-114070RB-I00), the Agencia Estatal de Investigación (CPP2021-009021), and AGUAR (2021PROD00064).

Data Availability Statement: All data generated or analyzed during this study are included in this published article or available from the corresponding author upon reasonable request.

Acknowledgments: We gratefully acknowledge ESPE (Universidad de las Fuerzas Armadas) for providing facilities and resources for Scanning Electron Microscopy (SEM) microscopy, particularly Nanomaterials Characterization Laboratory Center of Nanoscience and Nanotechnology (CENCI-NAT). Special thanks to Alexis Debut. We appreciate the technical assistance of Karla Vizuete for their guidance and support during the process.

Conflicts of Interest: The authors declare no conflict of interest.

Abbreviations

The following abbreviations are used in this manuscript:

SLA	Stereolithography
FDM	Fused Deposition Modeling
UV	Ultraviolet

References

1. Bacha, T.W.; Manuguerra, D.C.; Marano, R.A.; Stanzione, J.F. Hydrophilic modification of SLA 3D printed droplet generators by photochemical grafting. *RSC Adv.* **2021**, *11*, 21745–21753. [CrossRef] [PubMed]
2. Garcia-Cardosa, M.; Granados-Ortiz, F.J.; Ortega-Casanova, J. A Review on Additive Manufacturing of Micromixing Devices. *Micromachines* **2021**, *13*, 73. [CrossRef] [PubMed]
3. Rogers, C.I.; Qaderi, K.; Woolley, A.T.; Nordin, G.P. 3D printed microfluidic devices with integrated valves. *Biomicrofluidics* **2015**, *9*, 011502. [CrossRef] [PubMed]
4. Chen, C.; Mehl, B.T.; Munshi, A.S.; Townsend, A.D.; Spence, D.M.; Martin, R.S. 3D-printed microfluidic devices: Fabrication, advantages and limitations—A mini review. *Anal. Methods* **2016**, *8*, 6005–6012. [CrossRef] [PubMed]
5. Dixit, C.K.; Kadimisetty, K.; Rusling, J. 3D-printed miniaturized fluidic tools in chemistry and biology. *TrAC Trends Anal. Chem.* **2018**, *106*, 37–52. [CrossRef] [PubMed]
6. Chen, J.V.; Dang, A.B.C.; Dang, A. Comparing cost and print time estimates for six commercially-available 3D printers obtained through slicing software for clinically relevant anatomical models. *3D Print. Med.* **2021**, *7*, 1. [CrossRef] [PubMed]
7. Shallan, A.I.; Smejkal, P.; Corban, M.; Guijt, R.M.; Breadmore, M.C. Cost-Effective Three-Dimensional Printing of Visibly Transparent Microchips within Minutes. *Anal. Chem.* **2014**, *86*, 3124–3130. [CrossRef] [PubMed]
8. Tiboni, M.; Tiboni, M.; Pierro, A.; Del Papa, M.; Sparaventi, S.; Cespi, M.; Casettari, L. Microfluidics for nanomedicines manufacturing: An affordable and low-cost 3D printing approach. *Int. J. Pharm.* **2021**, *599*, 120464. [CrossRef] [PubMed]
9. Camarillo-Escobedo, R.M.; Flores-Nuñez, J.L.; Garcia-Torales, G.; Hernandez-Campos, E.; Camarillo-Escobedo, J.M. 3D printed opto-microfluidic autonomous analyzer for photometric applications. *Sens. Actuators A Phys.* **2022**, *337*, 113425. [CrossRef]
10. Gong, H.; Woolley, A.T.; Nordin, G.P. High density 3D printed microfluidic valves, pumps, and multiplexers. *Lab Chip* **2016**, *16*, 2450–2458. [CrossRef] [PubMed]
11. Zhang, J.M.; Ji, Q.; Duan, H. Three-Dimensional Printed Devices in Droplet Microfluidics. *Micromachines* **2019**, *10*, 754. [CrossRef] [PubMed]
12. Zhang, J.; Xu, W.; Xu, F.; Lu, W.; Hu, L.; Zhou, J.; Zhang, C.; Jiang, Z. Microfluidic droplet formation in co-flow devices fabricated by micro 3D printing. *J. Food Eng.* **2021**, *290*, 110212. [CrossRef]
13. Ding, X.; Li, Z.; Liu, C. Monolithic, 3D-printed lab-on-disc platform for multiplexed molecular detection of SARS-CoV-2. *Sens. Actuators B Chem.* **2022**, *351*, 130998. [CrossRef] [PubMed]
14. Vangunten, M.T.; Walker, U.J.; Do, H.G.; Knust, K.N. 3D-Printed Microfluidics for Hands-On Undergraduate Laboratory Experiments. *J. Chem. Educ.* **2020**, *97*, 178–183. [CrossRef]

15. Hwang, Y.H.; Um, T.; Hong, J.; Ahn, G.N.; Qiao, J.; Kang, I.S.; Qi, L.; Lee, H.; Kim, D.P. Robust Production of Well-Controlled Microdroplets in a 3D-Printed Chimney-Shaped Milli-Fluidic Device. *Adv. Mater. Technol.* **2019**, *4*, 1900457. [CrossRef]
16. Song, R.; Abbasi, M.S.; Lee, J. Fabrication of 3D printed modular microfluidic system for generating and manipulating complex emulsion droplets. *Microfluid. Nanofluidics* **2019**, *23*, 92. [CrossRef]
17. Der, O.; Bertola, V. An experimental investigation of oil-water flow in a serpentine channel. *Int. J. Multiph. Flow* **2020**, *129*, 103327. [CrossRef]
18. Verma, R.K.; Ghosh, S. Effect of phase properties on liquid-liquid two-phase flow patterns and pressure drop in serpentine mini geometry. *Chem. Eng. J.* **2020**, *397*, 125443. [CrossRef]
19. Deltombe, R.; Kubiak, K.J.; Bigerelle, M. How to select the most relevant 3D roughness parameters of a surface. *Scanning* **2014**, *36*, 150–160. [CrossRef] [PubMed]
20. Kabir, H.; Garg, N. Rapid prediction of cementitious initial sorptivity via surface wettability. *npj Mater. Degrad.* **2023**, *7*, 52. [CrossRef]

Article

Molecular Dynamics Simulation of Femtosecond Laser Ablation of Cu$_{50}$Zr$_{50}$ Metallic Glass Based on Two-Temperature Model

Jingxiang Xu [1,*], Dengke Xue [1], Oleg Gaidai [1], Yang Wang [2] and Shaolin Xu [3]

1   College of Engineering Science and Technology, Shanghai Ocean University, Shanghai 201306, China; m200601296@st.shou.edu.cn (D.X.); o_gaidai@just.edu.cn (O.G.)
2   Research Institute of Frontier Science, Southwest Jiaotong University, Chengdu 610031, China; yang.wang@swjtu.edu.cn
3   Department of Mechanical and Energy Engineering, Southern University of Science and Technology, Shenzhen 518055, China; xusl@sustech.edu.cn
*   Correspondence: jxxu@shou.edu.cn; Tel.: +86-21-61900801

Abstract: Femtosecond laser machining, characterized by a small heat-affected zone, high precision, and non-contact operation, is ideal for processing metallic glasses. In this study, we employed a simulation method that combines the two-temperature model with molecular dynamics to investigate the effects of fluence and pulse duration on the femtosecond laser ablation of Cu$_{50}$Zr$_{50}$ metallic glass. Our results showed that the ablation threshold of the target material was 84 mJ/cm^2 at a pulse duration of 100 fs. As the pulse durations increased, the maximum electron temperature at the same position on the target surface decreased, while the electron–lattice temperature coupling time showed no significant difference. As the absorbed fluence increased, the maximum electron temperature at the same position on the target surface increased, while the electron–lattice temperature coupling time became shorter. The surface ablation of the target material was mainly induced by phenomena such as melting, spallation, and phase explosion caused by femtosecond laser irradiation. Overall, our findings provide valuable insights for optimizing the femtosecond laser ablation process for metallic glasses.

Keywords: femtosecond laser; metallic glass; ablation; two-temperature model; molecular dynamics

Citation: Xu, J.; Xue, D.; Gaidai, O.; Wang, Y.; Xu, S. Molecular Dynamics Simulation of Femtosecond Laser Ablation of Cu$_{50}$Zr$_{50}$ Metallic Glass Based on Two-Temperature Model. *Processes* **2023**, *11*, 1704. https://doi.org/10.3390/pr11061704

Academic Editors: Zejia Zhao, Guoqing Zhang and Wai Sze YIP

Received: 6 May 2023
Revised: 26 May 2023
Accepted: 31 May 2023
Published: 2 June 2023

1. Introduction

Metallic glasses, also known as amorphous alloys or liquid metals, are a unique material characterized by a disorderly arrangement of atoms [1]. Due to their unique atomic structure, they generally exhibit excellent mechanical properties, high elasticity, super-plasticity in the supercooled liquid region, exceptional soft magnetic properties, corrosion resistance, and remarkable electrocatalytic performance [2]. Despite these favorable properties, metallic glasses become harder and more brittle under low-temperature conditions, show a highly viscous state in medium- to high-temperature conditions, and are susceptible to self-passivation, which results in the formation of a surface film that can easily denature due to crystallization or oxidation caused by thermal effects [3]. To overcome these limitations, Yao et al. [4] suggests that employing femtosecond laser ablation to process metallic glasses offers several benefits, such as a small heat-affected zone, perquisite precise ablation thresholds, and the ability to perform precise, non-crystallization processing of metallic glasses. The purpose of this article is to enhance our understanding of the mechanism behind the ablation process during the femtosecond laser machining of metallic glasses.

There is currently limited research on femtosecond laser ablation in metallic glass processing. Some research has been conducted experimentally. Sano et al. [5] first reported on the use of the femtosecond laser for ablating bulk metallic glass and studied the relationship between pulse energies and ablation depth in air. Wang et al. [6] conducted micro-drilling and trenching studies on amorphous alloys using femtosecond laser ablation in air, proposing that femtosecond laser ablation has the potential to achieve the amorphous processing of metallic glasses. Ma et al. [7] further investigated the femtosecond laser ablation of metallic glass under different laser shots and fluence in the atmospheric environment and provided guidance for controlling the evolution of surface morphology induced by femtosecond laser pulse.

Researchers have conducted some studies in the computational aspect as well. The two-temperature model (TTM) [8] has been used to describe the evolution of electronic and lattice temperatures in the short-pulse laser ablation of metals. Molecular dynamics (MD) simulation has been used to investigate the various mechanisms involved in the ablation process, such as ablation mechanisms [9–11], laser-induced shock waves [12], and melting [13]. However, due to the short period of femtosecond laser pulses and the complex effects resulting from the overlapping of multiple processes, it is difficult to observe the evolution process through experimental means. TTM cannot observe changes in atomic structure, while MD lacks descriptions of electrons. Thus, the mechanism between the femtosecond laser and metallic glasses is not yet fully understood by these methods.

To conduct research in this field, many researchers employ the TTM-MD hybrid model. Most simulations focus on the interaction between femtosecond lasers and monometallic materials such as Cu [14], Al [15], Ni [16], and Au [17]. For metallic glasses, Marinier et al. [18] studied the ablation dynamics of CuZr metallic glass and crystal structure materials using a MD simulation combined with the TTM under femtosecond laser irradiation. Iabbaden et al. [19] also employed a hybrid TTM-MD model to investigate the structural evolution of crystalline CuZr alloys and amorphous alloys of the same compositions under ultrashort laser irradiation. However, they overlooked the structural evolution of metallic glasses under different femtosecond laser irradiation conditions, which could significantly guide practical processing. The related mechanisms involved in the femtosecond laser ablation of metallic glasses remain unclear, especially the impact of different laser parameters on the ablation of metallic glasses.

Our study employed a hybrid TTM-MD model to investigate the ablation process of $Cu_{50}Zr_{50}$ metallic glass with a single-pulse femtosecond laser of different fluence and pulse durations. The study discussed the structural evolution of the target under different laser conditions, the distribution of the maximum electron temperature of the target surface, the electron–lattice temperature coupling time, and the propagation of pressure waves. The study also determined the ablation threshold of $Cu_{50}Zr_{50}$ metallic glass, which helps to understand the mechanism between the femtosecond laser and metallic glasses, as well as promotes the application of the femtosecond laser in processing metallic glasses.

2. Computational Modeling

2.1. TTM-MD Method

The simulation is studied by the LAMMPS (Large-scale Atomic/Molecular Massively Parallel Simulator) software [20]. The TTM-MD method combines two schemes where electrons of laser excitation are treated as a continuum on a regular grid [16], and ions are modeled on a classical MD.

$$F_{ion} = -\frac{\partial U}{\partial r_{ion}} + F_{langevin} - \frac{\nabla P_e}{n_{ion}} \tag{1}$$

$$C_e(T_e)\frac{\partial T_e}{\partial t} = \nabla(k_e(T_e)\nabla T_e) - g_p(T_e - T_{ion}) + \frac{I(t)e^{-\frac{x}{l_p}}}{l_p} \tag{2}$$

In Equation (1), F_{ion} is the total force acting on an ion, $\frac{\partial U}{\partial r_{ion}}$ is the force due to interatomic interactions, $F_{langevin}$ represents the force due to electron–phonon coupling, $\frac{\nabla P_e}{n_{ion}}$ is the blast force acting on ions because of electronic pressure gradient [21,22], and n_{ion} is the ionic density.

Equation (2) is a 1D heat diffusion equation with an external heat source in the x direction. In this equation, $C_e(T_e)$ is the electronic specific heat, $k_e(T_e)$ is the electronic thermal conductivity, and g_p is the coupling coefficient for the electron–ion interaction. T_e and T_{ion} are, respectively, the electronic and ionic temperature. For the $Cu_{50}Zr_{50}$ metallic glass, a free electron model is used as $C_e(T_e) = \gamma T_e$, where the electronic specific heat coefficient γ is available in [19]. When the simulation time t is shorter than the pulse duration τ, $I(t) = I_0$ and 0 otherwise. I_0 is the initial laser pulse intensity, and the absorbed laser fluence is defined as $F_{abs} = I_0 \cdot \tau \cdot l_p$ is the penetration depth of the laser energy deposition.

The electronic thermal conductivity $k_e(T_e)$ is determined from Equation (3), by assuming it is simply proportional to the electron heat capacity $C_e(T_e)$,

$$k_e(T_e) = D_e \rho_e C_e(T_e) \tag{3}$$

where D_e is the electronic thermal diffusion coefficient and ρ_e is the number density of electrons. Moreover, the coupling constant for the electron–ion interaction g_p is given by:

$$g_p = \frac{3Nk_B\gamma_p}{\Delta Vm} \tag{4}$$

where k_B is the Boltzmann's constant, m is the atomic mass, and ΔV and N represent the volume of the electronic grid cell and the total number of atoms inside each cell, respectively. g_p is considered as constant in this model.

The values of the parameters used in the TTM-MD method are shown in Table 1. The density of our prepared $Cu_{50}Zr_{50}$ sample was 7.14 g/cm^3, while that of the sample prepared by Iabbaden et al. [19] was 7.18 g/cm^3. This difference falls within an acceptable range.

Table 1. The input parameters for TTM-MD simulations of $Cu_{50}Zr_{50}$ metallic glass.

Physical Properties	Unit	Value
density ρ_0	g·cm^{-3}	7.14
electron–phonon factor g_p	10^{17} W·m^{-3}·K^{-1}	1.00 [18]
specific heat constant γ	J·m^{-3}·K^{-2}	321.30 [18]
electronic thermal conductivity k_e	W·m^{-1}·K^{-1}	3.85 [23]
electronic thermal diffusion D_e	10^{-5} m^2·s^{-1}	3.98 [19]
depth l_p	nm	14.30 [19]

2.2. Computational Details

This study focused on the $Cu_{50}Zr_{50}$ metallic glass due to its favorable amorphous-forming ability. The potential function utilized in this research was fitted by Mendelev et al. [24] based on the Finnis–Sinclair embedded atom model potential (EAM/FS) [25,26]. The NPT ensemble is a simulation technique where the number of particles, temperature, and pressure are kept constant, while the volume can vary. This constant pressure and temperature make the NPT ensemble suitable for simulating systems in contact with an environment with a fluctuating pressure and temperature. On the other hand, the NVE ensemble refers to a simulation technique where the number of particles, volume, and energy are kept constant while the temperature can vary. The NVE ensemble is particularly useful for systems that are isolated and do not exchange energy with their surroundings.

The $Cu_{50}Zr_{50}$ metallic glass sample was prepared using a melting and quenching simulation [27,28]. Initially, the $Cu_{50}Zr_{50}$ binary alloy sample with equal numbers of Cu and Zr atoms was generated using LAMMPS commands. Then, the sample was subjected to a 200 ps relaxation under the NPT ensemble with periodic boundary conditions and a temperature of 300 K. Subsequently, the sample was heated at a rate of 50 K/ps from 300 K to 2500 K to melt it. After heating the sample for 40 ps at 2500 K, the temperature was instantaneously quenched to 300 K within 25 ps. Finally, the sample was relaxed for approximately 100 ps under the NPT and NVE ensembles, with a relaxation temperature of

300 K, zero relaxation pressure, and a time step of 0.001 ps. As a result, a $Cu_{50}Zr_{50}$ metallic glass sample was obtained, with a size of 644.9 nm × 4.3 nm × 4.3 nm and composed of approximately 660,000 atoms, as displayed in Figure 1. The relaxation and laser ablation MD simulation processes both use microcanonical ensemble (NVE).

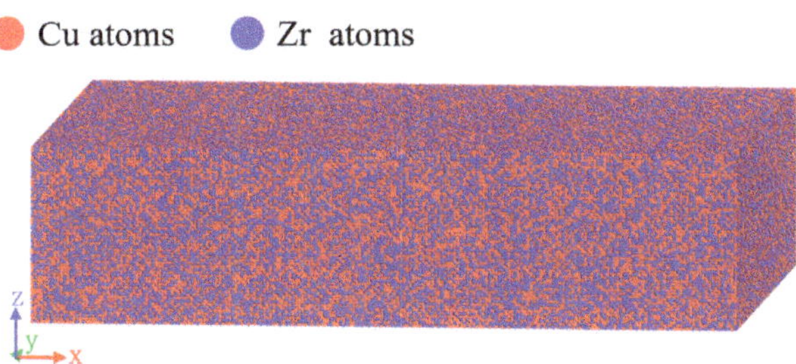

Figure 1. The substrate of the model (the Cu atoms are colored in red and the Zr atoms are colored in blue).

Figure 2 depicts a schematic model of the femtosecond laser ablation of $Cu_{50}Zr_{50}$ metallic glass. Upon being subjected to laser irradiation, the material absorbs the laser energy and undergoes thermal expansion or even ablation. Therefore, in constructing the simulation box, a vacuum layer was added to the upper surface of the target to prevent atoms from escaping the simulation box during the simulation process.

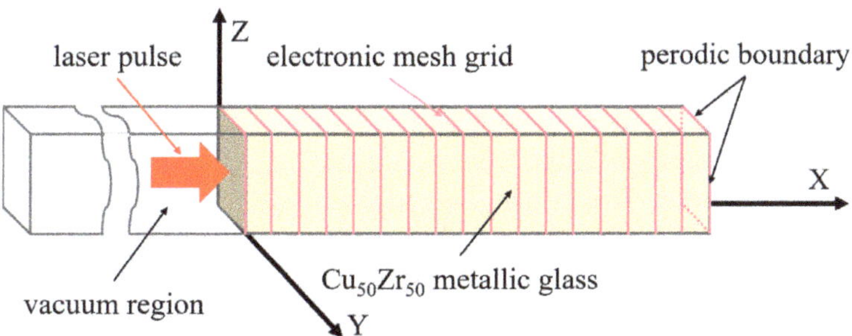

Figure 2. Simulation setup geometry used to model ultrafast laser pulse interaction in $Cu_{50}Zr_{50}$ metallic glass using TTM-MD simulations.

The heat diffusion Equation (2) is solved numerically on a 200 × 1 × 1 regular electronic cubic mesh, as shown in Figure 2. A step function $dx \approx 5$ nm is used for the T_e variation. Here, periodic boundary conditions are set in the y and z lateral directions and free boundaries are imposed following the laser shot $x = [0]$. The target irradiated by the femtosecond laser generates a pressure wave propagating towards the surface and into the target [29,30]. To avoid reflection of the pressure wave at the bottom of the target, we define non-reflecting boundary conditions composed of a viscous damping region [31–33] with a 100 nm in the x direction. A damping factor $\gamma_{damp} = 0.0025$ eV·ps·Å^{-2} [19] is used in this work in all simulations. The simulation is a regular MD run in the NVE ensemble after the laser pulse irradiation is completed.

To facilitate the simulation, we made some assumptions. These assumptions relate to the variables included in the simulation, the chosen force field, the simulation time step, and any other factors that may impact the accuracy and validity of the simulation results. Specifically, the assumptions of our proposed simulation include:

1. The model assumes that the interactions between atoms can be accurately represented by our chosen force field.

2. The model assumes the availability of accurate and reliable parameters to set up the force field.
3. The model assumes that the simulation time step is small enough to capture the dynamics of the system under investigation.
4. The model assumes that the simulation is conducted in a vacuum environment and heat exchange with the surroundings is neglected.
5. The model assumes that all of the laser energy is absorbed by the target and the reflectivity is zero.

3. Results

3.1. Ablation Threshold

When irradiated by the femtosecond laser, targets can undergo irreversible and permanent damage on their surfaces if the laser pulse parameters satisfy certain conditions, such as certain pulse duration and intensive enough laser fluence. Specifically, the surface of the target will be removed, and the minimum fluence that is required to achieve this kind of damage is known as the ablation threshold [18]. In our research, we determined the ablation threshold by gradually increasing the laser fluences. This method allowed for the determination of the specific energy requirements for ablation initiation in the target material and was chosen for its precision and reliability in determining the ablation threshold. Specifically, we determined the ablation threshold of the $Cu_{50}Zr_{50}$ metallic glass target material by incrementally increasing the laser fluence in 2 mJ/cm^2 intervals, with a baseline of 80 mJ/cm^2. We studied the results of the interaction between a femtosecond laser with a pulse duration of 100 fs and an absorbed fluence (80 + 2N, N = 0, 1, 2) mJ/cm^2 with the target material. The laser incidence direction was from left to right, and snapshots of the target at different times are shown in Figure 3.

From Figure 3a,b, it can be observed that the target surface expands significantly in the 0~40 ps time range after being irradiated by the laser pulse. During the 60–100 ps time range, the target surface remains in an expanded state but the target does not disintegrate, indicating that the ablation threshold of the target material is greater than 82 mJ/cm^2. In Figure 3c, at 20 ps, the target surface expands; and at 40 ps, a bubble appears at the position of x = 0 nm and the target expands. During the 60~100 ps time range, the target disintegrates due to the expansion of the bubble. The surface of the target is removed, and irreversible damage occurs. Therefore, the ablation threshold of the $Cu_{50}Zr_{50}$ metallic glass target material at a pulse duration of 100 fs is 84 mJ/cm^2. Hu et al. [34] predicted the damage threshold of the Cu50Zr50 amorphous alloy to be between 80 mJ/cm^2 and 120 mJ/cm^2. Further, we determined the ablation threshold to be within this range.

A comparison of Figure 4a,b indicates that the pressure within the target continuously propagates deeper, and its peak value gradually decreases over time. In Figure 4a, the pressure values remain relatively constant within the target material between x = 0~50 nm. Conversely, in Figure 4b, at the same position within the target, the pressure reaches its maximum value at 20 ps, subsequently decreasing. This variation suggests that there is a change in the stress state at the surface of the target material between 20~40 ps. The fracture mechanism of $Cu_{50}Zr_{50}$ metallic glass in Figure 3c was caused by the pressure induced by the femtosecond laser. At lower laser fluences (80 mJ/cm^2), the lattice temperature of the target material is not high enough and the tension near the surface is relatively small. As a result, only the expansion of the target material can be observed in the atomic structures shown in Figure 3a,b. When the laser fluence increases to 84 mJ/cm^2, more energy is absorbed at the surface of the target material, causing cavitation, which corresponds to breakage in Figure 3c. In Figure 4b, the tension near the surface decreases after 20 ps because it is "absorbed" by the cavities that ultimately lead to breakage (ablation).

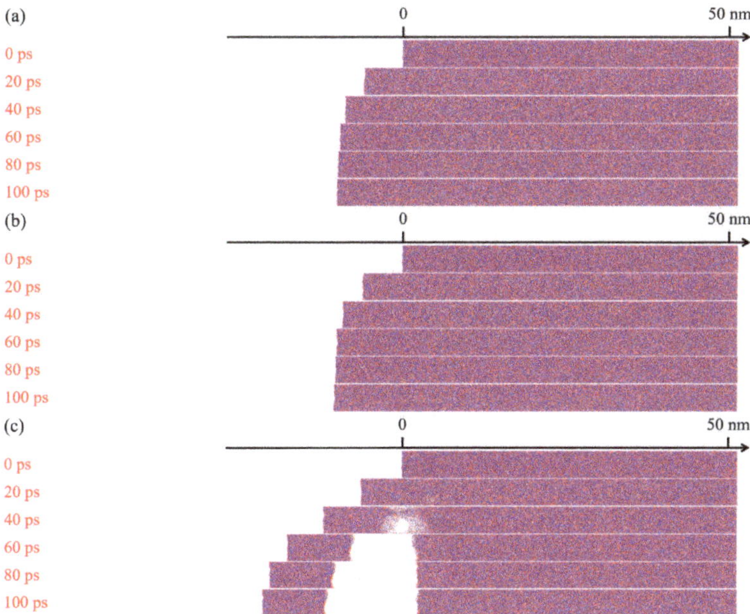

Figure 3. Snapshots (generated using the OVITO (Open Visualization Tool) software [35]) of $Cu_{50}Zr_{50}$ metallic glass irradiated by femtosecond laser with a pulse duration of 100 fs and absorbed fluence of (**a**) 80 mJ/cm^2, (**b**) 82 mJ/cm^2, and (**c**) 84 mJ/cm^2 from 0 to 100 ps. The laser pulse is incident from the left.

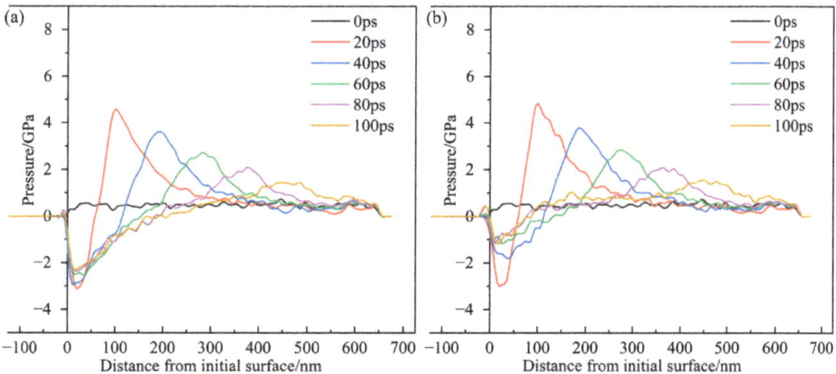

Figure 4. Spatial pressure profile for $Cu_{50}Zr_{50}$ metallic glass at different times and at absorbed fluence of (**a**) 80 mJ/cm^2 and (**b**) 84 mJ/cm^2. The initial surface is at position x = 0 nm.

3.2. Evolution of Femtosecond Laser Ablation at Different Fluences

To investigate the effects of different fluences on $Cu_{50}Zr_{50}$ metallic glass, laser pulses with a pulse duration of 100 fs and absorbed fluences of 90, 100, and 160 mJ/cm^2 were used to vertically irradiate the $Cu_{50}Zr_{50}$ metallic glass with a thickness of approximately 645 nm. The laser pulse was incident from left to right, and the snapshots of the target at different times are shown in Figure 5.

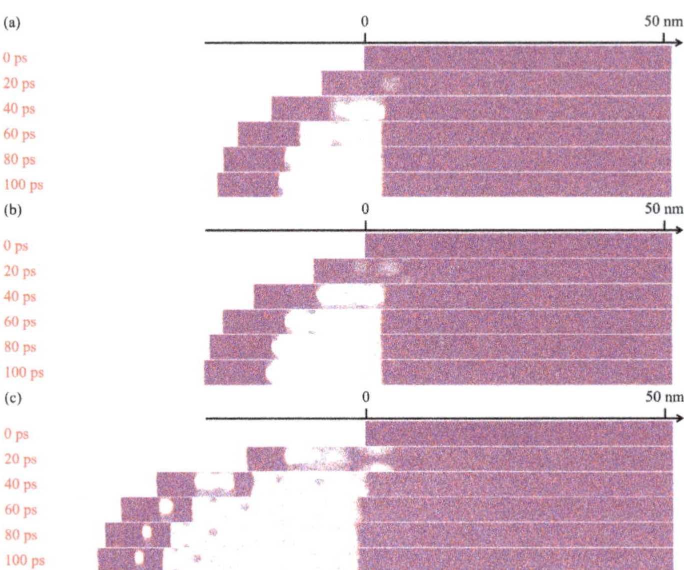

Figure 5. Snapshots of $Cu_{50}Zr_{50}$ metallic glass irradiated by femtosecond laser, with a pulse duration of 100 fs and absorbed fluences of (**a**) 90 mJ/cm^2, (**b**) 100 mJ/cm^2, and (**c**) 160 mJ/cm^2 from 0 to 100 ps. The laser pulse is incident from the left.

Figure 5 shows snapshots of the target after being irradiated by a single-pulse femtosecond laser with a pulse duration of 100 fs and absorbed fluence of 90~160 mJ/cm^2. At 20 ps, the targets at three laser conditions had already started to change. From 40 to 100 ps range, not only did the target material exhibit layer cracking, but large clusters of atoms were also produced, accompanied by violent phase explosions at the initial position of the target material. Moreover, it can be seen that an increasing number of single atoms and large clusters of atoms are ejected from the initial surface of the target material, causing the thickness of the ablated material to increase. If the pulse duration is the same, the target is more prone to ablation at the high absorbed fluence of the femtosecond laser.

According to Figure 6, the higher the absorbed fluence of the femtosecond laser, the more energy the target can absorb, and the electron temperature increases correspondingly. When the absorbed fluence is the same, after single-pulse femtosecond laser irradiation, the electron temperature in the absorption layer of the material decreases with the increase in depth, while the coupling time changes the opposite. The maximum electron temperatures of the initial target surface (x = 0 nm) after irradiation by femtosecond lasers with absorbed fluences of 90 mJ/cm^2, 100 mJ/cm^2, and 160 mJ/cm^2 were 8964 K, 9452 K, and 11,976 K, respectively. With the increase in the absorbed fluence, the maximum electron temperature gradually increases. This is mainly because under the condition of constant pulse duration, the increase in absorbed fluence represents the increase in peak power, thus causing the maximum electron temperature to rise. The reason why the highest electron temperature gradually increases with the increase in the absorbed fluence can be attributed to the following factors:

1. Increasing the absorbed fluence can increase the number of photons, which in turn improves the laser absorption rate and energy conversion efficiency. This promotes the absorption and conversion of energy into thermal energy in a more effective way.
2. More laser energy being absorbed into the material can lead to more excited electrons in a high-energy state, which results in an increase in electron temperature.
3. The increase in electron temperature can also lead to an increase in the thermal conductivity of the material, which accelerates the transfer of heat, causing the overall

material temperature to rise. This temperature increase may continue to occur during the reflux period, and the increase in electron temperature may be the result of complicated interactions between temperature and pressure.

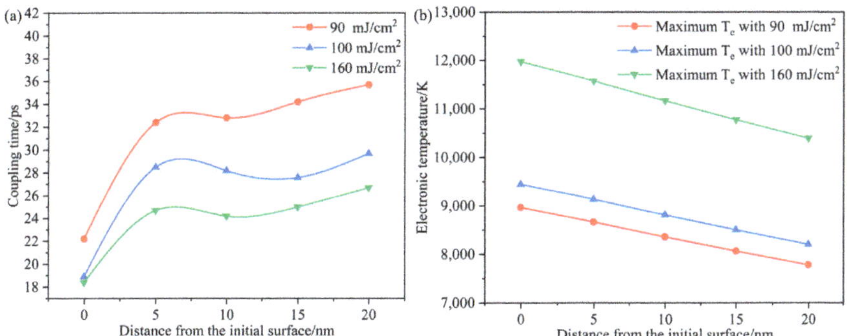

Figure 6. The variation of (**a**) the electron–lattice equilibrium temperature coupling time and (**b**) the maximum electronic temperature in the surface of $Cu_{50}Zr_{50}$ metallic glass irradiated by femtosecond laser irradiation, with a pulse duration of 100 fs and absorbed fluences of 90 mJ/cm^2, 100 mJ/cm^2, and 160 mJ/cm^2 depending on depth. The initial surface is at position x = 0 nm.

To investigate the cause of target fracture in $Cu_{50}Zr_{50}$ metallic glass under the loading of femtosecond lasers with a pulse duration of 100 fs and absorbed fluence of 90 mJ/cm^2, we analyzed the propagation of pressure inside the target material from 0 to 100 ps, as shown in Figure 7. The phenomenon of spallation is caused by the propagation of high-pressure compression waves generated near the surface of a target material. These waves reflect upon reaching the surface and create rarefaction waves that travel back into the bulk of the material. If the intensity of the rarefaction wave exceeds the strength of target, it can cause failure and the detachment of large pieces of matter near the surface. With the increase in time, the shock wave induced by the laser began to propagate deeper into the target material. The surface material underwent expansion and fracture between 0 ps and 100 ps due to the pressure on the surface of the target.

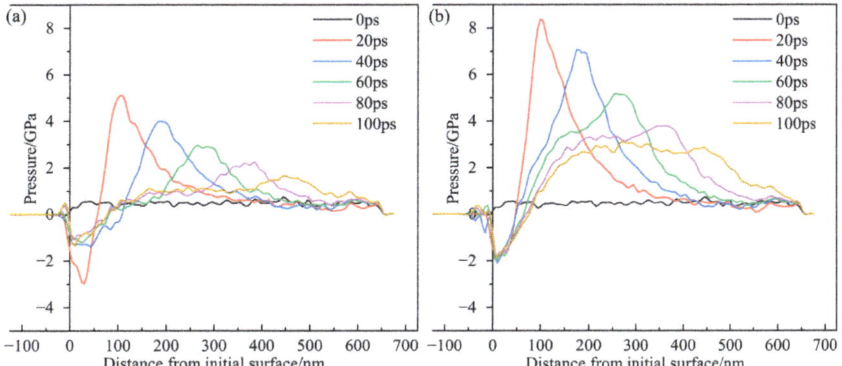

Figure 7. Spatial pressure profile for $Cu_{50}Zr_{50}$ metallic glass at different times and at absorbed fluences of (**a**) 90 mJ/cm^2 and (**b**) 160 mJ/cm^2. The initial surface is at position x = 0 nm.

3.3. Evolution of Femtosecond Laser Ablation at Different Pulse Durations

To investigate the effect of pulse durations on the femtosecond laser processing of metallic glass, $Cu_{50}Zr_{50}$ metallic glass with a thickness of approximately 645 nm was irradiated by single-pulse femtosecond lasers with an absorbed fluence of 160 mJ/cm^2 and pulse durations of 50 fs, 100 fs, 200 fs, and 500 fs. The snapshots of the target at different times are shown in Figure 8. The laser incidence direction is from left to right.

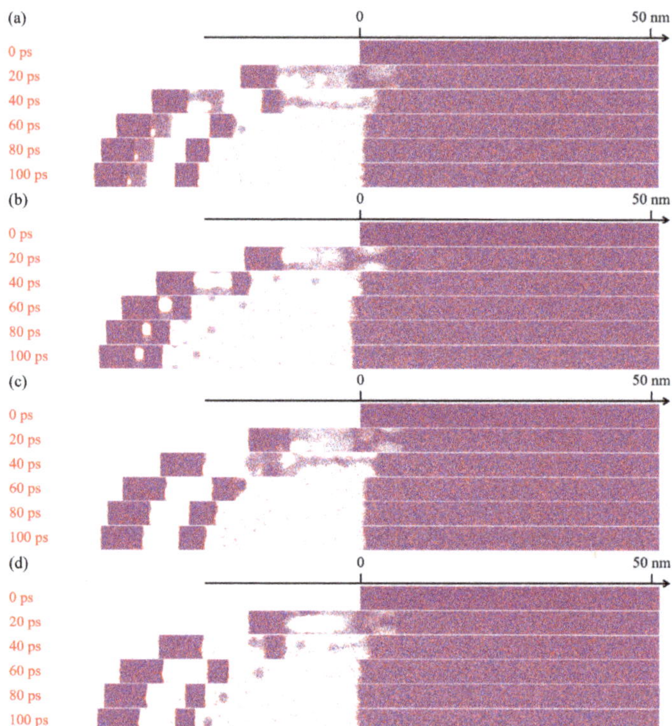

Figure 8. Snapshots of Cu$_{50}$Zr$_{50}$ metallic glass irradiated by femtosecond laser with absorbed fluence of 160 mJ/cm^2 and pulse durations of 50 fs (**a**), 100 fs (**b**), 200 fs (**c**), and 500 fs (**d**) from 0 to 100 ps. The laser pulse is incident from the left.

In Figure 8, when the absorbed fluence was 160 mJ/cm^2, the target material began to ablate at 20 ps, with the appearance of voids in the sub-surface of the target and the production of larger clusters of surface material dragged out by the expanding voids that led to target fracture. Between 40~60 ps, obvious laminations appeared in the target, with decreasing gaps between the layers; this was accompanied by more significant clusters of atoms being ablated from the target mother material. At 100 ps, ablation no longer occurred, and although the target material was partially removed, the subsurface region of the target material closer to the surface was occupied by thermal atoms that were emitted due to bottom-up thermal expansion.

Figure 9 shows the maximum electron temperature of the surface of the Cu$_{50}$Zr$_{50}$ metallic glass target, as well as the distribution of the electron–phonon coupling time with respect to space after irradiation with four pulse durations of 50 fs, 100 fs, 200 fs, and 500 fs, all with an absorbed fluence of 160 mJ/cm^2. From Figure 9a, it can be seen that increasing the pulse duration has no significant effect on the electron–lattice temperature coupling time of the target surface. From Figure 9b, it can be seen that the maximum electron temperature of the target surface decreases gradually with increasing depth. At the same position of the target, the maximum electron temperature decreases as the pulse duration increases. The maximum electron temperatures of the initial target surface (x = 0 nm) after irradiation by femtosecond lasers with pulse durations of 50 fs, 100 fs, 200 fs, and 500 fs were 12,116 K, 11,976 K, 11,893 K, and 11,758 K, respectively.

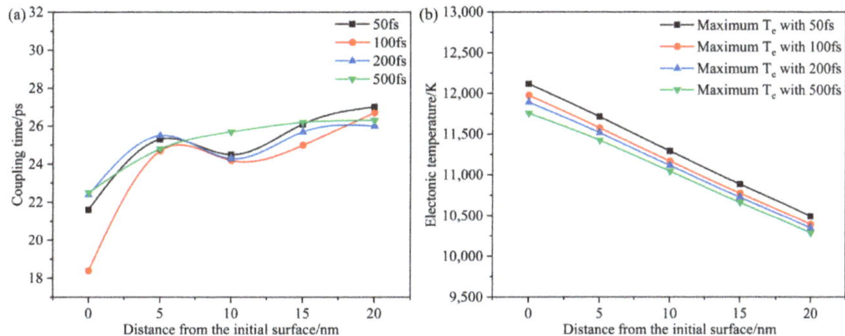

Figure 9. The variation of the electron–lattice equilibrium temperature coupling time (**a**) and the maximum electronic temperature (**b**) in the surface of $Cu_{50}Zr_{50}$ metallic glass irradiated by femtosecond laser irradiation, with absorbed fluence of 160 mJ/cm^2 and pulse durations of 50 fs, 100 fs, 200 fs, and 500 fs depending on depth. The initial surface is at position x = 0 nm.

As the pulse duration increases, the duration of femtosecond laser loading on $Cu_{50}Zr_{50}$ metallic glass tends to lengthen; thus, the time it takes for the electrons and lattice temperature to reach an equilibrium state also increases. With the increase in pulse duration, the highest electron temperature has a gradual downward trend. This is mainly because under the condition of constant pulse energy, the widening of pulse duration represents the decrease in peak power; thus, the highest electron temperature decreases.

4. Conclusions

In this paper, a hybrid TTM-MD approach was used to simulate the process of single-pulse femtosecond laser ablation of $Cu_{50}Zr_{50}$ metallic glass target material with a thickness of 644.9 nm, and the mechanisms underlying different absorbed fluences F_{abs} and pulse durations τ were analyzed.

1. It was determined that the ablation threshold of a target under $\tau = 100$ fs was 84 mJ/cm^2. Further, the tension near the surface decreased because it was "absorbed" by the cavities that finally caused ablation on the target material.
2. The process of the single-pulse femtosecond laser machining of $Cu_{50}Zr_{50}$ metallic glass was studied with $F_{abs} = 80, 82, 84, 90, 100, 160$ mJ/cm^2. As τ increased, the maximum T_e at the same position on the target surface decreased, while the $T_e - T_{ion}$ coupling time showed no significant difference. Further, it was found that the increase in F_{abs} leads to various forms of surface ablation on the target material, including melting, cavitation, spallation, material ejection, and phase explosion.
3. The mechanism between the femtosecond laser and metallic glass was researched by $\tau = 50, 100, 200, 500$ fs. It was found that the variation of τ had little effect on the electron–lattice temperature evolution, but it did have an impact on the structural changes during the ablation process. As the absorbed fluence increased, the maximum T_e at the same position on the target surface increased, while the $T_e - T_{ion}$ coupling time became shorter.

Author Contributions: Conceptualization, J.X. and S.X.; methodology, J.X. and Y.W.; validation, D.X. and O.G.; investigation, J.X. and D.X.; writing—original draft preparation, J.X. and D.X.; writing—review and editing, J.X., D.X., O.G., Y.W. and S.X.; supervision, J.X.; project administration, J.X.; funding acquisition, J.X. All authors have read and agreed to the published version of the manuscript.

Funding: This research was funded by the Young Eastern Scholar Program at Shanghai Institutions of Higher Learning and the Startup Foundation for Young Teachers of Shanghai Ocean University (No. A2-2006-00-200372).

Data Availability Statement: Data are contained within the article.

Acknowledgments: Support from the Young Eastern Scholar Program at Shanghai Institutions of Higher Learning is gratefully acknowledged. The project is also supported by the Startup Foundation for Young Teachers of Shanghai Ocean University (No. A2-2006-00-200372).

Conflicts of Interest: The authors declare no conflict of interest.

References

1. Wang, W. Brief History, Present and Future of Metallic Glasses. *Chin. J. Nat.* **2022**, *44*, 173–181.
2. Wu, Y.; Liu, X.; Lv, Z. A New Member of the Glass Family—Metallic Glass. *Physics* **2022**, *51*, 691–700.
3. Han, L.; Ming, P.; Kong, Z.; Zhang, X. Status and Research Progress of Metallic Glass Micro-Manufacturing Technology. *J. Netshape Form. Eng.* **2022**, *14*, 40–49.
4. Yao, Y.; Tang, J.; Zhang, Y.; Hu, Y.; Wu, D. Development of Laser Fabrication Technology for Amorphous Alloys. *Chin. J. Lasers* **2021**, *48*, 174–189.
5. Sano, T.; Takahashi, K.; Hirose, A.; Kobayashi, K.F. Femtosecond Laser Ablation of Zr55Al10Ni5Cu30 Bulk Metallic Glass. *Mater. Sci. Forum* **2007**, *539–543*, 1951–1954. [CrossRef]
6. Wang, X.; Lu, P.; Dai, N.; Li, Y.; Liao, C.; Zheng, Q.; Liu, L. Noncrystalline Micromachining of Amorphous Alloys Using Femtosecond Laser Pulses. *Mater. Lett.* **2007**, *61*, 4290–4293. [CrossRef]
7. Ma, F.; Yang, J.; Zhu, X.; Liang, C.; Wang, H. Femtosecond Laser-Induced Concentric Ring Microstructures on Zr-Based Metallic Glass. *Appl. Surf. Sci.* **2010**, *256*, 3653–3660. [CrossRef]
8. Anisimov, S.I.; Kapeliovich, B.L.; Perelman, T.L. Electron Emission from Metal Surfaces Exposed to Ultrashort Laser Pulses. *J. Exp. Theor. Phys.* **1974**, *66*, 375–377.
9. Perez, D.; Lewis, L.J. Ablation of Solids under Femtosecond Laser Pulses. *Phys. Rev. Lett.* **2002**, *89*, 255504. [CrossRef]
10. Lorazo, P.; Lewis, L.J.; Meunier, M. Short-Pulse Laser Ablation of Solids: From Phase Explosion to Fragmentation. *Phys. Rev. Lett.* **2003**, *91*, 225502. [CrossRef]
11. Perez, D.; Lewis, L.J. Molecular-Dynamics Study of Ablation of Solids under Femtosecond Laser Pulses. *Phys. Rev. B* **2003**, *67*, 184102. [CrossRef]
12. Smirnova, J.A.; Zhigilei, L.V.; Garrison, B.J. A Combined Molecular Dynamics and Finite Element Method Technique Applied to Laser Induced Pressure Wave Propagation. *Comput. Phys. Commun.* **1999**, *118*, 11–16. [CrossRef]
13. Kotake, S.; Kuroki, M. Molecular Dynamics Study of Solid Melting and Vaporization by Laser Irradiation. *Int. J. Heat Mass Transf.* **1993**, *36*, 2061–2067. [CrossRef]
14. Xie, J.; Yan, J.; Zhu, D. Atomic Simulation of Irradiation of Cu Film Using Femtosecond Laser with Different Pulse Durations. *J. Laser Appl.* **2020**, *32*, 022016. [CrossRef]
15. Förster, G.D.; Lewis, L.J. Numerical Study of Double-Pulse Laser Ablation of Al. *Phys. Rev. B* **2018**, *97*, 224301. [CrossRef]
16. Ivanov, D.S.; Zhigilei, L.V. Combined Atomistic-Continuum Modeling of Short-Pulse Laser Melting and Disintegration of Metal Films. *Phys. Rev. B* **2003**, *68*, 064114. [CrossRef]
17. Yuan, W.; Sizyuk, T. Ablation Study in Gold Irradiated by Single Femtosecond Laser Pulse with Electron Temperature Dependent Interatomic Potential and Electron–Phonon Coupling Factor. *Laser Phys.* **2021**, *31*, 036002. [CrossRef]
18. Marinier, S.; Lewis, L.J. Femtosecond Laser Ablation of CuxZr1-x Bulk Metallic Glasses: A Molecular Dynamics Study. *Phys. Rev. B Condens. Matter Mater. Phys.* **2015**, *92*, 184108. [CrossRef]
19. Iabbaden, D.; Amodeo, J.; Fusco, C.; Garrelie, F.; Colombier, J.P. Molecular Dynamics Simulation of Structural Evolution in Crystalline and Amorphous CuZr Alloys upon Ultrafast Laser Irradiation. *Phys. Rev. Mater.* **2022**, *6*, 126001. [CrossRef]
20. Plimpton, S. Fast Parallel Algorithms for Short-Range Molecular Dynamics. *J. Comput. Phys.* **1995**, *117*, 1–19. [CrossRef]
21. Chen, J.K.; Tzou, D.Y.; Beraun, J.E. A Semiclassical Two-Temperature Model for Ultrafast Laser Heating. *Int. J. Heat Mass Transf.* **2006**, *49*, 307–316. [CrossRef]
22. Falkovsky, L.A.; Mishchenko, E.G. Electron-Lattice Kinetics of Metals Heated by Ultrashort Laser Pulses. *J. Exp. Theor. Phys.* **1999**, *88*, 84–88. [CrossRef]
23. Choy, C.L.; Tong, K.W.; Wong, H.K.; Leung, W.P. Thermal Conductivity of Amorphous Alloys above Room Temperature. *J. Appl. Phys.* **1991**, *70*, 4919–4925. [CrossRef]
24. Mendelev, M.I.; Kramer, M.J.; Ott, R.T.; Sordelet, D.J.; Yagodin, D.; Popel, P. Development of Suitable Interatomic Potentials for Simulation of Liquid and Amorphous Cu–Zr Alloys. *Philos. Mag.* **2009**, *89*, 967–987. [CrossRef]
25. Daw, M.S.; Baskes, M.I. Embedded-Atom Method: Derivation and Application to Impurities, Surfaces, and Other Defects in Metals. *Phys. Rev. B* **1984**, *29*, 6443–6453. [CrossRef]
26. Finnis, M.W.; Sinclair, J.E. A Simple Empirical *N*-Body Potential for Transition Metals. *Philos. Mag. A* **1984**, *50*, 45–55. [CrossRef]

27. Zhu, P.; Fang, F. On the Mechanism of Material Removal in Nanometric Cutting of Metallic Glass. *Appl. Phys. A* **2014**, *116*, 605–610. [CrossRef]
28. Murali, P.; Guo, T.F.; Zhang, Y.W.; Narasimhan, R.; Li, Y.; Gao, H.J. Atomic Scale Fluctuations Govern Brittle Fracture and Cavitation Behavior in Metallic Glasses. *Phys. Rev. Lett.* **2011**, *107*, 215501. [CrossRef]
29. Shugaev, M.V.; He, M.; Levy, Y.; Mazzi, A.; Miotello, A.; Bulgakova, N.M.; Zhigilei, L.V. Laser-Induced Thermal Processes: Heat Transfer, Generation of Stresses, Melting and Solidification, Vaporization, and Phase Explosion. In *Handbook of Laser Micro- and Nano-Engineering*; Springer International Publishing: Cham, Switzerland, 2020; pp. 1–81.
30. Shugaev, M.V.; Zhigilei, L.V. Thermoelastic Modeling of Laser-Induced Generation of Strong Surface Acoustic Waves. *J. Appl. Phys.* **2021**, *130*, 185108. [CrossRef]
31. Available online: https://docs.lammps.org/fix_viscous.html (accessed on 1 March 2023).
32. Zhigilei, L.V.; Lin, Z.; Ivanov, D.S. Atomistic Modeling of Short Pulse Laser Ablation of Metals: Connections between Melting, Spallation, and Phase Explosion. *J. Phys. Chem. C* **2009**, *113*, 11892–11906. [CrossRef]
33. Schäfer, C.; Urbassek, H.M.; Zhigilei, L.V. Metal Ablation by Picosecond Laser Pulses: A Hybrid Simulation. *Phys. Rev. B* **2002**, *66*, 115404. [CrossRef]
34. Hu, Y.; Jiang, G. Study on the Damage Dynamics of CuZr Amorphous Alloy Nanofilm Irradiated by Femtosecond Laser. *J. Sichuan Univ. Nat. Sci. Ed.* **2019**, *56*, 897–902.
35. Stukowski, A. Visualization and Analysis of Atomistic Simulation Data with OVITO–the Open Visualization Tool. *Model. Simul. Mater. Sci. Eng.* **2010**, *18*, 015012. [CrossRef]

Article

High-Speed Laser Cutting Silicon-Glass Double Layer Wafer with Laser-Induced Thermal-Crack Propagation

Chunyang Zhao [1], Zhihui Yang [1], Shuo Kang [2], Xiuhong Qiu [1] and Bin Xu [1,*]

[1] Guangdong Provincial Key Laboratory of Micro/Nano Optomechatronics Engineering, College of Mechatronics and Control Engineering, Shenzhen University, Shenzhen 518060, China
[2] New Display Technology and Equipment Center, Jihua Laboratory, Foshan 528200, China
* Correspondence: binxu@szu.edu.cn

Abstract: This paper studied laser induced thermal-crack propagation (LITP) dicing of a glass-silicon double-layer wafer with high scanning speed. A defocusing continuous laser was used in the experimental system as the volumetric heat source for the glass layer and the surface heat source for the silicon layer. Based on the principle of thermal-crack propagation, the commercial software ABAQUS was used on the simulated analysis, and the results of temperature field and thermal stress field distribution with high and low speed were compared. The experiment was executed in accordance with the simulation parameters. The surface morphology of the cut section was described by optical microscopy and a profilometer, and combined with the results, the non-synchronous propagation process of the crack under high speed scanning was revealed. Most importantly, the scanning section with a nanoscale surface roughness was obtained. The surface roughness of the silicon layer was 19 nm, and that of glass layer was 9 nm.

Keywords: laser induced thermal-crack propagation; high speed scanning; glass-silicon double-layer wafer; nanoscale surface roughness

Citation: Zhao, C.; Yang, Z.; Kang, S.; Qiu, X.; Xu, B. High-Speed Laser Cutting Silicon-Glass Double Layer Wafer with Laser-Induced Thermal-Crack Propagation. *Processes* **2023**, *11*, 1177. https://doi.org/10.3390/pr11041177

Academic Editor: Antonino Recca

Received: 3 March 2023
Revised: 29 March 2023
Accepted: 4 April 2023
Published: 11 April 2023

1. Introduction

It has been reported that glass-silicon bilayers are widely used in precision devices such as integrated circuits (ICs) and micro-electro-mechanical systems (MEMS) [1,2].

Knowles et al. described a process that involves connecting glass and silicon wafers through anodic bonding to encapsulate micro devices. This process ensures good reliability, stability, and gas tightness. Monocrystalline silicon is preferred over polysilicon for the wafers due to its superior mechanical and electrical properties [3]. Esashi et al. proposed a first level of packaging for these devices based on the wafer-level chip scale packaging (WLCSP) technology [4]. After the circuit deposition procedure, this method enables the combination of patterned glass and silicon wafers in a single step, which can then be shredded. Until now, finding a suitable, high-quality, and efficient dicing technology has been a pressing issue [5–7]. Conventional scanning of brittle materials is performed by means of a rotating diamond scanning wheel mounted on a mechanical cutter. The process involves the injection of coolant to cool the scanning area. Miyake et al. noted that left-handed coolant with chips and microcracks may weaken the edge strength and trigger the impending failure of the microdevice [8]. The limitations above-mentioned have significant implications for the productivity and yield rates. In order to improve the edge quality, further treatments such as water grinding are required. Due to these drawbacks, researchers have explored non-mechanical scanning techniques. In this regard, Riveiro et al. demonstrated that well-designed laser systems provide a viable alternative for cutting soda glass [9]. Laser scanning is a non-contact process that prevents tool wear and eliminates contamination. However, studies by Lindroos et al. have shown that conventional laser processing involves exposing the substrate to a highly intense focused laser beam, leading to thermal defects such as heat-affected zones and microcracks, which can be severe in the cut area [10].

Compared to other scanning methods, laser-induced thermal-crack propagation (LITP) appears to be a highly promising technique for scanning brittle materials [11]. Its most significant features include a low operating temperature (below 500 °C) and a zero-width scanning path (in full-body scanning mode) [12]. Lumley et al. were the first to propose the LITP-based scanning technique [13], and significant progress has been made in the industrial application of this technology over the past few decades [14–16]. Laser induced thermal-crack propagation (LITP) is the conversion of the bonded wafer into a form of thermal energy absorbed by multiphotons and transferred to the interior and borders of the sheet. This leads to an uneven temperature rise inside the sheet, creating a temperature gradient [17]. The temperature rise leads to the thermal expansion of the material, while the temperature gradient and the differences in the thermophysical properties of glass and silicon lead to inhomogeneous thermal expansion of the material, which results in thermal stresses within the material [18]. Compressive stresses appear in the region near the laser point, while tensile stresses appear in the region near the laser point. Next, prefabricated cracks in the material expand in response to the tensile stress, and as the cracks expand step by step, the elastic energy is released as surface energy. As the laser beam moves, the energy is absorbed and converted, and the crack continues to grow, scanning the material [19,20].

CAI Yecheng et al. proposed an effective three-dimensional model to calculate the principle of the thermal action of laser-induced thermal-crack propagation in scanning glass/silicon bonded wafers at low velocities [21]. The results illustrate the crack propagation process in the laser-induced thermal-crack propagation (LITP) scanning of bilayer wafers at low velocities and apply the laser-induced thermal-crack propagation (LITP) technique to the field of silicon-glass bonded double layer wafer dicing.

The issue of processing efficiency is a challenge that the laser-induced thermal-crack propagation (LITP) technology cannot overcome [22,23]. Increasing the scanning speed is the most direct and effective method to improve processing efficiency. In this regard, finite element simulation using the commercial software Abaqus was performed to analyze the scanning mechanism of laser high-speed-induced thermal cracking in glass/silicon bonded wafers. Optical microscopy and profilometer were used to observe the surface morphology of the scanned path and determine the asynchronous expansion and crack propagation mode of the glass and silicon layers. Based on these results, a comparison between high-speed and low-speed laser-induced thermal cracking scanning revealed that better cross-sectional quality could be obtained at high laser speeds. This study provides a solution for improving production efficiency in laser-induced thermal-crack propagation (LITP) processing, which can be more widely used in various fields, but also has limitations: the technology is currently only available for processing with brittle materials.

2. Materials and Methods

2.1. Experimental Procedure

Figure 1 shows the schematic diagram of the experimental setup used for dicing a glass-silicon double layer wafer based on LITP. The experiments were carried out at an ambient temperature of 20 °C using a continuous fiber-coupled semiconductor laser source that emits at 1064 nm in the TEM00 mode, with a maximum output power of 300 [24,25]. The laser head emitted a highly concentrated beam of light that was directed vertically onto the surface of the specimen being processed, ensuring precision and accuracy during the processing. The specimen was firmly fixed onto the X-Y positioning table to guarantee its accurate alignment and positioning during the process. The operating system plays a crucial role in determining the processing position of the specimen, ensuring that the laser beam is directed precisely to the desired location. Additionally, the laser head can be adjusted to control the Z-axis direction, allowing for the processing of materials at varying depths and thicknesses.

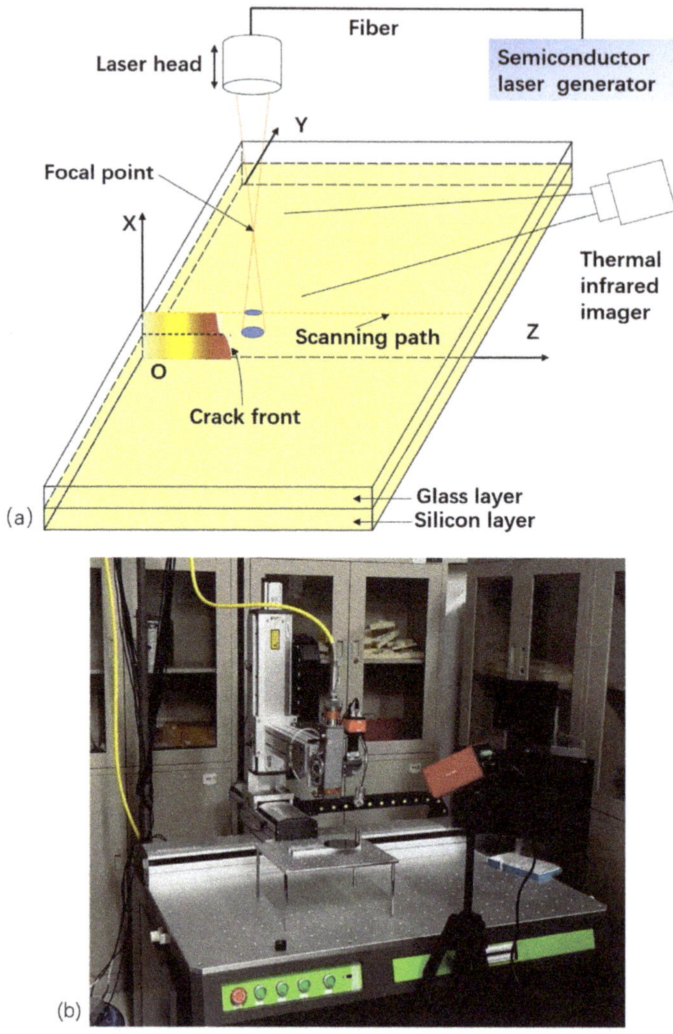

Figure 1. (**a**) Schematic of the experimental setup and the coordinate system of the specimen. (**b**) Physical image of the experimental equipment.

The samples were formed by combining a boron float 33 borosilicate glass wafer, manufactured by Schott, with an anode made of an n-type silicon wafer featuring (100) crystal planes. The prototype bonding and depletion layers measured between 2 and 20 nm and 1 µm, respectively. The silicon and glass wafers had a thickness of 0.5 mm and a diameter of 101.6 mm (4 inches). Prior to sample processing, an initial crack was made through the thickness, approximately 1 mm deep, using a diamond wire saw at the designated processing position.

The specimen was secured onto the X-Y stage with the glass side facing the laser head, and the scan path was programmed through the CNC operating system. The laser scanning parameters used during the experiment are detailed in Table 1. The cut surface was then examined under an optical microscope (Axiovert 200, Zeiss, Oberkochen, Germany), while the crack edge profile and surface roughness were measured using a surface profiler (PGI Dimension, Taylor Hobson, Leicester, UK).

Table 1. Processing parameters for the high speed LITP dicing of the silicon/glass double layer wafer.

Sample Size (mm^3)	Laser Power (W)	Scanning Speed (mm/s)	Laser Spot Diameter (mm)
$100 \times 100 \times (0.5 + 0.5)$	195	110	4.0

2.2. Theoretical Approach

The thermal stress field was meticulously computed, and the crack propagation process was simulated by utilizing the commercial software ABAQUS finite element analysis. To provide a comprehensive understanding of the numerical analysis, below are the fundamental assumptions and simplifications summarized.

1. The material distribution of glass and silicon layers was uniform and flawless; both the glass and silicon layers were isotropic in thermal analysis, while in the stress analysis, the glass layer was isotropic and the silicon layer was anisotropic, unlike other layers.
2. The glass-silicon double wafer anode bond layer was free of any defects; the friction problem and heat transfer between the sample and the fixture during the scanning process were ignored. Since the thickness of the bond layer was extremely small (2–20 nm) and can be considered zero, the gravitational and residual stresses after anode bonding could also be neglected.
3. All parameters were ignored or set to zero except for the process parameters, influencing factors, and material parameters considered in the simulation such as beam quality factor, etc.

On a macroscopic level, the process of low-intensity heat action aligns with Fourier's law of heat transfer. By applying Fourier's law and the first law of thermodynamics, the heat transfer issue can be analyzed. The 3D instantaneous heat conduction differential Equation (1), in the right-angle coordinate system, provides the 3D temperature distribution $T(x,y,z,t)$ based on the given initial and boundary conditions.

$$\frac{\partial}{\partial_x}\left(\lambda \cdot \frac{\partial_T}{\partial_x}\right) + \frac{\partial}{\partial_y}\left(\lambda \cdot \frac{\partial_T}{\partial_y}\right) + \frac{\partial}{\partial_z}\left(\lambda \cdot \frac{\partial_T}{\partial_z}\right) + \dot{q} = \rho c \cdot \frac{\partial_T}{\partial_t} \tag{1}$$

$$T|_{t=0} = T_0 \tag{2}$$

$$-k\frac{\partial T}{\partial_z}\bigg|_{\partial\Omega} = h(T - T_\infty) + \sigma\varepsilon\left(T^4 - T_\infty^4\right) \tag{3}$$

$$T|_{in-G} = T|_{in-Si} \tag{4}$$

where λ is the thermal conductivity; T is the temperature; $\dot{q}$ is the heat flow density of heat source; ρ is the density; c is the heat; t is the time. The initial temperature T_0 represents the outside room temperature; $\partial\Omega$ is the boundary surface of regional Ω; n is the outer normal direction of $\partial\Omega$; σ is the Steffen–Boltzmann constant with a value of $5.67 \times \frac{10^{-8}W}{m^2 \cdot K^4}$; ε is the emissivity or total energy emissivity, which takes values between 0 and 1; h is the convective heat transfer coefficient. The density of glass is 2230 (Kg/m^3) and the density of silicon is 2329 (Kg/m^3).

In order to solve Equations (7) and (8), it is necessary to establish a mathematical formula to describe the thermal source distribution when a silicon glass bilayer wafer is irradiated by laser. This process occurs simultaneously at the air–glass interface and the glass–silicon interface, where some energy is reflected, some is absorbed, and the rest is transmitted. Tables 2 and 3 provide the refractive indices of BF33 glass and silicon at 1064 nm, taking into account the refractive index of air at room temperature (20 °C). Using Equations (5) and (6), the normal incident reflectivity R_{AG} (refractive index from air to glass is 1.474) and R_{G-Si} (refractive index from glass to silicon is 3.550) were calculated to be 3.67% and 17.1%, respectively. It should be noted that the energy reflected from the silicon surface is absorbed by the glass layer. Therefore, when calculating the total absorption coefficients of the glass and silicon layers, this energy must be considered and corrected according to the experimental data. The corrected values were 0.392% and 79.61%, respectively.

$$R_{12} = (n_1 - n_2)^2 / (n_1 + n_2)^2 \tag{5}$$

$$I(z) = I_0 e^{\alpha z} \tag{6}$$

Table 2. Properties of Borofloat 33 glass material [21].

T	Thermal Conductivity	Specific Heat	Poisson's Ratio	Young's Modulus	Expansion Coefficient
T (°C)	λ_G (W/m·°C)	C_G (J/kg·°C)	ζ	E (GPa)	α (10^{-6}/°C)
25.0	1.08	758	0.200	64.0	3.20
125.0	1.25	1071	0.202	65.0	3.20
225.0	1.43	1175	0.204	66.0	3.26
325.0	1.60	1244	0.206	67.0	3.38
425.0	1.76	1290	0.208	68.0	3.61
525.0	1.92	1325	0.210	69.0	5.70

Table 3. Physical properties of silicon [21].

T	Specific Heat	Thermal Conductivity	Expansion Coefficient	Elastic Constant D_{ijmn} (GPa)		
T (°C)	C_G (J/kg·°C)	λ_G (W/m·°C)	α (10^{-6}/°C)	C_{11}	C_{12}	C_{44}
25.0	713	148.0	2.63	155.6	63.94	79.51
125.0	788	98.9	3.23	164.3	63.25	78.78
225.0	830	76.2	3.60	162.9	62.69	78.05
325.0	859	61.9	3.83	161.5	62.06	77.33
425.0	887	51.0	4.01	160.1	61.43	76.60
525.0	908	42.2	4.14	158.7	60.81	75.87

Therefore, the process of laser irradiation was utilized as a source of heat that spreads throughout the entire volume of the glass layer, while acting as a source of heat only on the surface of the silicon layer, as depicted in Figure 2. The mathematical expressions for the volume heat flux q_G and surface heat flux q_{Si} are described using Beer–Lambert's law and the Gaussian distribution of laser energy, which are presented as Equations (7) and (8), respectively.

$$q_G(x,y,z) = \frac{2(1 - R_{AG}) \cdot \alpha \cdot P_0}{\pi[r_G + (H_G + H_{Si} - z)tan_{\theta/2}]^2} \cdot e^{-\alpha \cdot (H_G + H_{Si} - z)} \cdot e^{-2\frac{(x-x_0-vt)^2-(y-y_0)^2}{[r_G+(H_G+H_{Si}-z)tan_{\theta/2}]^2}} \tag{7}$$

$$q_{Si}(x,y) = \frac{2\eta_{si}P_0}{\pi r_{si}^2} \cdot e^{-2\frac{(x-x_0-vt)^2-(y-y_0)^2}{r_{si}^2}} \tag{8}$$

where x_0, y_0 are the position coordinates of the scanning start beam center; q_G is the heat flux function of the bulk heat source in the glass layer; q_{Si} is the surface heat source heat flux function of the silicon layer; R_{AG} represents the reflectance of 1064 nm laser incident on the air-glass interface; P_0 is the laser power; H_G, H_{Si} are the thickness of the glass layer and silicon layer, respectively; r_G, r_{Si} are the spot radius of laser beam at the outer surface of glass layer and silicon layer, respectively; θ is the divergence angle of the laser beam; v is the scanning speed; η_{si} is the laser absorption rate on the silicon layer surface.

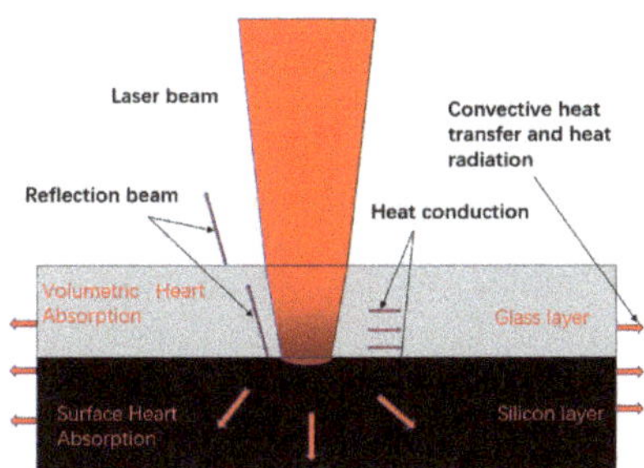

Figure 2. Schematic of the laser irradiates the glass-silicon double layer wafer.

Equations (7) and (8) are established based on the global coordinate system used in the finite element model, describing the thermal energy generated by any node in the finite element model at any time point in the form of heat flux. The introduction of these equations enables dynamic time steps in the laser scanning process and provides more accurate and detailed analytical methods for studying the thermal distribution and energy transfer in laser processing. Furthermore, by considering the heat flux produced by each node in the finite element model, a deeper understanding of the physical processes and phenomena in laser processing can be achieved, leading to a more effective design and control of laser processing systems.

Figure 3 showcases a detailed representation of the specimen's grid, which is composed of two distinct layers—a top layer of glass and a bottom layer of silicon. The unique properties of each layer necessitate a careful consideration of the mesh optimization strategy, as described in the subsequent text.

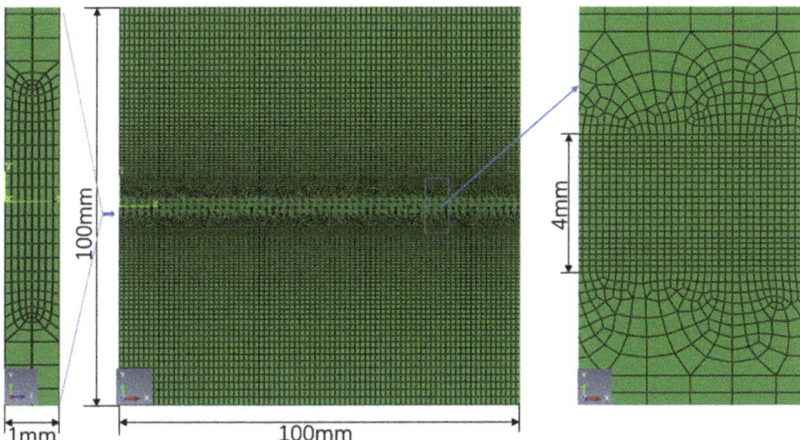

Figure 3. The mesh configuration of the three-dimensional model employed in the FEM analysis.

To optimize the mesh and ensure accurate simulation results, the two layers were divided into separate regions based on the laser scanning path and the maximum spot diameter, as detailed in [26–28]. The resulting mesh was finely tuned with a total of 140,211 nodes and 116,976 elements, providing a comprehensive and detailed analysis of the specimen's behavior under different laser processing conditions.

3. Results

Figure 4 shows the finite element simulation results of the maximum history temperature at each point on the simulated scan line during the high-speed scan. From the time the laser spot did not enter the workpiece to the time it completely left the workpiece area, the whole scanning process took only 945.45 ms. During the high-speed scanning process, the maximum historical temperature at each point in the stable crack expansion stage varied within 0.2 °C with time, which was almost unchanged. The maximum historical temperature at each point on the interface layer was 348.77 °C, and the temperature difference value between the silicon layer and the interface layer was 41.69 °C compared to the outer surface of the silicon layer, which was 307.08 °C. The outer surface of the glass layer was only 121.26 °C, and the temperature difference value between the glass layer and the interfacial layer was 227.49 °C. The highest historical temperature on the scan line of the interfacial layer was only 443.52 °C. In addition, the lag time of the points on the outer surface of the glass layer with the same x-coordinate was 159.5 ms and the lag distance was 17.54 mm compared to the interface layer. Compared to this, the points on the outer surface of the silicon layer only lagged 5.0 ms and the lag distance was only 0.55 mm.

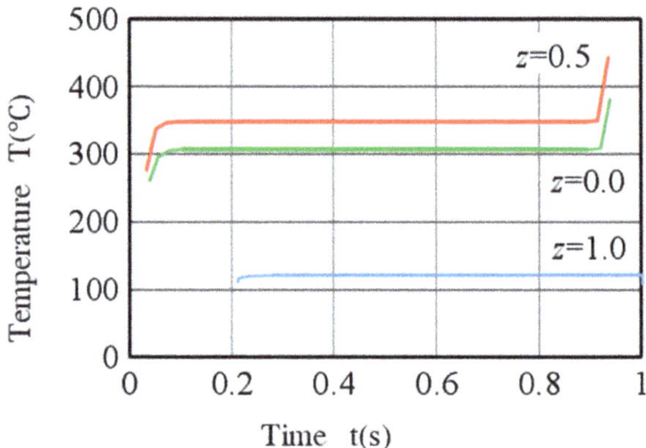

Figure 4. Historical maximum temperature curves of each point on the upper boundary layer, the outer surface of the glass layer, and the outer surface of the silicon layer of the scan line.

The simulation results of temperature fields on the upper and lower surfaces of the glass layer are presented in Figure 5a,b, respectively. Specifically, Figure 5a depicts the temperature distribution at low scanning speed (t = 13 s), while Figure 5c shows the temperature distribution at high scanning speed (t = 0.8 s). Upon the comparison of both figures, it was evident that the heat diffusion region caused by high-speed scanning was restricted to a small area near the bonding interface due to the low thermal conductivity of the glass. Furthermore, the temperature distribution on the upper surface of the glass layer lagged behind that of the interface layer and was relatively lower.

Figure 6c,d shows the temperature field and stress field distribution on the upper surface of the glass layer and the lower surface of the silicon layer at a high speed scanning time of 0.039 s. At this time, the laser spot center was located at the edge of the workpiece, and the maximum temperature of the workpiece was 322.6 °C, which had not yet reached the historical maximum temperature of 348.77 °C in the stable extension stage.

Figure 7c,d shows the temperature field and stress field distribution on the upper surface of the glass layer and the lower surface of the silicon layer at a high speed scanning time of 0.175 s. At this point, after the crack started at the cut entrance, the distance between the laser spot and the leading edge of the crack increased rapidly. The stress field distribution showed that the glass layer had the highest stress value near the interface, while the silicon layer had the highest stress value near the outer surface. This was basically consistent with the stress distribution and shape of the crack's leading edge in the low-speed scanning, but the crack's leading edge span was significantly increased.

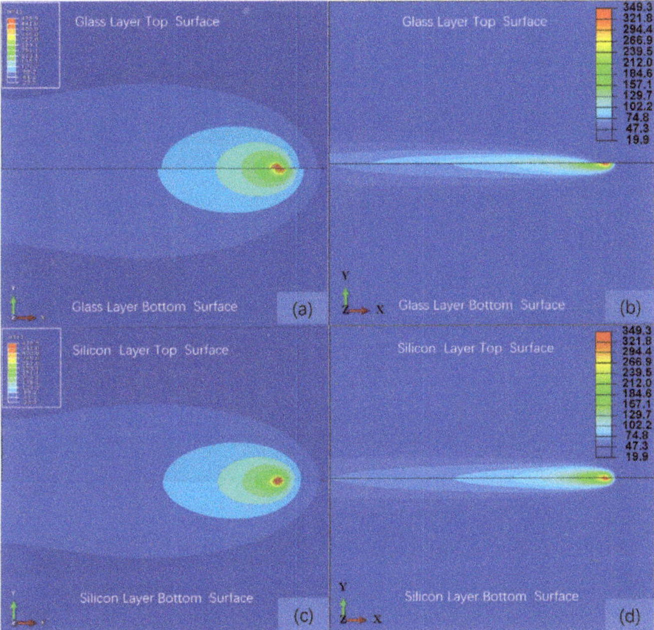

Figure 5. (**a**) Temperature field distribution on the upper and lower surfaces of the low speed glass layer. (**b**) Temperature field distribution on the upper and lower surfaces of the high speed glass layer. (**c**) Temperature field distribution on the upper and lower surface of the low speed silicon layer. (**d**) Temperature field distribution on the upper and lower surface of the high speed silicon layer.

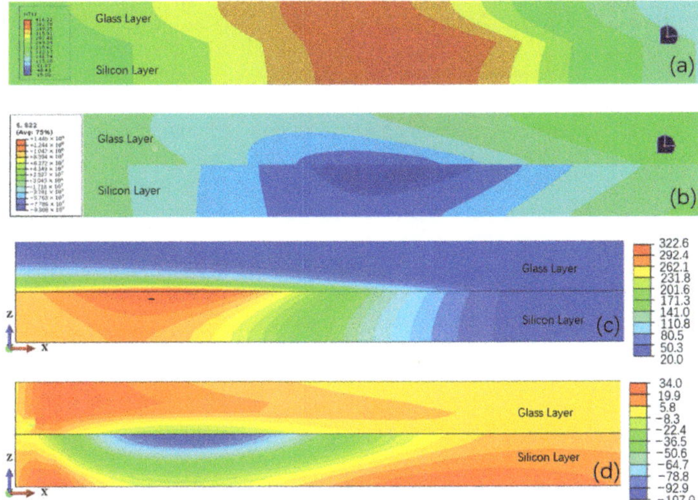

Figure 6. (**a**) Temperature field distribution at the low speed scanning inlet. (**b**) Distribution of the stress field at low speed incision. (**c**) Temperature field distribution at the high speed scanning inlet, (**d**) Distribution of the stress field at high speed incision.

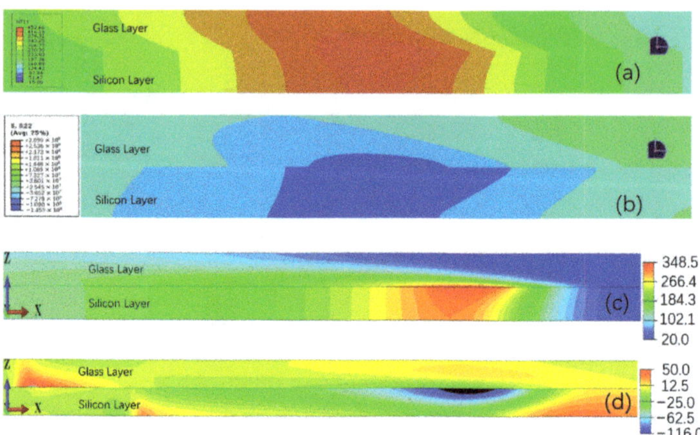

Figure 7. (**a**) Low speed temperature field distribution during steady expansion phase. (**b**) Distribution of the low speed stress field in the stable expansion stage. (**c**) High speed temperature field distribution during the steady expansion phase. (**d**) Distribution of the high speed stress field in the stable expansion stage.

According to Figures 6 and 7, the results show that the temperature gradient along the thickness of the glass layer at the center of the laser spot was very large, which led to tensile stresses near the upper surface of the glass layer in this region, and the maximum tensile stress could even exceed 40 MPa. This stress distribution situation was the opposite of the compressive stress in the glass layer at the center of the laser spot under low-speed scanning.

Figure 8 shows that a shallow groove prefabricated with a diamond wire saw was visible at the edge of the sample crack initiation. A scalloped ripple line could be observed on the upper surface of the glass layer near the shallow groove, with the upper right corner as the center area. This indicates that the crack in the glass layer at the entrance of the cut was not guided by the crack in the silica layer, but rather, the crack independently expanded from the corner point on the upper surface of the glass layer; a small area of "Γ" shaped corrugations was also visible at the entrance of the silica layer, extending up to 1.238 mm from the edge of the cut. The corrugation lines in this section were obviously different from the regularly distributed corrugation lines on the left side, and the cracks in the silicon layer in this section expanded independently. In the middle and left areas, the corrugation lines on the surface of the section had an inverted "S" shape, and the interval of the corrugation lines gradually stabilized. At this time, the glass layer and silicon layer cracked simultaneously. It can be seen that the crack initiation process at the entrance of the cut was smooth, the section was flat and no trajectory shift occurred, and after a short period of transition zone of the cut, the crack quickly entered the stable expansion stage.

Figure 9a shows that the cross-sectional photograph was taken at a location 7.123 mm from the edge of the crack, and the picture was roughly divided into two parts, left and right, with this location as the dividing line. It can be seen that there was a glass layer cross-sectional corrugation line at this location, and the trace was clearer than both the left and right sides. The ripple lines on the right side of the line were distributed in an obvious periodic pattern; while on the left side of the ripple line, the glass layer section was smooth and flat along the area in the scanning direction, with no obvious ripple traces left by the crack expansion. Further observation showed that the dividing ripple line did not continue to extend in the Z-negative direction until it intersected with the silicon layer, but started at a distance of about 459 μm from the upper surface of the glass layer and extended forward in the horizontal direction, forming a continuous horizontal ripple line. It was found that the distance from the horizontal ripple line to the interface layer increased and then decreased along the laser cutting direction.

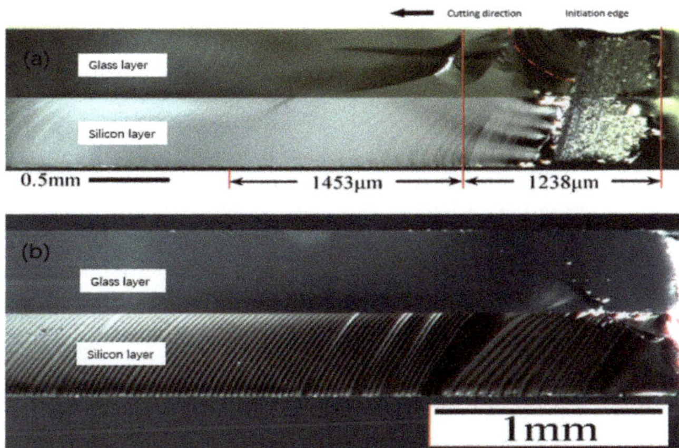

Figure 8. (**a**) High speed entrance profile optical photograph. (**b**) Low speed entrance profile optical photograph [21].

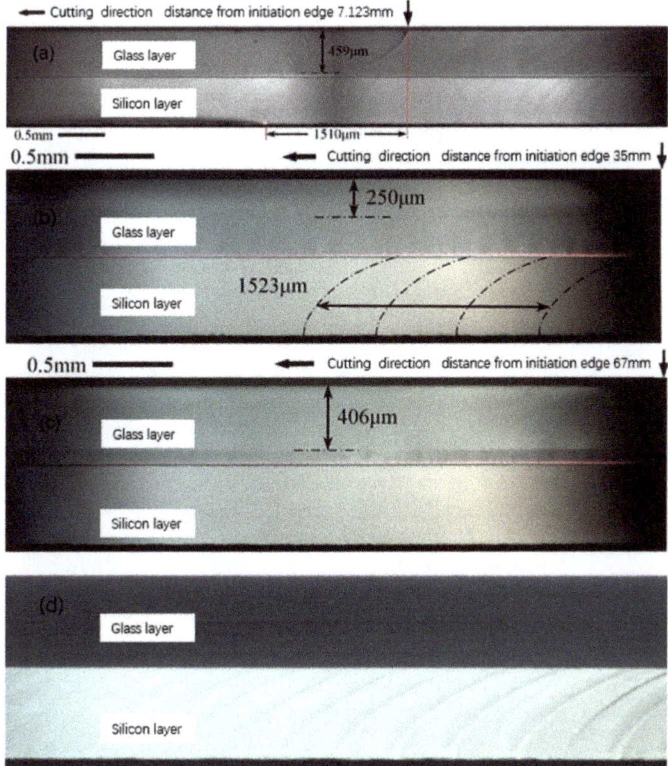

Figure 9. (**a**) High-speed optical photograph of the fracture surface about 7 mm from the entrance. (**b**) High-speed optical photograph of the fracture surface, about 35 mm from the entrance. (**c**) High-speed optical photograph of the fracture surface, about 67 mm from the entrance. (**d**) Low-speed optical photograph of the fracture surface [21].

From the section photos in Figure 9a–c, the section was smooth after 10 mm from the cracking edge, where the periodic ripple lines in the silicon layer section were barely visible. In contrast, the glass layer cross section only had obvious horizontal ripple lines, which visually indicates that the cross section quality had been greatly improved at high scanning speeds. According to the laser-induced thermal cracking scribing glass mechanism, the plate was not cut as a whole, the crack only extended in the shallow surface area of the plate, the stress formed a large gradient in the thickness direction, the compressive stress zone below the crack restricted the further expansion of the crack to the depth, and the subsequent mechanical method made the shallow-induced thermal cracking scribing process extend the crack to achieve the cutting of the plate. Therefore, the glass section processed by the induced thermal cracking scribing method will have a distinct horizontal corrugation line, which appears at the same location as the crack extension depth of the first step of the induced thermal cracking scribing. The horizontal ripple line also appeared during the high-speed cutting process, which indicates that the region of the glass layer near the upper surface was not cracked synchronously with the silicon layer, the crack expansion pattern inside the glass/silicon double bonded flat plate changed, and the crack expansion in the glass layer material was non-synchronous with the silicon layer.

Based on the simulation and measurement experimental results, the crack extension process of the glass/silicon double layer plate under high-speed scanning can be summarized. At the initial scan, the glass and silicon layers cracked independently near the upper surface, and the cracks did not extend to the whole material. Thereafter, the glass layer stopped cracking and the silicon layer cracked as a whole and drove the glass layer to crack. The crack entered a stable growth phase. After the crack expanded to a stable distance, the thermal conductivity of the glass layer was too low, the temperature difference value between the glass layer interface and the outer surface was too large, and the whole glass layer failed to form sufficient thermal expansion, so the crack expansion of the glass layer became a local expansion near the interface area and expanded simultaneously with the silicon layer crack, while the area near the outer surface was in an uncracked state. After the silicon and glass layers cracked near the interface, the glass layer did not crack as a whole until the laser continued to advance a sufficient distance, and this continued until the laser scanned the area close to the scan exit.

The results of the cross-sectional profile measurements at the stable crack stage are compared by combining laser scanning at low speed (as in Figure 10) with scanning at high speed (as in Figure 11). It can be seen that in the asynchronous cracking mode, the deviation of the crack location near the upper surface of the glass layer from the silicon layer section was about 15.6 µm, as shown in Figure 11a, and the surface cutting quality of the glass and silicon layer sections improved significantly compared to the low-speed scan condition (as shown in Figure 10a,b), as shown in Figure 11b,c. The surface roughness Ra of the silicon layer was 19 nm, while the surface roughness of the glass layer reached 9 nm. The reason for the periodic ripples in the silicon layer cross section was the same as in the low-speed scan, and was caused by the intermittent crack extension and the influence of the single-crystal silicon cracking surface.

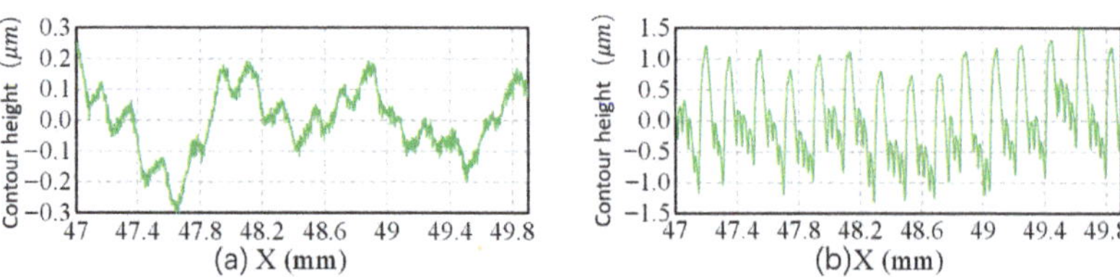

Figure 10. Low-speed profile of the fracture surface at the steady [21]. (a) Silicon layer; (b) glass layer.

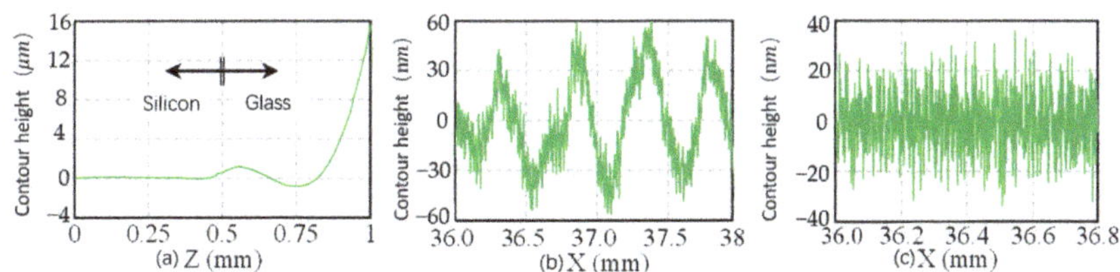

Figure 11. High-speed profile of the fracture surface at the steady crack propagation stage. (**a**) Cross section. (**b**) Silicon layer. (**c**) Glass layer.

4. Conclusions

In summary, the LITP technology was applied to high-speed laser-induced thermal-crack propagation for cutting glass/silicon bilayer bonded wafers. In this paper, the mechanism of high-speed induced thermal cracking of glass/silicon bilayer bonded wafers by a 1064 nm semiconductor laser was investigated in depth using both numerical simulation and experimental results and compared with low-speed induced thermal cracking. It can be concluded that the region near the upper surface of the glass layer cracked asynchronously with the silicon layer during the stable extension stage. The crack propagation pattern in the glass/silicon bilayer bonded wafer changed, and the crack propagation in the glass layer material was not synchronized with that in the silicon layer. The surface roughness of the silicon layer was 19 nm, while the surface roughness of the glass layer reached 9 nm. Obviously, the high-speed laser-induced thermal-crack propagation not only ensures excellent processing quality, but also significantly improves the processing efficiency.

Author Contributions: C.Z.: Conceptualization, Writing—original draft preparation, Funding acquisition. Z.Y.: Methodology, Software, Investigation. S.K.: Investigation, Supervision. X.Q.: Supervision, Writing—review and editing, Funding acquisition. B.X.: Data Curation, Funding acquisition. All authors have read and agreed to the published version of the manuscript.

Funding: This research was funded by the National Natural Science Foundation of China (Grant No. 62003216); the National Natural Science Foundation-Aerospace Joint Fund (Grant No. U203720-5); the Natural Science Foundation of Guangdong Province (Grant No. 2022A1515010600); the Research Committee of the Shenzhen University and Shenzhen Natural Science Foundation University Stable Support Project (Grant No. 20200826160002001); the Shenzhen Top Talents Start-Up Fund (Grant No. 000711).

Data Availability Statement: Not applicable.

Conflicts of Interest: No conflicts of interest exist in this submitted manuscript, and the manuscript was approved by all authors for publication. I would like to declare on behalf of my co-authors that the work described was original research that has not been published previously, and not under consideration for publication elsewhere, in whole, or in part.

References

1. Kang, H.S.; Hong, S.K.; Oh, S.C.; Choi, J.Y.; Song, M.G. Cutting glass by laser. In Proceedings of the Second International Symposium on Laser Precision Microfabrication, Singapore, 16–18 May 2002; SPIE: Bellingham, WA, USA; Volume 4426, pp. 367–370.
2. Wang, H.J.; Yang, T. A review on laser drilling and cutting of silicon. *J. Eur. Ceram. Soc.* **2021**, *41*, 4997–5015. [CrossRef]
3. Knowles, K.M.; Van Helvoort, A.T.J. Anodic bonding. *Int. Mater. Rev.* **2006**, *51*, 273–311. [CrossRef]
4. Esashi, M. Wafer level packaging of MEMS. *J. Micromech. Microeng.* **2008**, *18*, 073001. [CrossRef]
5. Zhang, X.H.; Chen, G.Y.; An, W.K.; Deng, Z.H.; Zhou, Z.X. Experimental investigations of machining characteristics of laser-induced thermal cracking in alumina ceramic wet grinding. *Int. J. Adv. Manuf. Technol.* **2014**, *72*, 1325–1331. [CrossRef]
6. Nagimova, A.; Perveen, A. A review on laser machining of hard to cut materials. *Mater. Today Proc.* **2019**, *18*, 2440–2447. [CrossRef]
7. Zhimalov, A.B.; Solinov, V.F.; Kondratenko, V.S.; Kaplina, T.V. Laser cutting of float glass during production. *Glass and ceramics.* **2006**, *63*, 319–321. [CrossRef]
8. Miyake, Y. Separation technology for FPD glass. *J. Jpn. Soc. Abras. Technol.* **2001**, *47*, 342–347.

9. Riveiro, A.; Quintero, F.; Lusquiños, F.; Comesaña, R.; Pou, J. Study of melt flow dynamics and influence on quality for CO_2 laser fusion cutting. *J. Phys. D Appl. Phys.* **2011**, *44*, 135501. [CrossRef]
10. Puurunen, R.L.; Kattelus, H.; Suntola, T. *Handbook of Silicon Based MEMs Materials and Technologies*; Lindroos, V., Tilli, M., Lehto, A., Motooka, T., Eds.; Elsevier: Oxford, UK, 2010.
11. Cheng, X.; Yang, L.; Wang, M.; Cai, Y.; Wang, Y.; Ren, Z. Laser beam induced thermal-crack propagation for asymmetric linear cutting of silicon wafer. *Opt. Laser Technol.* **2019**, *120*, 105765. [CrossRef]
12. Lumley, R.M. Controlled separation of brittle materials using a laser. *Ceram. Bull.* **1969**, *48*, 850.
13. Xu, J.; Hu, H.; Zhuang, C.; Ma, G.; Han, J.; Lei, Y. Controllable laser thermal cleavage of sapphire wafers. *Opt. Lasers Eng.* **2018**, *102*, 26–33. [CrossRef]
14. Kechagias, J.D.; Fountas, N.A.; Ninikas, K.; Petousis, M.; Vidakis, N.; Vaxevanidis, N. Surface characteristics investigation of 3D-printed PET-G plates during CO_2 laser cutting. *Mater. Manuf. Process.* **2022**, *37*, 1347–1357. [CrossRef]
15. Fountas, N.A.; Ninikas, K.; Chaidas, D.; Kechagias, J.; Vaxevanidis, N.M. Neural networks for predicting kerf characteristics of CO_2 laser-machined FFF PLA/WF plates. In Proceedings of the MATEC Web of Conferences, Rennes, France, 8–10 September 2022; EDP Sciences: Les Ulis, France; Volume 368, p. 01010.
16. Samieian, M.A.; Cormie, D.; Smith, D.; Wholey, W.; Blackman, B.R.; Dear, J.P.; Hooper, P.A. Temperature effects on laminated glass at high rate. *Int. J. Impact Eng.* **2018**, *111*, 177–186. [CrossRef]
17. Yahata, K.; Yamamoto, K.; Ohmura, E. Crack Propagation Analysis in Laser Scribing of Glass. *J. Laser Micro/Nanoeng.* **2010**, *5*, 109–114. [CrossRef]
18. Cai, Y.; Wang, M.; Zhang, H.; Yang, L.; Fu, X.; Wang, Y. Laser cutting sandwich structure glass–silicon–glass wafer with laser induced thermal–crack propagation. *Opt. Laser Technol.* **2017**, *93*, 49–59. [CrossRef]
19. Zhao, C.; Cai, Y.; Ding, Y.; Yang, L.; Wang, Z.; Wang, Y. Investigation on the crack fracture mode and edge quality in laser dicing of glass-anisotropic silicon double-layer wafer. *J. Mater. Process. Technol.* **2020**, *275*, 116356. [CrossRef]
20. Zhao, C.; Zhang, H.; Wang, Y. Semiconductor laser asymmetry cutting glass with laser induced thermal-crack propagation. *Opt. Lasers Eng.* **2014**, *63*, 43–52. [CrossRef]
21. Cai, Y.; Yang, L.; Zhang, H.; Wang, Y. Laser cutting silicon-glass double layer wafer with laser induced thermal-crack propagation. *Opt. Lasers Eng.* **2016**, *82*, 173–185. [CrossRef]
22. Weinhold, S.; Gruner, A.; Ebert, R.; Schille, J.; Exner, H. Study of fast laser induced cutting of silicon materials. In Proceedings of the Laser Applications in Microelectronic and Optoelectronic Manufacturing (LAMOM) XIX, San Francisco, CA, USA, 1–6 February 2014; SPIE: Bellingham, WA, USA; Volume 8967, pp. 333–339.
23. Haupt, O.; Schuetz, V.; Schoonderbeek, A.; Richter, L.; Kling, R. High quality laser cleaving process for mono-and polycrystalline silicon. In Proceedings of the Laser-based Micro-and Nanopackaging and Assembly III, San Jose, CA, USA, 24–29 January 2009; SPIE: Bellingham, WA, USA; Volume 7202, pp. 155–165.
24. Yahata, K.; Ohmura, E.; Shimizu, S.; Murakami, M. Boundary Element Analysis of Crack Propagation in Laser Scribing of Glass. *J. Laser Micro/Nanoeng.* **2013**, *8*, 102–109. [CrossRef]
25. Masolin, A.; Bouchard, P.-O.; Martini, R.; Bernacki, M. Thermo-mechanical and fracture properties in single-crystal silicon. *J. Mater. Sci.* **2013**, *48*, 979–988. [CrossRef]
26. Green, M.A. Self-consistent optical parameters of intrinsic silicon at 300 K including temperature coefficients. *Sol. Energy Mater. Sol. Cells* **2008**, *92*, 1305–1310. [CrossRef]
27. Watanabe, H.; Yamada, N.; Okaji, M. Linear thermal expansion coefficient of silicon from 293 to 1000 K. *Int. J. Thermophys.* **2004**, *25*, 221–236. [CrossRef]
28. Zhao, C.; Zhang, H.; Yang, L.; Wang, Y.; Ding, Y. Dual laser beam revising the separation path technology of laser induced thermal-crack propagation for asymmetric linear cutting glass. *Int. J. Mach. Tools Manuf.* **2016**, *106*, 43–55. [CrossRef]

Article

Low-Temperature Gas Cooling Correction Trajectory Offset Technology of Laser-Induced Thermal Crack Propagation for Asymmetric Linear Cutting Glass

Chunyang Zhao [1], Zhihui Yang [1], Xiuhong Qiu [1], Jiayan Sun [1] and Zejia Zhao [1,2,*]

[1] Guangdong Provincial Key Laboratory of Micro/Nano Optomechatronics Engineering, College of Mechatronics and Control Engineering, Shenzhen University, Shenzhen 518060, China; zcy724317@163.com (C.Z.); 15007960781@163.com (Z.Y.); 18916536921@163.com (X.Q.); sunjiayan730@163.com (J.S.)

[2] Institute of Semiconductor Manufacturing Research, College of Mechatronics and Control Engineering, Shenzhen University, Shenzhen 518060, China

* Correspondence: zhaozejia@szu.edu.cn

Abstract: Laser-induced thermal crack propagation (LITP) is a high-quality and efficient processing method that has been widely used in fields such as glass cutting. However, the problem of trajectory deviation often arises in actual cutting operations, especially in asymmetric cutting. To address this issue, a low-temperature gas cooling trajectory deviation correction technique was proposed in this study. This technique modifies the temperature and stress distribution by spraying low-temperature gas onto the processing surface and maintaining a relative position with the laser, thereby correcting the trajectory deviation. The finite element simulation software ABAQUS was employed to numerically simulate the dynamic propagation of temperature fields, thermal stress, and cracks in the asymmetric linear cutting and circular cutting of soda-lime glass with the proposed low-temperature gas cooling trajectory deviation correction technique, and the correction mechanism was elucidated. In the simulation results, the optimal relative distance (ΔX) between the low-temperature gas and scanning laser was obtained by analyzing the transverse tensile stress. Based on the analysis of the experimental and numerical simulation results, it is concluded that the cryogenic gas cooling technique can effectively correct the trajectory deviation phenomenon of asymmetric linear cutting of soda lime glass by LITP.

Keywords: low-temperature gas cooling technology; laser-induced thermal crack propagation (LITP); soda-lime glass; track correction

Citation: Zhao, C.; Yang, Z.; Qiu, X.; Sun, J.; Zhao, Z. Low-Temperature Gas Cooling Correction Trajectory Offset Technology of Laser-Induced Thermal Crack Propagation for Asymmetric Linear Cutting Glass. *Processes* **2023**, *11*, 1493. https:// doi.org/10.3390/pr11051493

Academic Editor: Hideki KITA

Received: 17 April 2023
Revised: 3 May 2023
Accepted: 10 May 2023
Published: 15 May 2023

1. Introduction

As a brittle material, glass has been widely utilized in various electronic device screens due to its high transparency, high mechanical strength, uniform texture, smooth surface, and corrosion resistance. With the popularity of smartphones, handheld devices, and tablet computers, the demand for high-quality, high-efficiency, and high-strength glass cutting has been increasing, which has become an important technological challenge in the glass manufacturing industry [1–4].

At present, the conventional mechanical cutting method is widely employed to cut glass, which involves the use of hard metal tools and physical contact force between them and the glass [5]. C.T. Pan et al. [6] proposed that the size of the median crack increases with the increase of cutting depth and cutting pressure, but an appropriate size of the median crack can eliminate edge chipping and obtain a smooth separation surface. M. Zhou et al. [7] demonstrated that adding ultrasonic vibration to diamond cutting during glass cutting can effectively reduce tool wear and improve the surface finish, which has been verified by experiments. Although this contact scratching method often generates many microcracks and small chippings on the cutting edge, these microcracks may propagate under continuous forces in the subsequent processing, leading to a rapid decrease in the number of products.

Therefore, traditional mechanical cutting methods are gradually being replaced by non-traditional cutting methods [8–10]. Finding a suitable cutting technology with high quality and efficiency has become an urgent issue [11,12]. Laser-induced thermal crack propagation (LITP) is a non-traditional and competitive glass processing method with features of non-contact, high-quality, high-efficiency, and high-strength processing [13]. Lumley [14] proposed LITP for cutting brittle materials. Laser-induced thermal crack propagation involves a laser instead of a tool in a non-contact form. The cut section is smooth, clean, and straight, free of contamination and defects, and extremely strong. The material only undergoes separation but is not removed during the crack propagation process, which not only avoids material damage but also greatly improves the material utilization efficiency. Therefore, laser-induced thermal crack propagation is widely recognized as the method with the highest cross-section quality for glass cutting [15,16].

H.S. Kang et al. conducted a study on laser-induced thermal crack propagation for liquid crystal displays (LCDs), plasma displays (PDPs), and flat panel displays (FPDs). The results showed that the uneven distribution of the liquid after being sprayed onto the glass surface resulted in poorer edge quality when using liquid cooling for glass cutting compared to gas cooling [17]. C.H. Tsai et al. [18] used a diamond tool to scribe and scanned along the scribe with a continuous CO_2 laser. They found that the uncertainty of the size and direction of cracks at the bottom of the scribe resulted in the separated surface after processing not being perpendicular to the surface of the flat glass. Salman Nisar et al. [19] conducted a study on symmetrically cutting glass using diode laser-induced thermal crack propagation. The results showed that there was a serious trajectory deviation phenomenon at the entry and exit points of the material. During the stable cutting stage, the crack could extend along the trajectory of the laser movement.

However, asymmetric linear cutting (where the cut line does not coincide with the material centerline, including asymmetric straight lines and circular curves) is more common in practical processing. The asymmetry of the material on both sides of the scan line causes the stress distribution asymmetry and the location of the shear minimum does not appear on the scan line, causing the crack trajectory to be shifted; while the trajectory shifting mechanism of circular curve cutting also includes the effect of the asymmetry of the temperature field distribution and the hysteresis of crack expansion, causing the crack trajectory to be shifted. C.Y. Zhao et al. [20] proposed a dual-beam laser trajectory offset correction technique that can effectively correct the trajectory offset phenomenon that occurs when cutting asymmetric glass by laser-induced thermal crack propagation. However, this technique increases the complexity and cost of the equipment and raises the maximum temperature of the material during the cutting process, which may lead to defects in the material separation surface. In addition, the technique is better for the correction of asymmetric straight lines but is less effective for the correction of trajectory offsets with small cutting arc radii.

In this study, a cryogenic gas cooling trajectory deviation correction technique is proposed to solve the trajectory deviation problem in asymmetric cutting. The technique involves directing the cryogenic gas to the side where the temperature gradient needs to be increased in order to change the temperature and stress distribution. Numerical simulations were performed to investigate the temperature field, thermal stresses, and dynamic crack expansion during the correction of a cryogenic gas cooling trajectory deviation for the asymmetric linear cutting of glass. The optimal relative distance (ΔX) between the cryogenic gas and the scanning laser was obtained by combining the experimental results and simulation data. The correction mechanism of the cryogenic gas cooling trajectory deviation correction technique is further explained.

2. Theoretical and Simulation Models

The principle of the low-temperature gas cooling trajectory deviation correction technique is to direct low-temperature gas towards the side that requires an increase in the temperature gradient to form a larger temperature gradient and reduce asymmetric stress, thereby reducing the trajectory deviation. As the gas jet generates a significant amount of pressure, the same arrangement strategy is employed to symmetrically place the nozzle on the upper and lower sides of the specimen to shield the effect of the jet pressure on processing, ensuring that the pressure on both sides of the material is consistent during jetting (as shown in Figure 1).

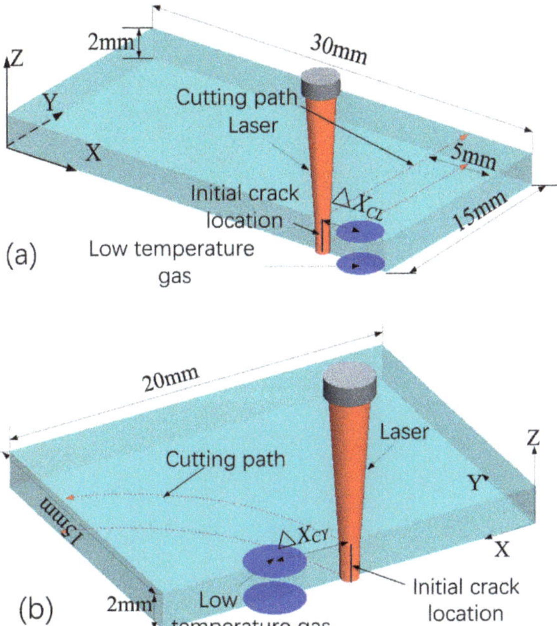

Figure 1. Low-temperature gas cooled revising the trajectory deviation technology. (**a**) Asymmetric line. (**b**) Circular arc curve.

Upon the laser entering the scanning area, the material temperature rapidly drops, leading to material shrinkage and tensile stress. In the case of laser scanning on asymmetric glass, the asymmetric thermal stress can cause a trajectory deviation. Therefore, low-temperature gas jetting to the side that requires an increased temperature gradient can improve the symmetry of the cutting process and reduce the trajectory deviation. For asymmetric linear cutting, low-temperature gas jetting should be applied to the side with less material to correct the trajectory. For circular arc cutting, low-temperature gas jetting should be applied to the side that requires an increased temperature gradient on the inner side of the arc to correct the trajectory.

The absorption mode of sodium–calcium flat glass for a continuous semiconductor laser with a wavelength of 1064 nm is in the bulk absorption mode, as shown in Figure 2. The expression for the planar Gaussian distribution cone divergent bulk absorption heat source with uniform linear motion can be represented by Equation (1) [21,22].

$$q_G(x,y,z) = \frac{2(1 - R_G)\cdot\alpha\cdot P_0}{\pi[r_G + (H_G - z)tan_{\theta/2}]^2}\cdot e^{-\alpha\cdot(H_G - z)}\cdot e^{-2\frac{(x-x_0-vt)^2-(y-y_0)^2}{[r_G+(H_G-z)tan_{\theta/2}]^2}} \tag{1}$$

where x_0, y_0 are the position coordinates of the scanning start beam center; q_G is the heat flux function of the heat source in the glass; R_G is the reflectance of the 1064 nm laser incident on the air–glass interface; P_0 is the laser power; H_G is the thickness of the glass layer; r_G is the spot radius of the laser beam at the outer surface of the glass layer; θ is the divergence angle of the laser beam; and v is the scanning speed.

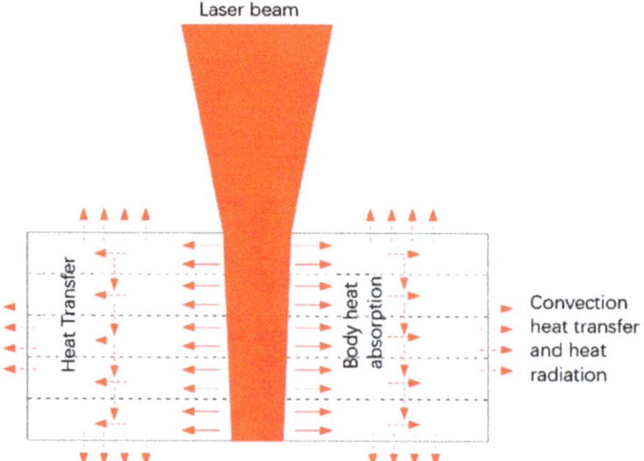

Figure 2. The schematic of the laser irradiates the glass–silicon double layer wafer.

For isotropic and homogeneous materials, the temperature field $T(x,y,z,T)$ is obtained using a transient solver program controlled by the heat diffusion in Equations (2)–(4) based on the conservation of energy and Fourier's law in the Cartesian coordinate system.

$$k\left(\frac{\partial^2 T}{\partial^2 x} + \frac{\partial^2 T}{\partial^2 y} + \frac{\partial^2 T}{\partial^2 z}\right) + Q = \rho C \cdot \frac{\partial T}{\partial_t} \tag{2}$$

$$T(x,y,z,0) = 20\,^{\circ}C \tag{3}$$

$$-k\left(\frac{\partial_t}{\partial_n}\right)_W = h(t_W - t_f) \tag{4}$$

where k, Q, ρ are the thermal conductivity, heat generated per unit volume of laser irradiation and the specific heat. It is assumed that the initial temperature of the glass is at room temperature of 20 °C, W is the thermal convection surface between the outside air and the glass, and h is the glass, the convection coefficient between the outside air and the glass.

To investigate the effect of low-temperature gas cooling on trajectory correction, the flow field distribution of the gas jet injected onto a flat plate was simulated using the FLUENT simulation software. The numerical solution of the heat transfer coefficient distribution on the flat plate was obtained through the simulation, and an approximate analytical solution for the heat transfer coefficient distribution was obtained by fitting the numerical solution using MATLAB.

During the FLUENT simulation process, the RNG k-ε model was used to investigate the impinging flow of a single circular jet [23–25]. The nozzle was perpendicular to the material surface, and the model and mesh division is shown in Figure 3, where $H/D = 10$. The boundary condition was $Re = 14{,}000$, the indoor temperature was T = 300 K, the solid wall was a non-slip boundary, and the injection method was a fixed-point injection. The heat transfer coefficient distribution cloud map on the solid wall was extracted (as shown in Figure 4), and the heat transfer coefficient on the centerline was obtained and curve-fitted using MATLAB (as shown in Figure 5). The R-square value after fitting was 0.951, indicating that a good fitting curve can be obtained through Gaussian fitting. The final analytical solution for the fitted curve is Equation (5).

$$H = 443.4\exp\left(-\frac{x^2}{0.00145^2}\right) \tag{5}$$

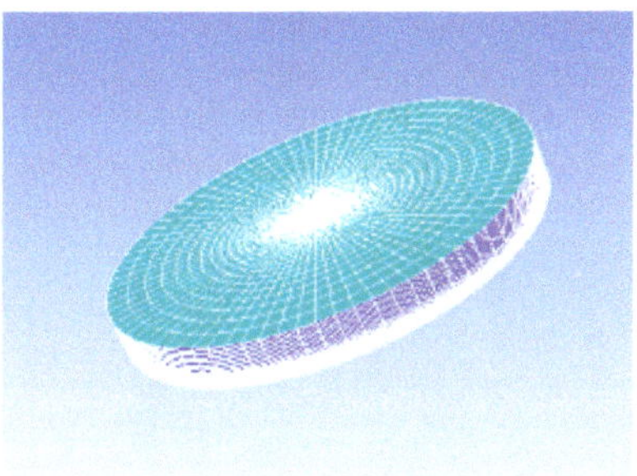

Figure 3. Model and mesh generation.

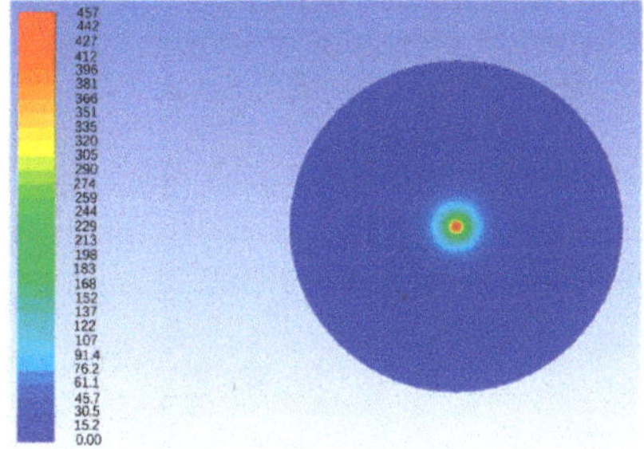

Figure 4. Heat transfer coefficient distribution.

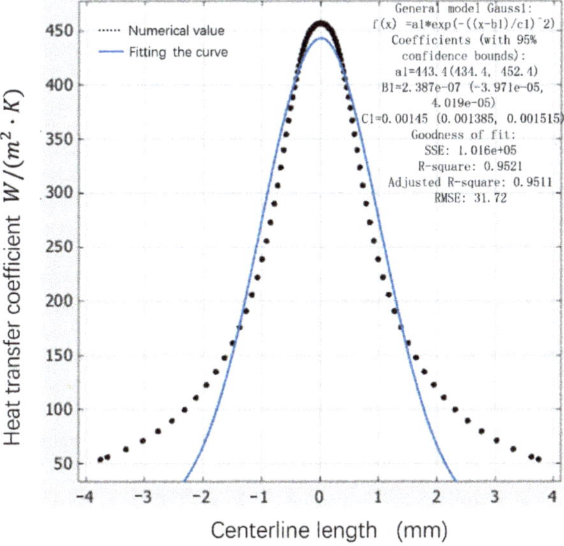

Figure 5. The fitted curve of heat transfer coefficient.

By injecting cryogenic gas, the local convective heat transfer coefficient can be increased and the local temperature of the material can be reduced, thus changing the distribution of the temperature field. Using the approximate analytical solution of the heat transfer coefficient distribution obtained above, it is introduced into the ABAQUS user subroutine for solving the analytical model of the temperature field and studying the effect of the relative position between the center position of the jet and the center position of the laser spot on the temperature field distribution.

The sequential coupled thermoelastic and extended finite element method (XFEM) analysis was conducted using the ABAQUS software, with a modeled size of 30 mm × 15 mm × 2 mm. The mesh and initial crack state (200 μm) of the specimen are illustrated in Figure 6. The total number of elements used was 63,720.

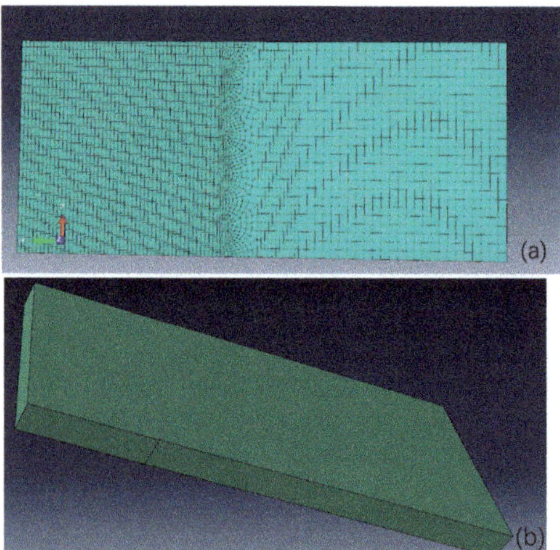

Figure 6. Mesh of the specimen and the initial crack state. (**a**) Mesh of the specimen; (**b**) initial crack state.

In asymmetric cutting, both the crack extension direction and the crack plane may be deflected, and these two types of deflection (as shown in Figure 7) are tilt type and twist type, respectively. The tilt type means that the crack plane is rotated around the O_z axis by θ (tilt angle), and the rotation of the crack face in the θ direction is achieved by continuous adjustment of the crack leading edge; the twist type means that the crack plane is rotated around the O_x axis by ϕ (twist angle), and the rotation of the crack face in the ϕ direction is achieved by "step" splitting of the crack leading edge. The main type of deflection is tilted, and it is the transverse shear stress that affects the direction of crack expansion in asymmetric cuts.

$$\left.\begin{aligned}
\sigma_{y'y'} &= \sigma_{\theta\theta}{}^I + \sigma_{\theta\theta}{}^{II} = [K_I/(2\pi r)^{1/2}]f_{\theta\theta}{}^I + [K_{II}/(2\pi r)^{1/2}]f_{\theta\theta}{}^{II} = K_I'(\theta)/(2\pi r)^{1/2} \\
\tau_{x'y'} &= \sigma_{r\theta}{}^I + \sigma_{r\theta}{}^{II} = [K_I/(2\pi r)^{1/2}]f_{r\theta}{}^I + [K_{II}/(2\pi r)^{1/2}]f_{r\theta}{}^{II} = K_{II}'(\theta)/(2\pi r)^{1/2} \\
\tau_{x'z'} &= 0
\end{aligned}\right\} \tag{6}$$

$$\left.\begin{aligned}
\sigma_{y'y'} &= \sigma_{\phi\phi}{}^I + \sigma_{\phi\phi}{}^{III} = [K_I/(2\pi r)^{1/2}]g_{\phi\phi}{}^I = K_I'(\phi)/(2\pi r)^{1/2} \\
\tau_{x'y'} &= 0 \\
\tau_{x'z'} &= \tau_{z'\phi}{}^I + \tau_{z'\phi}{}^{III} = [K_{III}/(2\pi r)^{1/2}]g_{z'\phi}{}^{III} = K_{III}'(\phi)/(2\pi r)^{1/2}
\end{aligned}\right\} \tag{7}$$

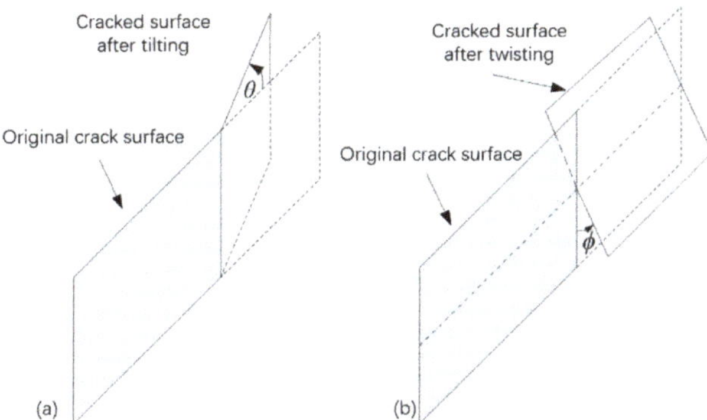

Figure 7. Model of crack nonplanar extension (**a**) sloping type; (**b**) distortion type.

3. Experimental Steps

In the trajectory offset correction technique with cryogenic gas cooling, the schematic diagram of the trajectory correction system is shown in Figure 8. The equipment used in this study is a fiber-coupled 300 W continuous-wave, diode-pumped, solid-state laser with a wavelength of 1064 nm, and the specifications of the main lasers used in the experiment are shown in Table 1. The gas pump (The gas flow rate is 5 L/min) is placed in a space with an ambient temperature of −20 °C. The gas was fed through a tee and a gas guide tube, which was fixed to the machine through a snakeskin tube, and the relative positions of the laser and the jet were adjusted by an XY trimmer table. To ensure that the two lasers irradiate vertically onto the material surface, the two lasers are arranged on either side of the material. The material is placed above the focal point and the distance to the focusing lens is varied to obtain the desired laser beam size on the material surface.

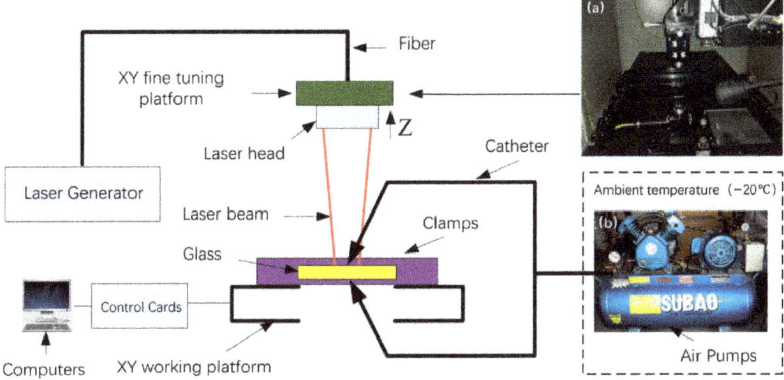

Figure 8. Experimental setup for cryogenic gas cooling trajectory correction technology during asymmetric cutting with LITP technology. (**a**) Diagram of laser and experimental setup equipment, (**b**) diagram of air pump equipment.

Table 1. The main parameters of the laser used in the experiment.

Laser beam wavelength	(nm)	1064
Range of output power	(W)	0–300
Launch angle	(mrad)	≤10
Beam mode		TEM_{00}
Output mode		Continuous
Output power stability	(RMS)	≤2%

The specimens used were soda-lime flat glass with a sample size of $30 \times 15 \times 2 \text{ mm}^3$. The physical properties are shown in Table 2. An initial crack of more than 200 µm was prefabricated on the edge of the glass sample with a diamond wire saw. The best cutting quality was achieved at this time when the laser scanning trajectory was at a distance $\Delta L = 5$ mm from the nearest edge of the specimen, and the processing parameters were laser power $P = 30$ W, scanning speed $V = 4$ mm/s, and spot diameter $D = 2$ mm, while in the circular curve cutting process, the specimen size was $20 \times 15 \times 2 \text{ mm}^3$ and the cutting arc radius $R = 10$ mm, and the processing parameters were laser power $P = 30$ W, scanning speed $V = 3$ mm/s, and spot diameter $D = 3$ mm.

Table 2. Physical properties of soda-lime glass.

Density	$D\ \left(\text{kg}/m^3\right)$	2480
Thermal Conductivity	$\lambda\ (\text{W}/\text{m}\cdot^{\circ}\text{C})$	0.8
Specific Heat	$C\ (\text{J}/\text{kg}\cdot^{\circ}\text{C})$	836
Poisson's Ratio	ξ	0.23
Young's Modulus	(GPa)	74
Expansion coefficient	$\alpha\ \left(10^{-6}/^{\circ}\text{C}\right)$	9.1
Fracture toughness	$\left(\text{MPa m}^{1/2}\right)$	30

4. Results

Figure 9 shows the temperature field distribution during asymmetric cutting. When the relative distance between the laser spot and the jet center is $\Delta X_{CL} = \Delta X_{CY} = 0.7$ mm, the temperature distribution of the low-temperature gas-cooling trajectory correction technique is shown in Figure 10. It can be observed that introducing low-temperature gas on the right side of the scanning line during the trajectory correction process of the asymmetric straight cutting path using low-temperature gas cooling reduces the temperature in that area, and the highest temperature of the specimen is lower than that without correction. The same effect is achieved when introducing low-temperature gas on the inner side of the scanning line for trajectory correction during circular arc cutting with low-temperature gas cooling.

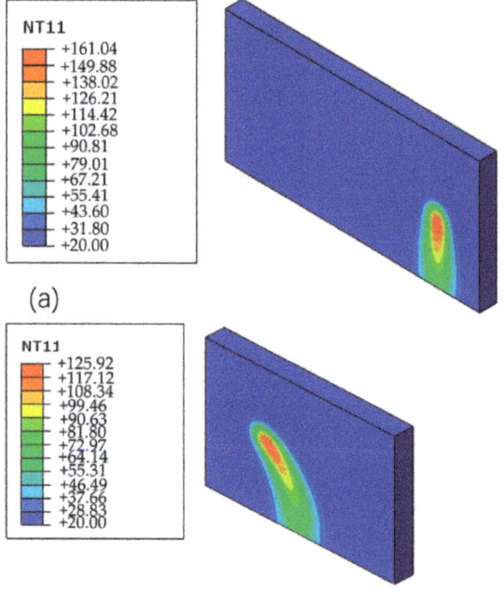

Figure 9. Temperature field distribution of asymmetric cutting. (**a**) Asymmetric line; (**b**) circular arc curve.

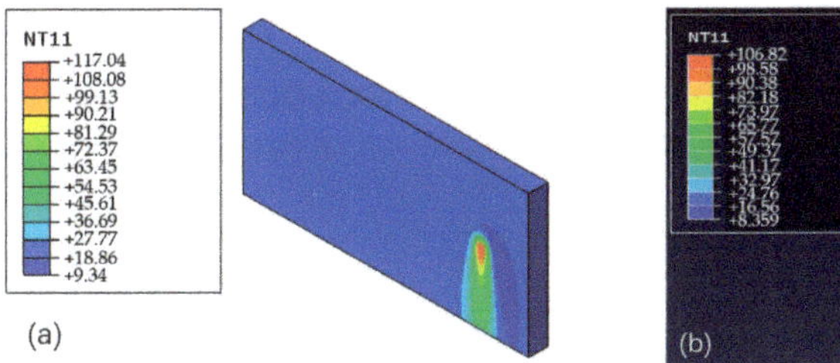
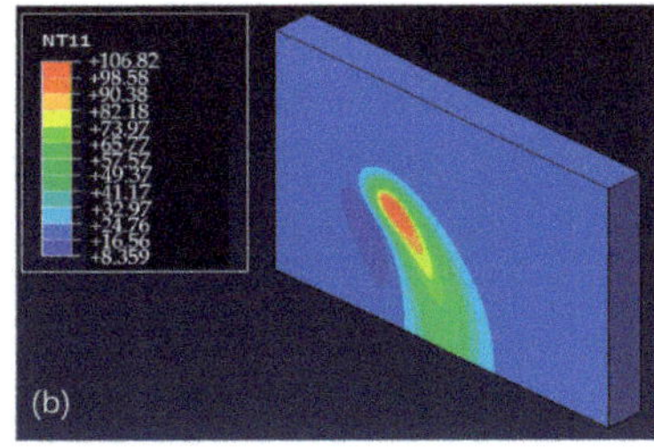

Figure 10. Temperature field distributions of low-temperature gas cooled revising the trajectory deviation. (**a**) Asymmetric line; (**b**) circular arc curve.

Figure 11a shows the influence of the relative distance ΔX_{CL} between the laser spot center and the jet center on the temperature distribution perpendicular to the scanning direction at the crack front during the process of correcting the asymmetric straight cutting path using low-temperature gas cooling when the laser is scanned to 10 mm. It can be observed that the highest material temperature always appears on the laser scanning line, and as ΔX_{CL} increases, the highest temperature gradually increases, and the temperature gradient on the left side of the scanning line changes less than that on the right side. However, when ΔX_{CL} further increases to 1 mm, the rate of change of temperature gradient on both sides of the scanning line decreases compared to that when $\Delta X_{CL} = 0.7$ mm, and the symmetry of the temperature distribution increases. Figure 11b shows the influence of the relative distance ΔX_{CY} between the laser spot center and the jet center on the temperature distribution perpendicular to the scanning direction at the crack front during the process of correcting the circular arc cutting path using low-temperature gas cooling when the laser scanning time is T = 3.2 s. It can be observed that as ΔX_{CY} increases, the temperature gradient on the inner side (center of the circle side) of the scanning line is significantly larger than that on the outer side (away from the center side of the circle), and when $\Delta X_{CY} = 1$ mm, the position of the highest temperature deviates to the outside of the scanning line, causing a change in the direction of the temperature gradient on the outer side of the scanning line.

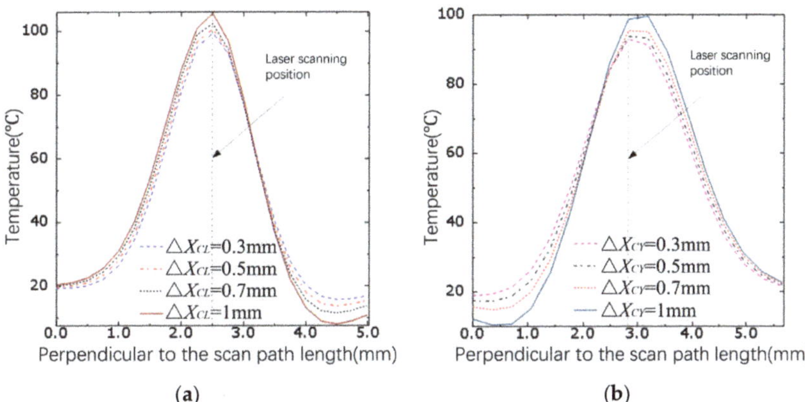

Figure 11. Temperature distribution graphs perpendicular to the scan path at the crack tip with the different transverse distance between the jet center and laser spot. (**a**) Asymmetric line; (**b**) circular arc curve.

The temperature field obtained from the low-temperature gas cooling trajectory correction technique described above is loaded into an extended finite element model to obtain the corresponding stress field distribution and crack propagation status. Figure 12 shows the stress field distribution of asymmetric cutting. Figure 13 shows the stress field distribution during crack propagation when the relative distance between the laser spot center and the jet center is $\Delta X_{CL} = \Delta X_{CY} = 0.7$ mm. Compared with the stress field distribution before correction, low-temperature gas cooling increases the symmetry of the cutting process, and the compressive stress distribution on both sides of the crack is more symmetrical during the stable propagation stage.

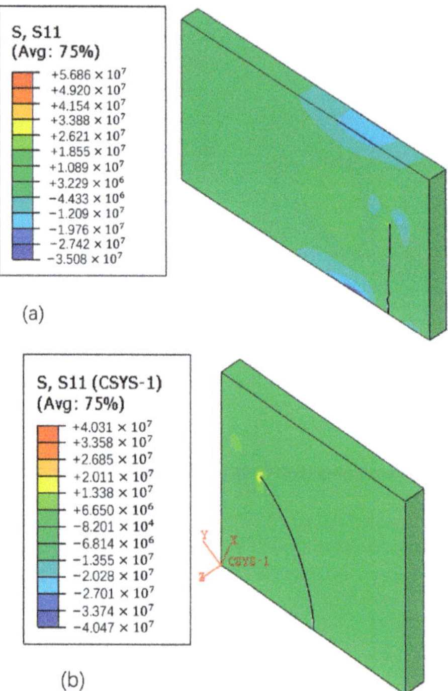

Figure 12. Stress field distributions of asymmetric cutting. (**a**) Asymmetric line; (**b**) circular arc curve.

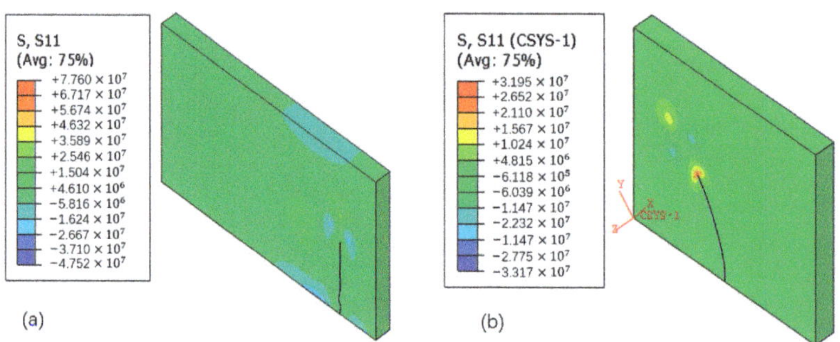

Figure 13. Stress field distributions of low-temperature gas cooled revising the trajectory deviation. (**a**) Asymmetric line; (**b**) circular arc curve.

Figure 14 shows the simulation results of the crack trajectory offset after low-temperature gas cooling trajectory correction. It can be seen that the crack propagation trajectory almost overlaps with the separated laser scanning trajectory, but in the initial stage of trajectory correction, the crack propagation is unstable due to the edge effect, and the crack trajectory shows some deviation. As the processing continues, low-temperature gas cooling can effectively correct the trajectory deviation, and the overlap between the crack propagation trajectory in the stable stage and the laser scanning trajectory is higher.

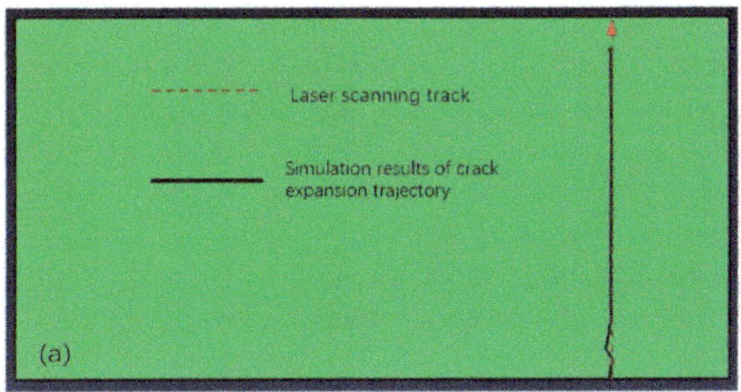

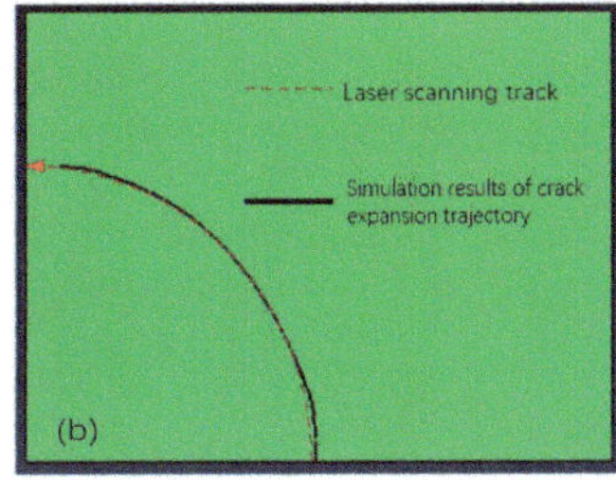

Figure 14. Simulation result of low-temperature gas cooled revising the trajectory deviation. (**a**) Asymmetric line; (**b**) circular arc curve.

The stress distribution curve of the crack front perpendicular to the scanning direction during the correction of the asymmetric linear cutting trajectory by low-temperature gas cooling is shown in Figure 15. It can be seen that the compressive stress on the left side of the scan line is greater than that on the right side due to the edge effect at the beginning and end of processing, while the compressive stress on the right side is greater than that on the left side at the stabilization section, and the difference between the compressive stresses on both sides of the scan line is smaller, which is beneficial to the trajectory correction. During the whole processing stage, the position of the minimum value of transverse shear force is basically maintained on the laser scan line, but the direction of transverse shear force changes at the end stage, which is due to the introduction of low-temperature cold air, which makes the temperature gradient in the scanning direction larger and the distance of the crack front lagging behind the laser spot smaller, and the crack front is closer to the edge of the material at the end section, and the effect of the edge is greater, thus changing the direction of the transverse shear force. Combining the distribution characteristics of tensile stress and transverse shear force, it can be seen that low-temperature air cooling can effectively correct the crack trajectory of asymmetric linear cutting in the stable extension section, but still cannot eliminate the effect of the edge effect, and the crack extension in the first and last sections is not stable.

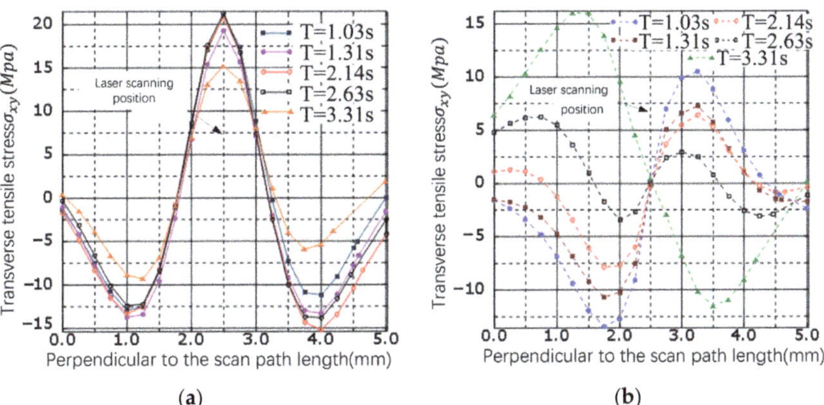

Figure 15. The stress distribution curve perpendicular to the scan path at the crack tip in low-temperature gas cooling revising the trajectory deviation for asymmetric linear cutting glass. (**a**) Tensile stress distribution curve. (**b**) Transverse shear stress distribution curve.

Figure 16 shows the stress distribution curve perpendicular to the scanning direction at the crack front during the process of correcting the circular arc cutting trajectory using low-temperature gas cooling. From the figure, it can be seen that in the initial stage, the compressive stress on the inner side of the laser scanning line is greater than that on the outer side of the arc, and the position of the minimum value of lateral shear force appears on the laser scanning line, which suppresses the unstable state of the crack in the initial stage to some extent. In the stable and final stages of trajectory correction, the compressive stress on both sides of the laser scanning line always shows that the inner side of the arc is smaller than the outer side, which is beneficial for trajectory correction. The position of the minimum value of lateral shear force appears on the separated laser scanning line in the stable stage and deviates from the laser scanning line position in the final stage.

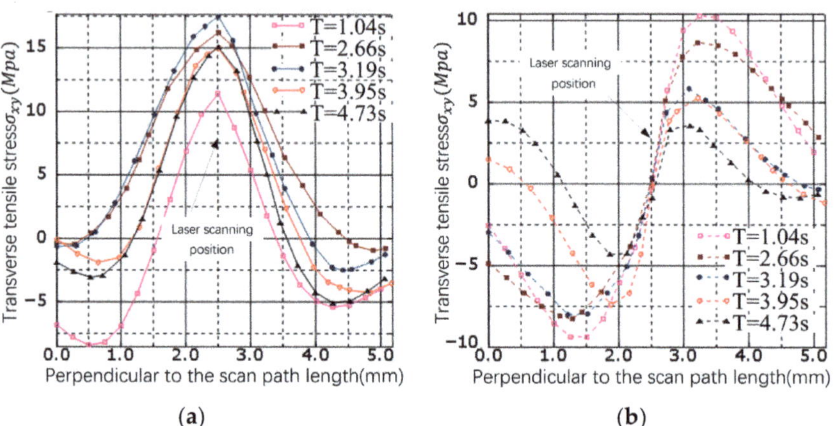

Figure 16. The stress distribution curve perpendicular to the scan path at the crack tip in low-temperature gas cooling revising the trajectory deviation for circular arc cutting glass. (**a**) Tensile stress distribution curve; (**b**) transverse shear stress distribution curve.

Based on the simulation results obtained, experimental research on trajectory correction using low-temperature gas cooling was conducted by changing the relative position relationship between the jet center and the laser spot center.

Figure 17 shows the correction effect of the deviation of the asymmetric straight cutting trajectory with respect to the relative distance ΔX_{CL} is between the laser spot center and the jet center. It can be seen that the trajectory deviation decreases first and then increases with the increase of ΔX_{CL}. Low-temperature gas cooling is applied to the right side of the scanning line, which increases the temperature gradient on the right side of the scanning line and promotes the position of the minimum value of lateral shear force to approach the scanning line. When ΔX_{CL} is 0.7 mm, the temperature gradient on both sides of the scanning line is the largest, the temperature distribution is the most asymmetrical, and the trajectory deviation is the smallest, achieving the best trajectory correction. At the same time, when ΔX_{CL} is 0.6 mm and 0.8 mm, the trajectory deviations are both below 50 μm, which can effectively reduce the deviation of the cutting trajectory.

Figure 18 shows the best correction effect of the low-temperature gas cooling trajectory correction technology, with a minimum trajectory deviation of 27 μm and high-quality separation surface. This is because the low-temperature gas cooling trajectory correction technology introduces a cooling mechanism based on the original separation laser, making the stress change during the crack propagation process smoother, resulting in a better-quality separation surface.

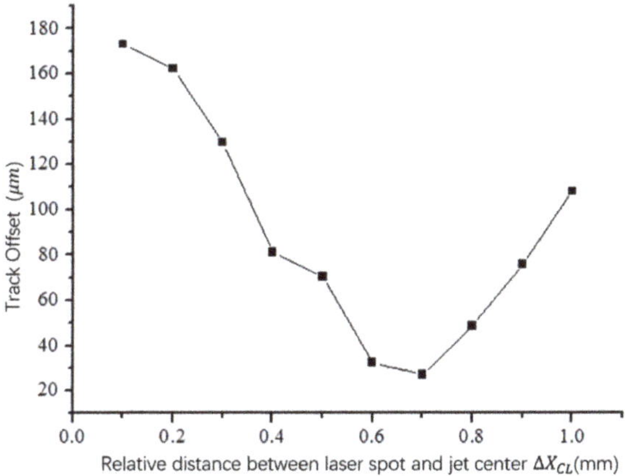

Figure 17. Deviations of separation path in low-temperature gas cooling revising the trajectory deviation for asymmetric linear cutting glass.

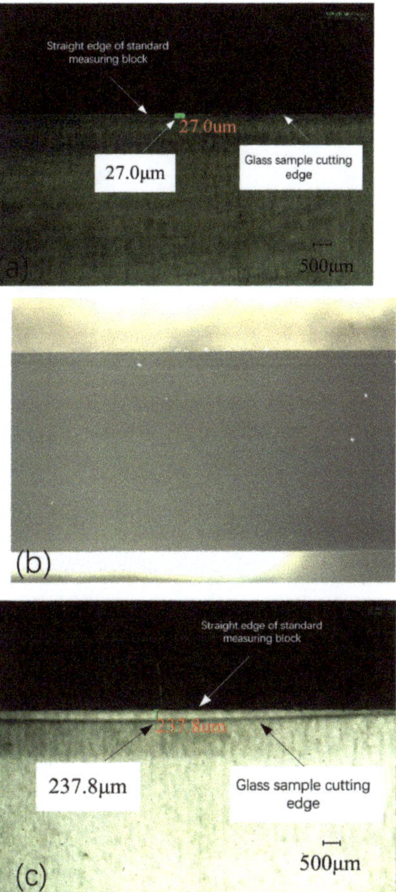

Figure 18. Best revised result in low-temperature gas-cooled revising technology. (**a**) Minimum trajectory deviation; (**b**) Cutting surface quality; (**c**) Asymmetric cutting.

Figure 19 shows the correction effect of the different relative distances of the laser optical axis and jet center ΔX_{CY} on the cutting trajectory of the circular arc curve. It can be seen that the minimum trajectory offset after correction appears at $\Delta X_{CY} = 0.7$ mm, when $\Delta X_{CY} < 0.7$ mm, with the increase of ΔX_{CY}, the inner temperature gradient of the scan line becomes larger and the outer temperature gradient becomes smaller, the asymmetry of temperature distribution increases and the trajectory offset decreases, but when $\Delta X_{CY} > 0.7$mm, with the further increase of ΔX_{CY}, the outer temperature gradient of the scan line direction changes and the trajectory offset becomes larger.

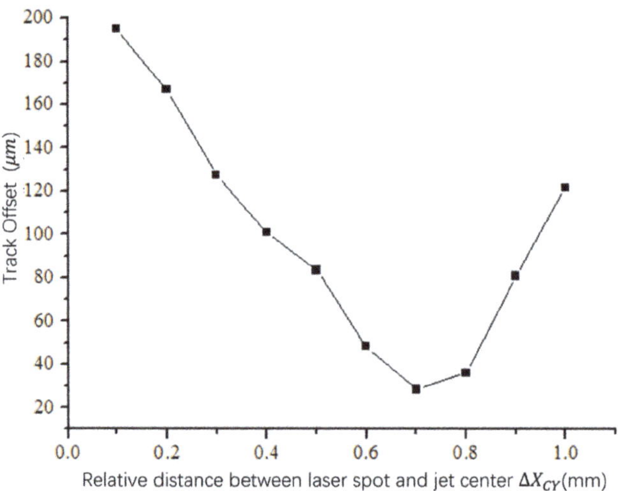

Figure 19. Deviations of separation path in low-temperature gas cooling revising the trajectory deviation for circular arc cutting glass.

Figure 20 shows the correction effect of the low-temperature gas cooling trajectory correction technology on a cutting arc with a radii of R = 5 mm and R = 10 mm. The trajectory deviation after correction for the R = 5 mm arc radius is 27.0 μm, and for the R = 10 mm arc radius is 28.5 μm, indicating a better correction effect for a smaller cutting arc radius. This is because the introduction of low-temperature gas increases the temperature gradient in the scanning direction, reducing the distance between the crack front and the laser spot. Without considering the influence of other correction effects, the smaller the distance between the crack and the laser, the closer the crack extension trajectory is to the laser scanning trajectory during the process of machining the circular arc curve, reducing the trajectory deviation compared to the asymmetric case (show in the Figure 21).

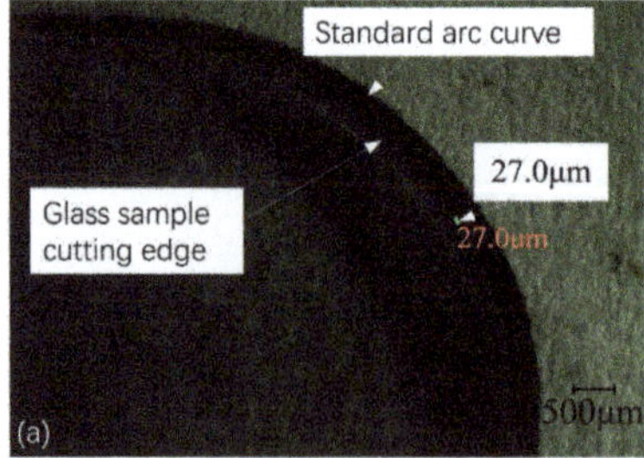

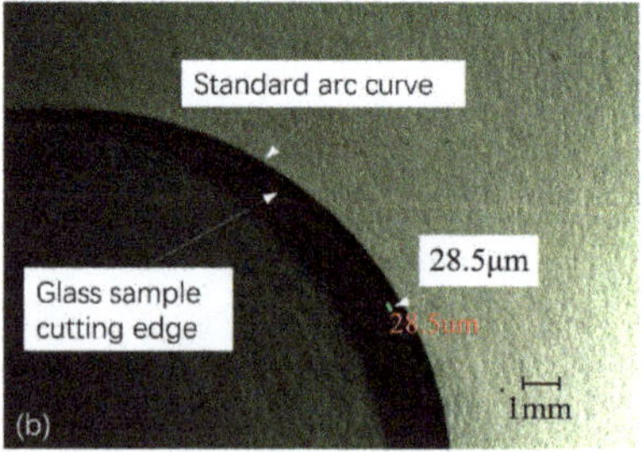

Figure 20. Best revised results with different cutting arc radius in low-temperature gas cooling revising the trajectory deviation technology. (**a**) R = 5 mm; (**b**) R = 10 mm.

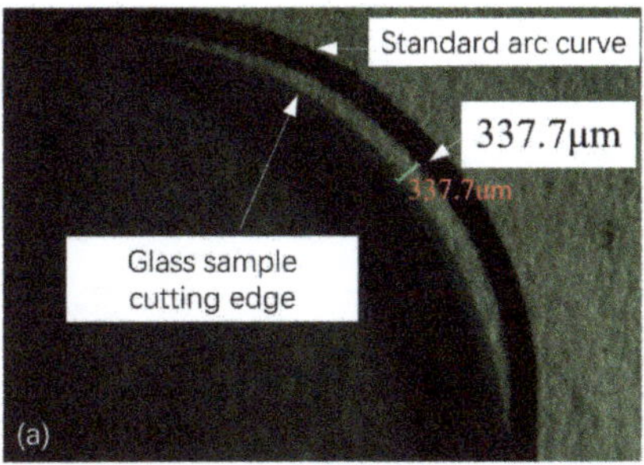
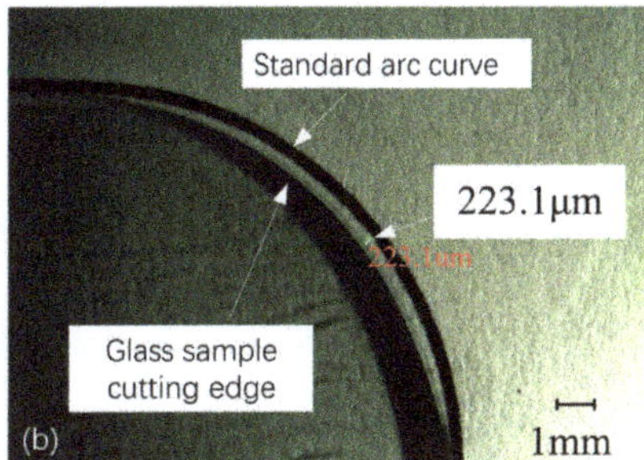

Figure 21. Trajectory offsets for different arc radius cuts under asymmetry. (**a**) R = 5 mm; (**b**) R = 10 mm.

5. Conclusions

To conclude, the trajectory correction technique of cryogenic gas cooling is proposed for the trajectory shift problem that occurs in the stable extension stage in the asymmetric cutting commonly processed in laser-induced thermal crack propagation (LITP). Through a finite element simulation of the effect of cryogenic gas cooling on the temperature field distribution during the trajectory correction process, it is concluded that during the trajectory of asymmetric linear cutting, cryogenic gas cooling loaded on the side with less material in the stable expansion section can effectively correct the crack trajectory of asymmetric linear cutting, and the compressive stress on both sides of the laser scan line in circular arc cutting always shows that the inner side of the arc is smaller than the outer side of the arc, and cryogenic gas cooling. In the stable extension section, loading on the inner side of the arc can effectively correct the crack trajectory of asymmetric linear cutting. The test was carried out by the constructed low-temperature gas cooling system. The results show that in the trajectory correction technique for asymmetric linear cutting, cryogenic gas is injected on the side with less material in the scan line, while in the trajectory correction technique for circular arc curve cutting, cryogenic gas is injected on the inner side of the circular arc curve. The optimal relative distance between the center of its cooling gas and the center of the laser spot are both $\Delta X_{CL} = \Delta X_{CL} = 0.7$ mm. For asymmetric linear cutting, the corrected minimum offset is 27 μm, while for the circular arc curve cutting the smaller the radius is, the better the correction small effect is; at R = 5 mm, the offset is 27 μm. The analysis of experimental results and numerical simulation results shows that the low-temperature gas cooling correction trajectory offset technology can effectively revise the deviation of the separation path in asymmetry linear cutting glass with LITP.

The results of this paper have important implications for improving the application rate of LITP technology in practical glass processing. However, in order to expand the application of this technology, more work is needed to investigate closed shaped cutting with LITP.

Author Contributions: C.Z.: Conceptualization, writing—original draft preparation, funding acquisition. Z.Y.: methodology, software, investigation. X.Q.: Investigation, supervision. J.S.: Supervision, writing—review and editing, funding acquisition. Z.Z.: Data curation, funding acquisition. All authors have read and agreed to the published version of the manuscript.

Funding: This research was funded by the National Natural Science Foundation of China (Grant No. 62003216); the National Natural Science Foundation–Aerospace Joint Fund (Grant No. U2037205); the Natural Science Foundation of Guangdong Province (Grant No. 2022 A1515010600); the Guangdong Basic and Applied Basic Research Foundation (2020A1515111011); Shenzhen University stability support plan (Grant No. 20200814105908002, 20220809212220001); and the Shenzhen Top Talents Start-Up Fund (Grant No. 000711).

Institutional Review Board Statement: Not applicable.

Informed Consent Statement: Not applicable.

Data Availability Statement: Not applicable.

Conflicts of Interest: No conflicts of interest exist in this submitted manuscript, and the manuscript was approved by all authors for publication. I would like to declare on behalf of my co-authors that the work described was original research that has not been published previously, and not under consideration for publication elsewhere, in whole, or in part.

References

1. Zheng, H.Y.; Lee, T. Studies of CO_2 laser peeling of glass substrates. *J. Micromech. Microeng.* **2005**, *15*, 2093. [CrossRef]
2. Solinov, E.F.; Solinov, V.F.; Buchanov, V.V.; Kustov, M.E.; Murav'ev, É.N. Application of laser technology for cutting glass articles. *Glass Ceram.* **2015**, *71*, 306–308. [CrossRef]
3. Kuo, Y.L.; Lin, J. Laser cleaving on glass sheets with multiple laser beams. *Opt. Lasers Eng.* **2008**, *46*, 388–395. [CrossRef]
4. Nisar, S.; Li, L.; Sheikh, M.A. Laser glass cutting techniques—A review. *J. Laser Appl.* **2013**, *25*, 042010. [CrossRef]
5. Zhimalov, A.B.; Solinov, V.F.; Kondratenko, V.S.; Kaplina, T.V. Laser cutting of float glass during production. *Glass Ceram.* **2006**, *63*, 319–321. [CrossRef]
6. Tsai, C.H.; Huang, B.W. Diamond scribing and laser breaking for LCD glass substrates. *J. Mater. Process. Technol.* **2008**, *198*, 350–358. [CrossRef]
7. Pan, C.T.; Hsieh, C.C.; Su, C.Y.; Liu, Z.S. Study of cutting quality for TFT-LCD glass substrate. *Int. J. Adv. Manuf. Technol.* **2008**, *39*, 1071–1079. [CrossRef]
8. Zhou, M.; Ngoi BK, A.; Yusoff, M.N.; Wang, X.J. Tool wear and surface finish in diamond cutting of optical glass. *J. Mater. Process. Technol.* **2006**, *174*, 29–33. [CrossRef]
9. Jia, X.; Li, K.; Li, Z.; Wang, C.; Chen, J.; Cui, S. Multi-scan picosecond laser welding of non-optical contact soda lime glass. *Opt. Laser Technol.* **2023**, *161*, 109164. [CrossRef]
10. Lai, S.; Ehrhardt, M.; Lorenz, P.; Zajadacz, J.; Han, B.; Lotnyk, A.; Zimmer, K. Ultrashort pulse laser-induced submicron bubbles generation due to the near-surface material modification of soda-lime glass. *Opt. Laser Technol.* **2022**, *146*, 107573. [CrossRef]
11. Mingareev, I.; Bautista, N.; Yadav, D.; Bhanderi, A. Ultrafast laser deposition of powder materials on glass. *Opt. Laser Technol.* **2023**, *160*, 109078. [CrossRef]
12. Finucane, M.A.; Black, I. CO_2 laser cutting of stained glass. *Int. J. Adv. Manuf. Technol.* **1996**, *12*, 47–59. [CrossRef]
13. Liang, L.; He, L.; Jiang, Z.; Tan, H.; Jiang, C.; Li, X. Experimental study on the direct planar metallization on glass by the particle sputtering in laser-induced plasma-assisted ablation. *J. Manuf. Process.* **2022**, *75*, 573–583. [CrossRef]
14. Cai, Y.; Yang, L.; Zhang, H.; Wang, Y. Laser cutting silicon-glass double layer wafer with laser induced thermal-crack propagation. *Opt. Lasers Eng.* **2016**, *82*, 173–185. [CrossRef]
15. Lumley, R.M. Controlled separation of brittle materials using a laser. *Ceram. Bull.* **1969**, *48*, 850.
16. Zhao, C.; Zhang, H.; Wang, Y. Semiconductor laser asymmetry cutting glass with laser induced thermal-crack propagation. *Opt. Lasers Eng.* **2014**, *63*, 43–52. [CrossRef]
17. Zhao, C.; Cai, Y.; Ding, Y.; Yang, L.; Wang, Z.; Wang, Y. Investigation on the crack fracture mode and edge quality in laser dicing of glass-anisotropic silicon double-layer wafer. *J. Mater. Process. Technol.* **2020**, *275*, 116356. [CrossRef]
18. Kang, H.S.; Hong, S.K.; Oh, S.C.; Choi, J.-Y.; Song, M.-G. Cutting glass by laser. In Proceedings of the Second International Symposium on Laser Precision Microfabrication, Singapore, 16–18 May 2001; Volume 4426, pp. 367–370.
19. Nisar, S.; Sheikh, M.A.; Li, L.; Safdar, S. Effect of thermal stresses on chip-free diode laser cutting of glass. *Opt. Laser Technol.* **2009**, *41*, 318–327. [CrossRef]
20. Zhao, C.; Zhang, H.; Yang, L.; Wang, Y.; Ding, Y. Dual laser beam revising the separation path technology of laser induced thermal-crack propagation for asymmetric linear cutting glass. *Int. J. Mach. Tools Manuf.* **2016**, *106*, 43–55. [CrossRef]
21. Ogata, K.; Nagato, K.; Ito, Y.; Nakano, H.; Hamaguchi, T.; Saito, I.; Fujiwara, T.; Nagata, T.; Ito, Y.; Nakao, M. Real-time observation of crack propagation and stress analysis during laser cutting of glass. *J. Laser Appl.* **2019**, *31*, 042008. [CrossRef]
22. Yang, L.J.; Wang, Y.; Tian, Z.G.; Cai, N. YAG laser cutting soda-lime glass with controlled fracture and volumetric heat absorption. *Int. J. Mach. Tools Manuf.* **2010**, *50*, 849–859. [CrossRef]
23. Singh, D.; Premachandran, B.; Kohli, S. Experimental and numerical investigation of jet impingement cooling of a circular cylinder. *Int. J. Heat Mass Transf.* **2013**, *60*, 672–688. [CrossRef]
24. Nouri-Bidgoli, H.; Ashjaee, M.; Yousefi, T. Experimental and numerical investigations on slot air jet impingement cooling of a heated cylindrical concave surface. *Heat Mass Transf.* **2014**, *50*, 437–448. [CrossRef]
25. Jacob, A.; Leena, R.; Krishnakumar, T.S. Numerical and Experimental Investigations on Jet Impingement Cooling. *Int. J. Mech. Mechatron. Eng.* **2013**, *7*, 516–520.

 *processes*

MDPI

Article

Scheduling Optimization of Printed Circuit Board Micro-Hole Drilling Production Line Based on Complex Events

Qian Zhou, Xupeng Hu, Siyuan Peng, Yonghui Li,

State Key Laboratory of Radio Frequency Heterogeneous Integration, Shenzhen University, Shenzhen 518060, China
* Correspondence: shy-no.1@163.com

Abstract: The interdependence between the scheduling method and the production efficiency of a micro-hole drilling production line for printed circuit boards (PCBs) holds significant importance, necessitating the optimization of such a production line's scheduling. Consequently, this research paper presents a scheduling optimization approach for the micro-hole drilling production line of a PCB, utilizing complex events as its foundation. Initially, a complex event model was constructed to establish correlations among extensive production line data. Subsequently, the typical complex events associated with the micro-hole drilling production line of a PCB were defined, thereby enabling the all-around monitoring of the operation state of such a production line. Furthermore, this study presents the establishment of a production scheduling model for PCB micro-hole drilling. With the goal of minimizing the maximum completion time, the catastrophe genetic algorithm was used to solve the initial scheduling scheme of the printed circuit board micro-hole drilling production line. The reliability and effectiveness of the catastrophe genetic algorithm in solving the hybrid-driven production scheduling problem of complex events were verified. Dynamic scheduling was performed when three complex events occurred in the production line: emergency order insertion, abnormal equipment operation, and tool failure. The scheduling optimization rate after identifying the emergency insertion event could reach 25.1%. The scheduling optimization rate of the production equipment operation event was related to the specific failure time of the equipment. The scheduling optimization rate after identifying the tool failure event could reach 25%. Rescheduling immediately after identifying the tool failure event could exert no effect on the initial scheduling process. It was proven that the identification and rescheduling of complex events can improve the production efficiency of a PCB micro-hole drilling production line.

Keywords: micro-hole drilling; production line; complex event; emergency insertion; tool failure

Citation: Zhou, Q.; Hu, X.; Peng, S.; Li, Y.; Zhu, T.; Shi, H. Scheduling Optimization of Printed Circuit Board Micro-Hole Drilling Production Line Based on Complex Events. *Processes* **2023**, *11*, 3073. https://doi.org/10.3390/pr11113073

Academic Editors: Yanzhen Zhang and Raul D. S. G. Campilho

Received: 8 September 2023
Revised: 12 October 2023
Accepted: 24 October 2023
Published: 26 October 2023

1. Introduction

As key components of communication systems, printed circuit boards (PCBs) support the rapid development of the new generation of information and communication technologies, such as the Internet of Things, big data, and artificial intelligence [1]. Micro-drilling production on a PCB is the basic requirement for the information interconnection of communication products [2]. How to improve the fluency and intelligence level of a PCB micro-drilling production line is one of the key challenges faced by PCB manufacturing enterprises [3]. Production scheduling is a decision-making process concerning the production line. As the nerve center of production process control, it allocates resources to tasks in a specific order within a given time [4]. Because the amount of data generated in the manufacturing process is too large and complex, it is difficult to perceive, transmit, and process, which easily leads to untimely decision-making, thereby reducing production efficiency and increasing resource losses [5]. Therefore, it is of great significance to optimize the production scheduling of PCB micro-hole drilling production lines.

At present, the optimization of production scheduling mainly focuses on two aspects: massive data processing and the optimization algorithm [6]. In terms of data processing, since the sensing of real-time data in a production line is completed via a large number of sensor nodes, there is a large number of redundant and invalid data in the sensing process [7]. However, due to the lack of effective automatic identification and acquisition system solutions for this real-time, multi-source information, there are phenomena such as time-consuming collection without added value, serious lag, and error-proneness when acquiring multi-source information [8]. Complex event processing technology can quickly process large amounts of data from various sources according to the consistency of the data, thereby generating accurate results to guide the production process [7]. In order to monitor an abnormal situation in the workplace, Lu et al. [9] used complex event-processing technology to correlate context information, accurately extracted data to monitor abnormal situations, and then established an abnormal complex event model. Li Zhe et al. [10] combined the complex event model and RFID technology to realize the comprehensive monitoring of a production line via monitoring the abnormal data in the production process in real time. In order to improve the scheduling ability of the production line with higher precision and faster efficiency, Ding et al. [11] proposed an analysis method using RFID-generated data based on complex-event-driven information, which realized the accurate processing of massive data. Wang et al. [12] associated RFID data with event information to form an original event and standardized the RFID event model, combined it with a detection engine, and more effectively and quickly dealt with complex events composed of RFID data. In addition, Mehdiyev N et al. [13] established a model to predict business processes and standardize control, and they proposed a standardized process control framework based on complex-event-driven information, which laid a theoretical foundation for achieving accurate production scheduling. Govindasamy et al. [14] proposed a probabilistic, complex event processing method based on an RFID automobile manufacturing environment, which uses complex event processing technology to process continuous flow probability data and uncertain data, effectively reduces the processing time and throughput of the system, and optimizes the scheduling ability of the production line. The production process of PCB micro-hole drilling is relatively complex, the machining process is obvious in stages, the process is discrete, and there are abnormal events, such as emergency insertion, abnormal equipment operation, and tool fracture failure [15]. The arrival of emergency orders in the production process will affect the existing scheduling scheme and even cause the order to delay delivery [16,17]; the abnormal operation of equipment in production will interrupt the production process and reduce production efficiency [18]; in the process of micro-hole drilling, tool fracture failure can easily lead to the scrapping of the sheet, reducing production efficiency and wasting too much resources [19,20]. Therefore, it is very important to find and identify the emergency insertion orders, abnormal equipment operation, and tool fracture failure in the PCB micro-hole drilling production process in time so as to rearrange production scheduling. According to the above review, it can be seen that complex event processing technology can quickly analyze and locate requirements from a continuous event flow, accurately identify abnormal conditions in the production process, and apply them to a PCB micro-hole drilling production line to optimize production scheduling when abnormal conditions occur in a timely and effective manner.

In terms of algorithms, heuristic algorithms are widely used to solve production scheduling problems [21], and they have achieved remarkable results. Common algorithms include the discrete whale algorithm, migratory bird optimization algorithm, reinforcement learning algorithm, and genetic algorithm. In order to solve the single-objective, flexible job shop scheduling problem with a minimum makespan, Caldeira et al. [22] proposed an improved Jaya algorithm that can efficiently balance the exploration and utilization of the search space. Yan Xu et al. [23] proposed the quantum whale optimization algorithm, which improves the shortcomings of the traditional whale optimization algorithm in solving flexible job shop scheduling. Jiang et al. [24] proposed the discrete grey wolf optimization algorithm, which can maintain the diversity in the population and solve the problem of the premature convergence of the grey wolf optimization algorithm. In order to solve HFSP, HAN et al. [25] improved the migratory bird optimization algorithm, and they adopted new acceptance criteria and competition mechanisms to ensure the diversity of the population and the exploration ability of the algorithm. Zhang Jie et al. [26] used the wolf algorithm in FJSP for the first time, and they completed the process of solving the job shop scheduling problem based on the wolf algorithm. However, the scope of application of the above algorithm based on the characteristics of animal behavior in nature will be limited, it is not applicable to all problems, ref. [27] and the reinforcement learning algorithm requires a lot of training data and time. For complex job shop scheduling problems, it takes a longer training time to get better results [28]. The genetic algorithm is more suitable for the optimization of complex problems because of its characteristics of the independence of the problem model, global optimality, random transfer, rather than certainty, implicit parallelism, etc. [29]. It is the best choice to solve the scheduling problem of a PCB micro-hole drilling production line. Liu et al. [30] improved the framework of the traditional genetic algorithm and effectively improved the convergence and accuracy of the traditional genetic algorithm. In order to make up for the deficiency of the genetic algorithm, Wu Shujing et al. [31] designed a mechanism to preserve excellent individuals and improve the search ability of the genetic algorithm. Zhou et al. [32] proposed an ant colony algorithm with the goal of minimizing the maximum completion time for the two problems of prematurity and instability in the genetic algorithm to solve the job shop scheduling problem. The catastrophe operator can randomly select some genes from the individual genes to mutate, thereby generating new individuals. This can increase the diversity of the population and help to avoid the algorithm's falling into the local optimal solution. In addition, the catastrophe operator can also increase the global search ability of the algorithm, thereby improving the convergence speed and convergence accuracy of the algorithm. By introducing the catastrophe operator, the algorithm can search the solution space more comprehensively so as to find a better solution.

This paper firstly uses complex event processing technology to correlate complex events with production data, forming complex events represented by emergency order insertion events and production line equipment operation events. At the same time, the real-time data of the tool and the remaining useful life model are correlated to form the tool failure event so as to establish the complex event model of the production line. Then, the production scheduling model based on the experimental platform of a PCB micro-hole drilling production line is established, and the solution process of the traditional genetic algorithm is processed to solve the initial scheduling scheme. Then, the emergency insertion order, production line equipment operation, and prediction of the tool's remaining useful life are taken as abnormal events, and a complex-event-driven production scheduling model is formed. Finally, the reliability and effectiveness of the catastrophe genetic algorithm in solving the complex-event–hybrid-driven production scheduling problem are verified using an example simulation. It is proven that the identification and rescheduling of complex events are very important to improve the production efficiency of a PCB micro-hole drilling production line.

2. Establishment of a Complex Event Model

In the process of processing massive production data, the related concepts and definitions of events are different. This paper abstracts the definitions of various literatures and sets the following definitions:

Original event: This refers to the original data collected during the production process, which is simple and repetitive. The data are fragmented, and there is a large number of label data, also known as label events;

Complex event: This refers to the combination of original events according to certain logical rules to generate events with higher levels of guiding significance, also known as complex events.

In the production process of a PCB micro-hole drilling production line, there are many processes, such as outbounds, AGV transfer, marking, PCB drilling, detection, and so on. The equipment includes an intelligent warehouse, an AGV trolley, a transfer robot, a conveyor belt, a marking machine, a drilling machine, and detection equipment. The processing of a PCB will produce a lot of RFID data stream and equipment running state data. Through complex event processing, these data can be processed in real time. In this study, complex event processing was carried out according to the data of processing equipment and sensing equipment. The representative complex events are summarized as follows: an emergency order insertion event based on RFID data, a production line equipment operation event based on equipment operation data, and a tool failure event. They provide a massive source of complex events for the production scheduling of a PCB micro-hole drilling production line.

2.1. Production Equipment Operation Event

In order to better perceive the running state of the equipment, in this study, we collected the operation state data of the equipment for the experimental platform of a PCB micro-hole drilling production line and processed the complex events on their basis.

For the detection equipment running state event, we described the event for which the visual inspection equipment detected the specified object, which included information such as the type of item, test results, and test time. Part of the encapsulation code is shown in Box 1.

Box 1. Equipment operation event encapsulation code.

```
<CEvent name="VisionDetection" type="complex">
<eid value="event6"/>
<timestamp value="2023-03-17T19:00:00Z"/>
<equipmentId value="TEST-01"/>
<equipmentStatus value="active"/>
<testType value="pressure"/>
<testValue value="10.5"/>
<testUnit value="kPa"/>
<error value="none"/>
</PEvent>
```

2.2. Emergency Insertion Event

Through the complex event processing system, the data stream can be processed in real time to form various complex RFID events in the production process. Through the analysis and processing of complex events, dynamic production scheduling can be realized using emergency insertion events.

The emergency order insertion event (IEC) refers to the insertion of a higher-priority work plan in the original production plan due to the system's scheduling during the PCB production process, resulting in an overall processing quantity greater than the original planned processing quantity because, after the arrival event (type = 'RFID_arrival'), the RFID departure event (type = 'RFID_departure') will occur. And both events occur on a workstation (location = 'workstation'). Then, if multiple such events occur continuously over a period of time (here, set to 10 s), and the total number exceeds the planned number of processing, then an emergency order insertion event can be considered to have occurred. The state monitoring sentences of EPL are shown in Box 2.

Box 2. Monitoring sentences of EPL.

```
SELECT count(*) as cnt
FROM RFIDEvent(type='RFID_departure', location='workstation').win:time(10 sec) as e1,
RFIDEvent(type='RFID_arrival', location='workstation', AID=e1.AID).win:time(10 sec) as e2
HAVING count(*) > original_plan_num
```

2.3. Tool Failure Event

In the actual production process of a PCB micro-hole drilling production line, the status and monitoring of the production line can be reflected in real time by effectively using the data collected in real time. At the same time, through the processing of a large number of historical data, the mechanism model of the tool and equipment can be constructed. These models can be combined with real-time data to achieve the active prediction of the production process. The main research content of this section was to construct a tool with a remaining useful life prediction model based on historical data and combine the model with real-time collected data to form a complex event of the tool's remaining useful life prediction. In order to collect a large amount of historical data, in this study, we first built a force measurement platform. In this study, the high-precision micro-force measurement system produced by Kistler Company was selected for data acquisition. The system is composed of a dynamometer, a charge amplifier, a data acquisition card, the computer-side force measurement software DynoWare (2825A-02–2), and other parts.

Figure 1 is the tool's remaining useful life prediction model system. The role of the force measurement platform is to collect real-time data and save historical data. The real-time data are denoised, extracted, and selected via noise reduction, feature extraction, and selection and then input into the tool's remaining useful life prediction based on the similarity principle. The improved model predicts the remaining useful life of the tool, forms a complex event for predicting the remaining useful life of the tool, and finally realizes active prediction dynamic scheduling.

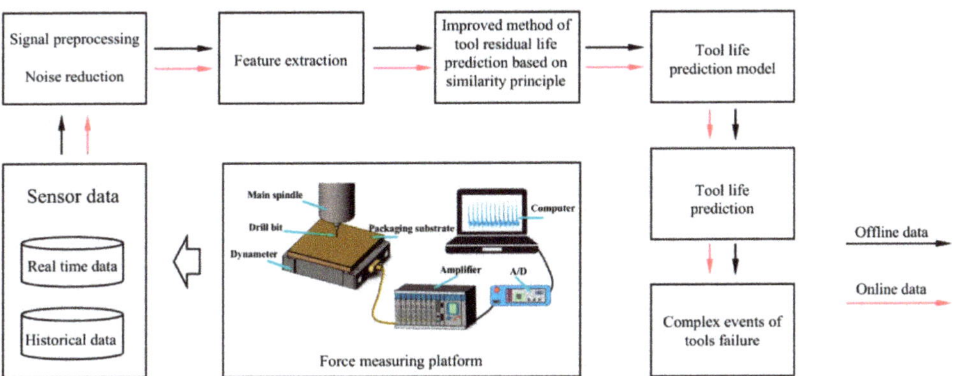

Figure 1. Tool's remaining useful life prediction model system.

Tool failure events were determined as follows:

$$CE_{\text{Tool life prediction}} = e_b(DM_{id}, DT_{id}, RUL, e) \tag{1}$$

where DM_{id} is the drilling equipment ID, DT_{id} is the tool ID, RUL_{id} is the tool's remaining useful life, and e is the sub-events of tool failure.

The remaining useful life of the tool can be predicted using the complex event of the tool's remaining useful life model. The event mainly includes the number of the tool, the remaining useful life of the tool, the measured axial force value and time, and the predicted time. When the remaining service life of the tool is predicted to be lower than a certain threshold, the remaining service life state of the tool becomes 1, indicating that it needs to be replaced actively. Part of the encapsulation code is shown in Box 3.

Box 3. Tool failure event encapsulation code.

```
<CEvent name="Tool life prediction" type="complex">
<eid value="Tool1"/> <!-- the number of tools -->
<operator name="Tool life status" status="1"/> <!-- The remaining service life state of the tool, 1
indicates the need for replacement -->
<operand>
<PEvent name="Axial force measurement" type="primitive">
<eid value="Machine1"/> <!-- the number of drilling machine -->
<force value="150"/> <!-- value of thrust force -->
<time value="2023-03-20T10:30:00"/> <!-- measuring time -->
</PEvent>
</operand>
<time value="2023-03-20T10:31:00"/> <!-- measuring time -->
</CEvent>
```

The tool's remaining useful life prediction model system and real-time sensor data are integrated, that is, the tool's improved remaining useful life prediction model based on the similarity principle can be realized to actively predict the remaining useful life of the tool. Assuming that the current RULb = eb (2, PCB002,267,0) is monitored in real time, the event indicates that the remaining useful life of the drill bit with an ID of PCB002 on the drilling machine is expected to be broken by 267 drilling holes. In order to facilitate the continuous processing timing of PCB micro-hole drilling using the experimental platform and RUL to participate in dynamic processing operation scheduling, RUL is expressed by time. According to the PCB002 drill bit, the drilling time is 3S, and the processing time is 800S. Therefore, according to the actual processing situation, the monitoring event of the processing state of the above RULb = eb (2, PCB002,267,0) drill bit can be expressed as RULb = eb (2, PCB002,800,0), which can be used to provide conditional input for subsequent dynamic production scheduling.

3. Production Scheduling Optimization Based on the Catastrophe Genetic Algorithm

3.1. Establishment of a Scheduling Model

The scheduling problem of a PCB micro-hole drilling production line can be described as an $n \times m$ scheduling problem; that is, the PCB micro-hole-drilling production line has n workpieces to be processed (denoted as workpiece set $J = \{ J_1, J_2, \ldots, J_n\}$), and it can perform different processing on m different equipment (denoted as equipment set $M = \{ M_1, M_2, \ldots, M_m\}$). Each workpiece J_i contains n_i processes (denoted as process set $O_{ij} = \{ O_{i1}, O_{i2}, \ldots, O_{ini}\}$). Each process can choose to be processed on the candidate equipment set $M (O_{ij})$ $(M (O_{ij}) \subseteq M)$ with processing capabilities. The processing indicator O_{ijk} indicates that the jth process of the workpiece i is processed on the machine k, and the value is 1 or 0, indicating whether it is processed on it or not, $i \in [1, n]$, $j \in [1, m]$, and $k \in [1, m]$. The processing time of the jth process of the workpiece i to be processed on machine k is T_{ijk}, and the starting time of the jth process of the workpiece i to be processed on machine k is S_{ijk}. Then, the completion time of the jth process of job i on machine k is $C_{ijk} = T_{ijk} + S_{ijk}$, and the maximum completion time of job i is $C_{imax} = \Sigma C_{ijk}$.

Production line scheduling constraints:

1. The same workpiece can only be processed by the same equipment once;
2. The processing sequence is fixed;
3. The equipment can only complete a single processing task per unit time;
4. In the initial scheduling, the processing tasks have no priority order;
5. Once processed, the current task must be completed before processing other workpieces;
6. The processing time is greater than zero;
7. The processing time is fixed.

For enterprise managers, the processing time, production cost, and energy consumption of equipment are all problems that need to be considered in production scheduling optimization. How to shorten the processing time, reduce the production cost, reduce the machine load, and improve equipment utilization as much as possible are all problems that need to be considered to improve enterprise efficiency. This paper establishes a scheduling problem of a PCB micro-hole drilling production line with the goal of minimizing total completion time.

Objective function:

$$f_1 = \min(C_{\max}) \tag{2}$$

Optimization rate:

$$f_2 = \frac{C_{\max} - f_1}{C_{\max}} \times 100\% \tag{3}$$

Constraint condition:

$$S_{jk} - (S_{ik} + T_{jk}) + L \times (1 - X_{ijk}) \geq 0 \tag{4}$$

$$S_{ik} - (S_{jk} + T_{jk}) + L \times X_{ijk} \geq 0 \tag{5}$$

$$\sum^{m} O_{ijk} = 1 \tag{6}$$

$$S_{i(j+1)} - C_{ijk} \geq 0 \tag{7}$$

Among them, the goal of Equation (2) is to minimize the maximum completion time. Equation (3) represents the processing time optimization rate of rescheduling immediately after identifying critical events. Equations (4) and (5) show that a machine can only process one job at a time. Equation (6) indicates that each process can only be carried out on one machine. Equation (7) indicates that the next process of the same workpiece must wait for the completion of the previous process.

Firstly, the operator catastrophe processing is carried out for the traditional genetic algorithm. With the goal of minimizing the maximum completion processing time, the initial scheduling scheme and dynamic scheduling scheme are solved for the emergency insertion event, the production line equipment operation event, and the tool failure event as abnormal events. Finally, the feasibility of the production scheduling model based on a data-model–complex-event hybrid drive is verified.

3.2. Catastrophe Genetic Optimization Algorithm

The genetic algorithm is a population search algorithm. Its group search has large coverage, which makes the genetic algorithm face a small risk of falling into the local optimal solution, which is more conducive to global optimization, and the genetic algorithm programming is simpler. The genetic algorithm can select the best individual of the population through fitness function science, and the fitness function does not need to meet the continuous or derivative criteria, so it is easy to build the fitness function and algorithm framework of the research problem. The genetic algorithm achieves strong optimization performance. There is a lot of interference information in the process of seeking the optimal solution in the initial population, which will affect the efficiency and accuracy of the algorithm. The genetic algorithm eliminates invalid interference information through gene selection, recombination, crossover, mutation, and other processes to ensure its high efficiency and high quality. The genetic algorithm achieves strong robustness under different problems. Flow chart as shown in the Figure 2.

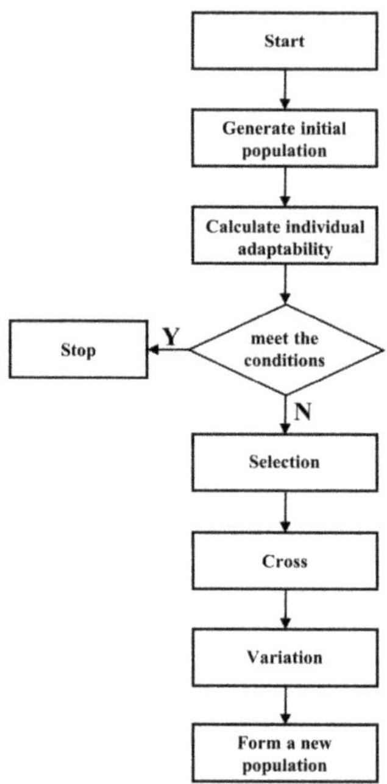

Figure 2. Flow chart of the genetic algorithm.

4. Complex-Event-Driven Dynamic Scheduling Mechanism

As the core equipment of network transmission, PCB is widely used in the communication industry. It is a typical discrete manufacturing and processing product. It includes a wide variety of products and complex forms. Its production has the characteristics of standardization and a large batch. The production process is relatively complex, the processing process is obvious, and the process is discrete. Therefore, the selection of PCB as the processing object to design the physical production line can meet the needs of this paper for flexible manufacturing research. Therefore, it is divided into a physical equipment layer, a data perception layer, a data processing layer, and an application service layer to design the overall architecture of the PCB micro-hole drilling production line experimental platform, as shown in Figure 3.

Events such as emergency order insertion events, production line equipment operation events, and tool failure events directly affect the production of PCBs, resulting in dynamic changes to production scheduling schemes. Complex-event-driven dynamic job scheduling is based on the catastrophe genetic algorithm as the core algorithm to solve the dynamic change in a production line job scheduling scheme caused by a complex event anomaly.

The dynamic scheduling driver mechanism used in this paper is shown in Figure 4. Firstly, it judges whether there are complex events in the PCB drilling production line. If there are complex events, the complex events are judged to determine whether it is a common complex event or a typical complex event on the PCB drilling production line. If it is a common complex event, it will determine whether the event affects the time of the next process according to the minimum processing time as the objective function. If the time exceeds the threshold set by the experimental platform, it will be dynamically adjusted. If the time is short, it will be periodically adjusted. If it is a typical complex event of a PCB drilling production line, it is necessary to judge the type of specific complex event. For abnormal equipment events, it is necessary to judge the processing delay caused by equipment abnormalities. If the delay accounts for less than 10% of the total processing time, one must wait for the equipment to restart. If it is greater than 10%, a new dynamic scheduling scheme is immediately formed via dynamic scheduling. For the emergency insertion event, it is necessary to form a new scheduling scheme after judging the priority of the job type of the insertion order; for the tool failure event, after the system pushes the predicted remaining useful life of the tool, it is necessary to select the appropriate tool change time. In this case, there is no need to reschedule.

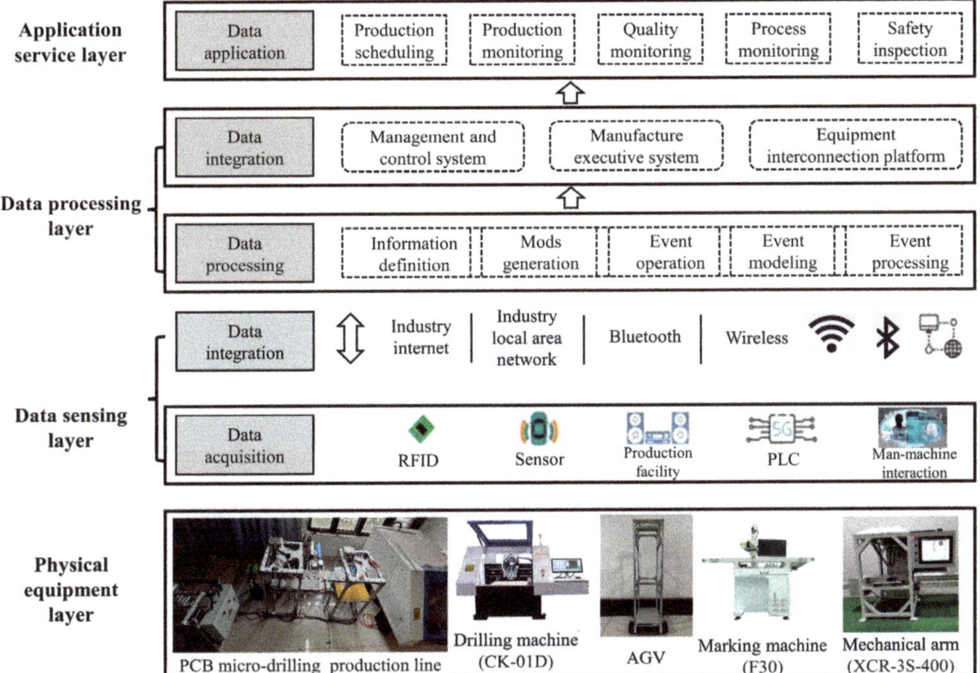

Figure 3. Overall architecture of the intelligent perception platform.

4.1. Processing Job Scheduling without Disturbance Events

For the production line without any disturbance event, the traditional genetic algorithm and the catastrophe genetic algorithm were used to solve the scheduling problem and compared. Specifically, according to the built PCB processing production line as the experimental object, for this paper, we set up six different processing workpieces and used five kinds of machine equipment in the production from process to process. The specific data are shown in Table 1. In the order of the first action process, each processing part was processed according to the fixed five processes. The corresponding column under each process indicates that a certain process of the workpiece is processed on machine tool equipment, and the time corresponding to the right side is the time required for the process to complete the processing on the equipment. This paper uses workpiece one (1, 92, 2, 147, 3, 210, 4, 126, 5, 240) as an example for illustration. This means that the workpiece PCB No. 1 board process 1 is processed at equipment 1, the processing time is 92 s, process 2 is processed at equipment 2, the processing time is 147 s, process 3 is processed at equipment 3, the processing time is 210 s, process 4 is processed at equipment 4, the processing time is 126 s, process 5 is processed at equipment 5, and the processing time is 240 s.

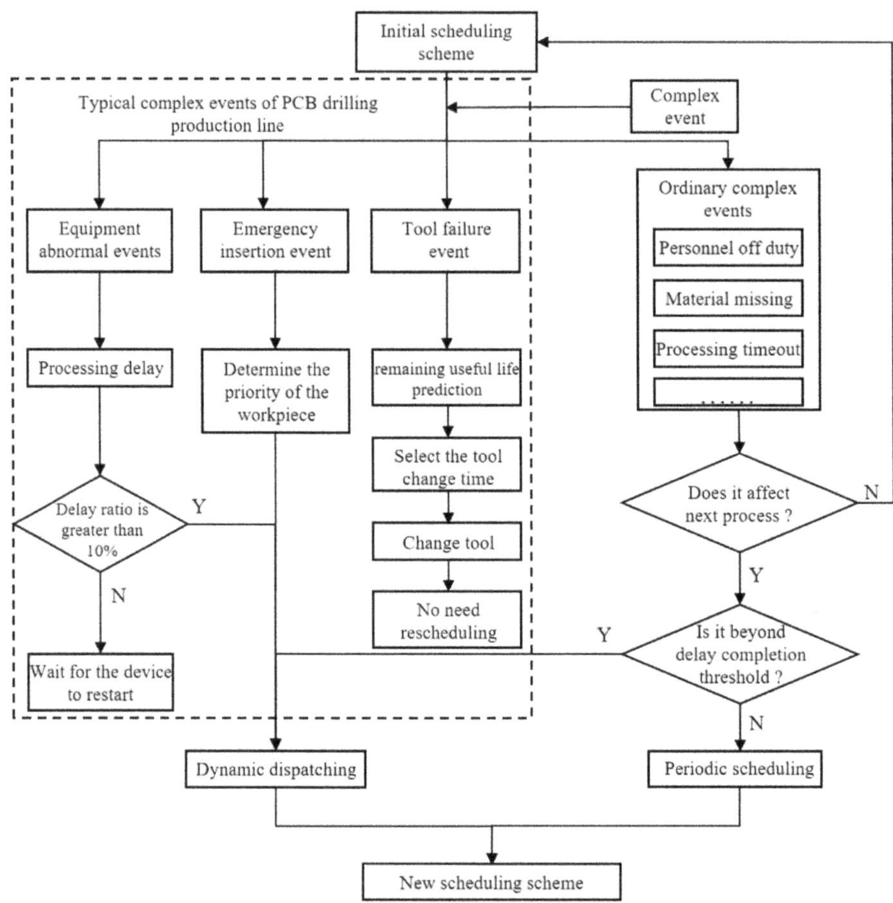

Figure 4. Scheduling flow chart.

Table 1. Process table.

Workpiece	Process 1	Time(s)	Process 2	Time(s)	Process 3	Time(s)	Process 4	Time(s)
1	1	210	2	294	3	420	4	252
2	2	420	3	210	4	336	1	210
3	3	294	1	462	2	336	4	378
4	4	420	3	252	1	210	2	294
5	1	252	4	336	3	294	2	378
6	2	336	1	378	4	210	3	420

Through MATLAB, the catastrophe genetic algorithm was used to solve the problem and compared with the traditional genetic algorithm to obtain an iterative diagram, as follows. It can be seen from the Figure 5 that the traditional genetic algorithm tends to be gentle after 20 iterations, indicating that the traditional genetic algorithm has the defect of 'precocity' and easily falls into the local optimal solution. The catastrophic genetic algorithm undergoes multiple disasters in the iterative process, which makes it easier to obtain better results by increasing the diversity of the population and jumping out of the local optimal solution.

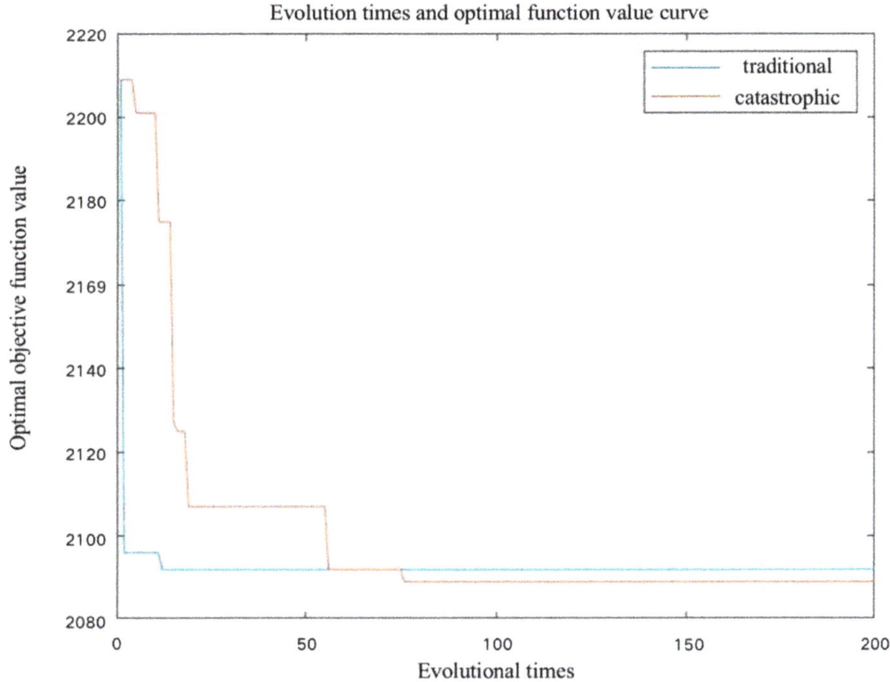

Figure 5. Evolution times and optimal function value curve.

From the results of a Gantt chart Figure 6 and program operation, it can be understood that the optimal result time was 2142 s. The processing sequence of specific equipment was illustrated by taking the No. 1 equipment as an example: the No. 1 equipment (4-1, 5-2, 2-3, 6-3, 3-4, 1-4), that is, at the beginning of production, process 1 of the No. 4 plate was first processed with the No. 1 equipment, followed by process 2 of the No. 5 plate, process 3 of the No. 2 plate, process 3 of the No. 6 plate, process 4 of the No. 3 plate, process 4 of the No. 1 plate, and so on.

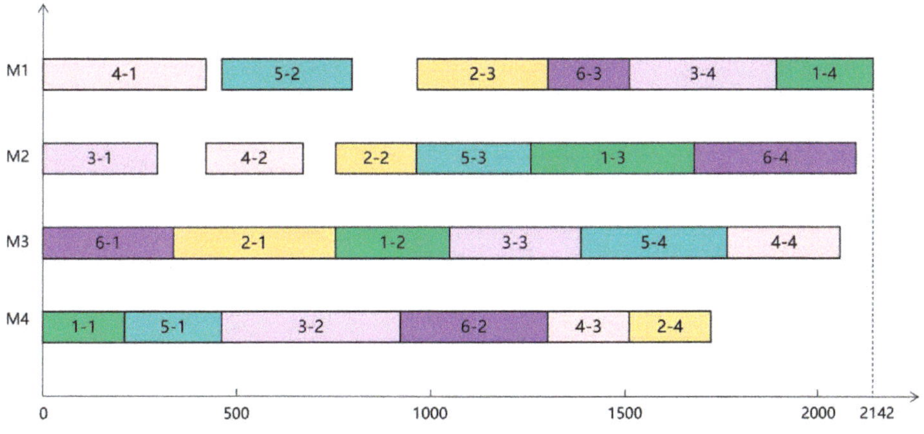

Figure 6. Initial scheduling scheme Gantt chart.

4.2. Event-Driven Dynamic Scheduling of Production Equipment Operation

From the complex event processing in Section 3, it can be understood that the production line can use the complex events of the equipment state and the RFID complex events to obtain the matrix formula (Formula (8)) of the workpiece and equipment processing state, as shown below.

$$S_{6\times4} = \begin{bmatrix} S_{11} & S_{12} & S_{13} & S_{14} \\ S_{21} & S_{22} & S_{23} & S_{24} \\ S_{31} & S_{32} & S_{33} & S_{34} \\ S_{41} & S_{42} & S_{43} & S_{44} \\ S_{51} & S_{52} & S_{52} & S_{54} \\ S_{61} & S_{62} & S_{63} & S_{64} \end{bmatrix} = \begin{bmatrix} 1 & 1 & 1 & 1 \\ 0 & 1 & 1 & 1 \\ 1 & 1 & 1 & 1 \\ 1 & 1 & 1 & 1 \\ 1 & 1 & 1 & 1 \\ 1 & 1 & 1 & 1 \end{bmatrix} \tag{8}$$

According to the matrix formula, the processing status of the workpiece and the processing status of the equipment can be monitored in real time. According to the above formula, it can be understood that workpiece 2 exhibited an abnormal event on equipment 1. The following analysis is based on this abnormal event.

When the PCB No. 2 board is abnormal on the marking machine, it means that the marking machine has no way to further process the subsequent plate. In view of this situation, the first scheme is to wait for the marking machine to restart and correct before processing. The second scheme is to reschedule the unprocessed plate. In this paper, the selection of the two schemes is investigated. Finally, the parameters are set according to the actual experience. Assuming that the time from the equipment shutdown to restarting accounts for more than 10% of the total processing time, the second scheme is selected. If the time ratio is less than 10%, the first scheme is selected. Using an example for analysis, assuming that the delay caused by the marking machine is 70 s, the total processing time is 2142 s, and the proportion of scheduling optimization can be deemed to be 3.3%. Therefore, the first scheme was used for continuous processing, and the workpiece delay processing scheduling Gantt chart was obtained, as shown in Figure 7. It can be seen that the impact of the equipment operation event on the initial scheduling depends on the length of the delay time caused by the abnormal event.

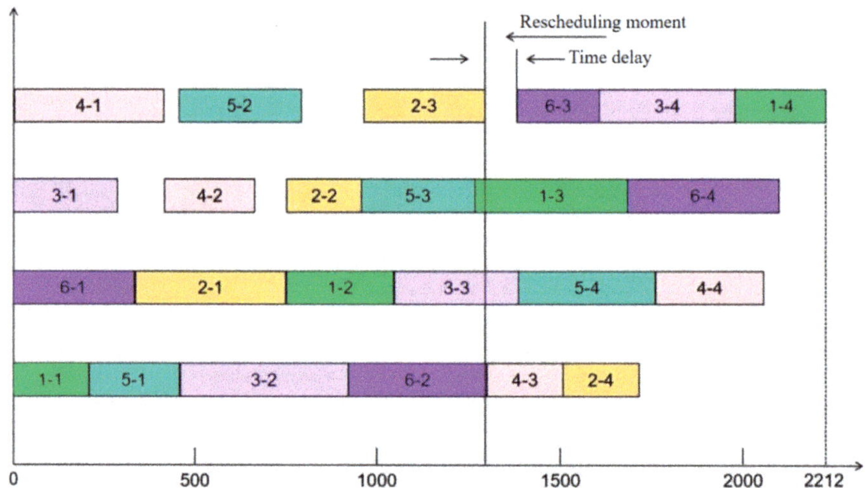

Figure 7. Gantt chart of workpiece delay processing scheduling.

4.3. Dynamic Scheduling of Event-Driven Emergency Insertion

As it is shown in Figure 8, When the PCB No. 2 board is completed by the marking machine, the experimental platform of the PCB micro-hole drilling production line monitors the occurrence of an emergency insertion event and judges that it is a new insertion of workpiece 7: J71 (M1, 480), J72 (M2, 200), J73 (M4, 380), and J74 (M3, 380). In the case of not identifying the emergency insertion event, job 7 will be processed after all jobs are processed, and the resulting scheduling Gantt chart is shown in the figure. At this time, the entire processing time is 3582 s.

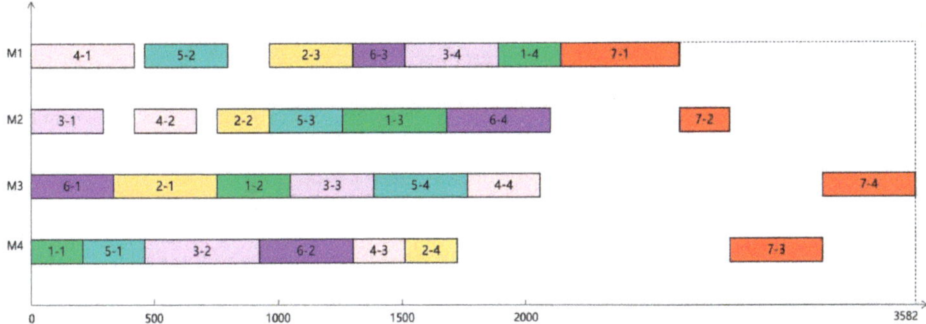

Figure 8. Dynamic scheduling Gantt chart of unidentified emergency order insertion.

If the system recognizes the emergency order insertion event and judges the priority of the plate for the emergency order insertion, it is judged that the priority of the newly inserted plate is the highest. At this time, the optimal production scheduling scheme will be obtained, and then the plate with a lower priority will be processed. In this process, it is also necessary to make full use of the idle time of the replaced plate for local dynamic optimization to obtain the optimization rate $f_2 \approx 25.1\%$, indicating that the emergency order insertion exerts a great disturbance of the initial scheduling. The minimum time for rescheduling completion was 2682 s, and the dynamic scheduling Gantt chart based on the emergency order insertion in Figure 9 was obtained. It can be seen that the newly inserted workpiece is preferentially processed on each machine. Then the unprocessed jobs in the original scheduling are processed.

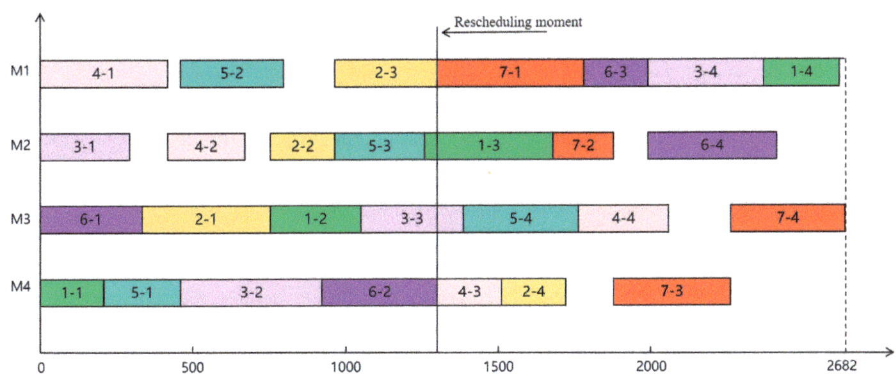

Figure 9. Dynamic scheduling Gantt chart for identifying complex events in emergency order insertion.

4.4. Tool-Failure-Event-Driven Dynamic Scheduling

In the production process of PCB in this paper, because there was only one drilling machine, only the remaining service life of the drilling machine was predicted and analyzed. In the actual production process, if multiple drilling machines are encountered, only the service life of other tools needs to be described according to the method of complex events in this paper.

According to the table and the initial production scheduling scheme, the processing time of each equipment is expressed as a matrix, and the row number is the equipment number:

$$
A = \begin{bmatrix} J_{41} & J_{52} & J_{23} & J_{63} & J_{34} & J_{14} \\ J_{31} & J_{42} & J_{22} & J_{53} & J_{13} & J_{64} \\ J_{61} & J_{21} & J_{12} & J_{33} & J_{54} & J_{44} \\ J_{11} & J_{51} & J_{32} & J_{62} & J_{43} & J_{24} \end{bmatrix} = \begin{bmatrix} 420 & 336 & 336 & 210 & 378 & 252 \\ 294 & 252 & 210 & 294 & 420 & 420 \\ 336 & 420 & 294 & 336 & 378 & 294 \\ 210 & 252 & 462 & 378 & 210 & 210 \end{bmatrix} \tag{9}
$$

From the tool failure event in the third chapter, it can be understood that the monitoring event of the current tool processing state can be expressed as $RUL_b = e_b$ (2, PCB002,800,0). This event indicates that the remaining service life of the drill bit with the ID of PCB002 on the drilling machine was expected to be 800 s, and the tool belonged to the normal processing situation. However, in order to ensure smooth processing, the micro drill can be replaced in advance to avoid the interruption of the tool during the machining process.

This paper assumes that the time to replace the micro drill is 20 s. At present, the PCB micro-hole drilling production line monitors the drilling machine in the processing of PCB No. 6 according to the complex events of the production state and judges that the drilling machine has been processed for 185 s. At this time, the remaining tool life is 800 s. The system actively predicts the time to replace the micro drill. From the matrix, we can confirm that the drilling machine enters a new processing cycle after the completion of J64, and the processing time is 294 s, 252 s, 210 s, 294 s, 420 s, and 420 s.

In the case of not identifying the remaining service life of the tool, Figure 10 is obtained, when the J22 processing is carried out to 61 s, the tool fails, and the workpiece 2 is scrapped. At this time, the tool change operation is performed. If rescheduling is not performed, the J21 processing is performed again after the J12 processing is completed. In this case, the scheduling time is 2856 s.

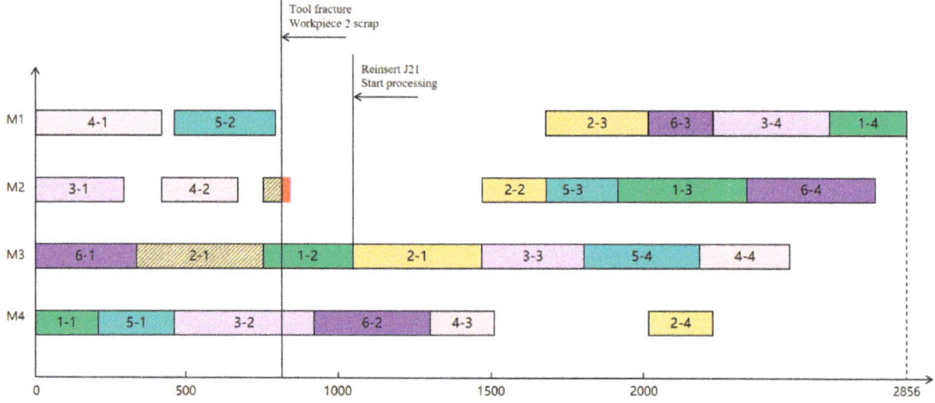

Figure 10. Scheduling Gantt chart of unidentified tool failure events.

If dynamic rescheduling is performed immediately after the tool change is completed, the new scheduling Gantt chart is obtained without waiting to process other workpieces directly. As it is shown in Figure 11, at this time, the whole machining process takes 2646 s.

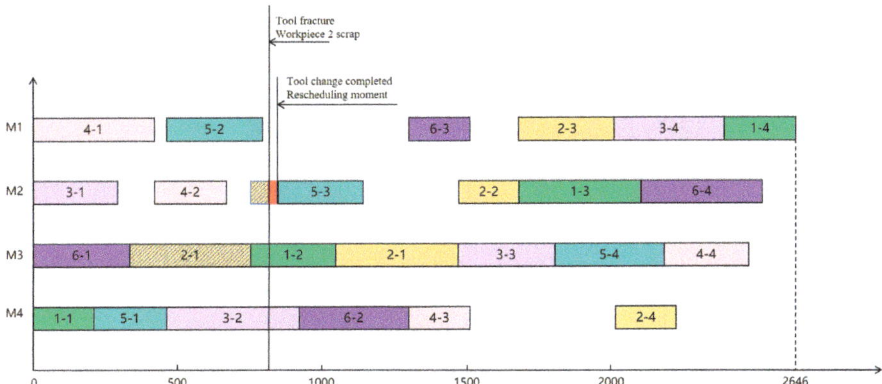

Figure 11. Dynamic rescheduling Gantt chart of unidentified tool failure events.

If the remaining service life event of the tool is known, it can be concluded that the time to change the drill is after the completion of J64, before the start of processing of J31, before the start of processing of J42, and after the completion of processing of J42. The corresponding remaining life is 565S, 313S, 313S, and 61S, and the result of calculating the idle time is greater than the drill change time of 20S. Therefore, it is best to choose to change the drill in these four time periods, which means that it is most suitable for the drill to change the tool after processing this cycle or in the middle of the next cycle, and it will not have any impact on production scheduling. If other time periods are selected for changing the drill, it will affect the normal operation of production scheduling. Therefore, the optimal scheduling Gantt chart is shown in Figure 12. Compared to the processing time when complex events are not identified, the optimization rate at this time is $f_2 = 25\%$, and the disturbance has no effect on the initial scheduling. Therefore, the prediction of the remaining useful life of the tool can change the tool before the tool's failure, which can reduce the time of the whole production scheduling and improve processing efficiency while avoiding the scrapping of the workpiece.

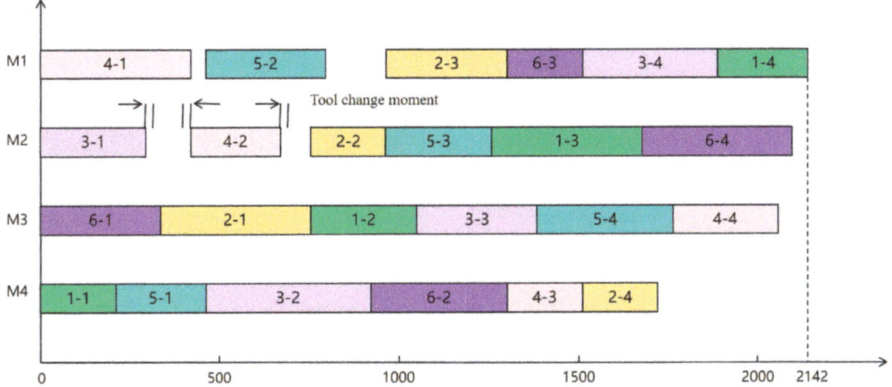

Figure 12. Dynamic scheduling Gantt chart for identifying tool failure events.

The scheduling time of the equipment operation event needs to be calculated according to the time spent waiting for the equipment to restart, which does not reflect the specific value in Table 2. However, it can be seen from whether the emergency insertion event and the tool failure event in Table 3 identify the scheduling time of the complex event that the dynamic scheduling time after identifying the complex event is significantly less than the scheduling time when the complex event is not identified. The identification of complex events can improve the production efficiency of a PCB micro-hole drilling production line. In addition, in the tool failure event, the production time of rescheduling immediately after a tool change is shorter than that of continuing processing after waiting for a tool change.

Table 2. Comparison table of scheduling times and completion times under different conditions.

	Whether Complex Events Are Identified	Time to Finish Work (s)
Equipment operation events	No	2142 + device recovery time
	Yes (delay proportion less than 10%)	
	Yes (delay proportion more than 10%)	Reschedule immediately

Table 3. Comparison table of scheduling times and completion times under different conditions.

Event Name	Whether Complex Events Are Identified	Time (s)	Optimization Rate
Emergency insertion event	No	3582	25.1%
	Yes	2682	
Tool failure event	No (continue processing after tool change)	2856	25% (continue processing after tool change) 19% (rescheduling immediately after tool change)
	No (rescheduling immediately after tool change)	2646	
	Yes	2142	

5. Conclusions

Aiming at the key challenges in PCB micro-hole drilling production lines, this paper has established three complex event models: an emergency insertion event, a production line equipment operation event, and a tool failure event. Based on complex events, combined with the initial scheduling optimization results of the catastrophic genetic algorithm, the dynamic scheduling of the production line was realized. The conclusions are as follows.

The simulation results of the initial scheduling of a PCB micro-hole drilling production line show that the catastrophic genetic algorithm has a higher number of iterations than the traditional genetic algorithm, can avoid falling into the local optimal solution, and has better recognition and better decision-making ability. It is more accurate and effective than the traditional genetic algorithm in solving complex-event-driven production scheduling problems, and it provides accurate and effective solutions for various production line scheduling problems.

Three complex event models of an emergency insertion event, a production line equipment operation event, and a tool failure event have been established. Dynamic scheduling optimization is carried out when the above complex events occur in a production line. The results show that the scheduling optimization rate can reach 25.1% after the emergency insertion event is identified. The scheduling optimization rate of a production equipment operation event is related to the specific failure time of equipment. The scheduling optimization rate can reach 25% after the tool failure event is identified. Rescheduling immediately after the tool failure event is identified can have no effect on the initial scheduling process.

There are multiple events that occur at the same time in the micro-hole drilling production line of a PCB, so the dynamic scheduling of multiple events can be studied in the future.

Author Contributions: Conceptualization, H.S. and Q.Z.; methodology, Q.Z. and X.H.; validation, Q.Z. and X.H.; investigation, X.H. and Q.Z.; resources, H.S.; writing—original draft preparation, Q.Z.; writing—review and editing, T.Z., S.P. and Y.L.; visualization, H.S.; supervision, H.S.; funding acquisition, H.S. All authors have read and agreed to the published version of the manuscript.

Funding: The research proposed in this paper was supported by the National Key Research and Development Program of China (No. 2020YFB1708500) and the National Natural Science Foundation of China (No. 52075344).

Data Availability Statement: Not applicable.

Conflicts of Interest: The authors declare no conflict of interest.

References

1. He, L.F.; Wen, W.F.; Zhu, G.E.; Zha, H.P.; Guo, D.W. New challenge of 5G communication to PCB technology for mobile communication terminal. *Electron. Co. Inform. Technol.* **2019**, *3*, 43–45.
2. Shi, H.Y.; Liu, X.W.; Lou, Y. Materials and micro drilling of high frequency and high speed printed circuit board: A review. *Int. J. Adv. Manuf. Technol.* **2019**, *100*, 823–841. [CrossRef]
3. Balderas, D.; Ortiz, A.; Méndez, E.; Ponce, P.; Molina, A. Empowering Digital Twin for Industry 4.0 using metaheuristic optimization algorithms: Case study PCB drilling optimization. *Int. J. Adv. Manuf. Technol.* **2021**, *113*, 1295–1306. [CrossRef]
4. Liu, M. *Proactive-Aware-Event-Driven Job-Shop-Scheduling Problems in the Internet of Manufacturing Things*; South China University of Technology: Guangdong, China, 2021.
5. Parente, M.; Figueira, G.; Amorim, P.; Marques, A. Production scheduling in the context of Industry 4.0: Review and trends. *Int. J. Prod. Res.* **2020**, *58*, 5401–5431. [CrossRef]
6. Miettinen, K. *Nonlinear Multiobjective Optimization*; Springer Science & Business Media: Berlin/Heidelberg, Germany, 2012.
7. Yang, Z.Y.; Wang, S.L. Introducing A Novel Data Organization and Access Model for Real-Time Monitoring System. *Int. J. Adv. Comput. Technol.* **2012**, *4*, 236–243.
8. Huang, G.Q.; Zhang, Y.F.; Chen, X.; Newman, S.T. RFID-enabled real-time wireless manufacturing for adaptive assembly planning and control. *J. Intell. Manuf.* **2008**, *19*, 701–713. [CrossRef]
9. Lu, T.; Zha, X.; Zhao, X. Multi-stage monitoring of abnormal situation based on complex event processing. *Procedia Comput. Sci.* **2016**, *96*, 1361–1370. [CrossRef]
10. Li, Z. *Research on the Application of RFID Technology Event Processing Framework in MES*; Hunan University: Changsha, China, 2014.
11. Ding, K.; Jiang, P. RFID-based production data analysis in an IoT-enabled smart job-shop. *J. Autom. Sin.* **2017**, *5*, 128–138. [CrossRef]
12. Wang, F.; Liu, S.; Liu, P. Complex RFID event processing. *VLDB J.* **2009**, *18*, 913–931. [CrossRef]
13. Mehdiyev, N.; Krumeich, J.; Werth, D.; Loos, P. Determination of Event Patterns for Complex Event Processing Using Fuzzy Unordered Rule Induction Algorithm with Multi.objective Evolutionary Feature Subset Selection. In Proceedings of the 2016 49th Hawaii International Conference on System Sciences, Koloa, HI, USA, 5–8 January 2016; pp. 1719–1728.
14. Govindasamy, V.; Thambidurai, P. RFID probabilistic complex event processing in a real-time product manufacturing system. *Nature* **2013**, *2*, 139–144.
15. Lin, C.J.; Huang, M.L. Modified artificial bee colony algorithm for scheduling optimization for printed circuit board production. *J. Manuf. Syst.* **2017**, *44*, 1–11. [CrossRef]
16. He, X.M.; Dong, S.H.; Zhao, N. Research on rush order insertion rescheduling problem under hybrid flow shop based on NSGA III. *Int. J. Prod. Res.* **2020**, *58*, 1161–1177. [CrossRef]
17. Aliweza, E.; Stefan, H.; Tallal, J.; Jorg, F. A Lean Based Overview on Sustainability of Printed Circuit Board Production Assembly. *Procedia CIRP* **2015**, *26*, 305–310.
18. Kim, J.; Seo, D.; Moon, J.; Kim, J.; Kim, H.; Jeong, J. Design and Implementation of an HCPS-Based PCB Smart Factory System for Next-Generation Intelligent Manufacturing. *Appl. Sci.* **2022**, *12*, 7645. [CrossRef]
19. Tao, S.; Gao, Z.S.; SHI, H.Y. Characterization of Printed Circuit Board Micro-Holes Drilling Process by Accurate Analysis of Drilling Force Signal. *Int. J. Precis. Eng. Manuf.* **2022**, *23*, 131–138. [CrossRef]
20. Huang, G.; Wan, Z.; Yang, S. Mechanism investigation of micro-drill fracture in PCB large aspect ratio micro-hole drilling. *J. Mater. Process. Technol.* **2023**, *316*, 117962. [CrossRef]
21. Wang, J.Y.; Pan, R.L.; Qin, F. Improved genetic algorithm to solve flexible job-shop scheduling problem. *Manuf. Autom.* **2022**, *44*, 91–94+106.
22. Caldeira, R.H.; Gnanavelbabu, A. Solving the flexible job shop scheduling problem using an improved Jaya algorithm. *Comput. Ind. Eng.* **2019**, *137*, 106064. [CrossRef]
23. Yan, X.; Ye, C.; Yao, Y.Y. Solving Job-Shop scheduling problem by quantum whale optimization algorithm. *Appl. Res. Comput.* **2019**, *36*, 975–979.
24. Jiang, T.; Zhang, C. Application of Grey Wolf Optimization for Solving Combinatorial Problems: Job Shop and Flexible Job Shop Scheduling Cases. *IEEE Access* **2018**, *6*, 26231–26240. [CrossRef]
25. Han, D.; Tang, Q.; Zhang, Z.; Li, Z. An improved migrating birds optimization algorithm for a hybrid flow shop scheduling within steel plants. *Mathematics* **2020**, *8*, 1661. [CrossRef]
26. Zhang, J.; Li, B.; Jiang, J.; Gu, H. Flexible job-shop scheduling problem based on coyote optimization algorithm. *Manuf. Eng.* **2020**, *10*, 39–44.

27. Tang, H.B.; Ye, C.M. Application of Bionic Swarm Intelligence Algorithm to Production Scheduling: A Review. *Ind. Eng. J.* **2010**, *13*, 1–5+42.
28. Kayhan, B.M.; Yildiz, G. Reinforcement learning applications to machine scheduling problems: A comprehensive literature review. *J. Intell. Manuf.* **2023**, *34*, 905–929. [CrossRef]
29. Ge, J.K.; Qiu, Y.H.; Wu, C.M.; Pu, G.L. Summary of genetic algorithms research. *Appl. Res. Comput.* **2008**, *25*, 2911–2916.
30. Liu, Z.; Wang, J.; Zhang, C.; Chu, H.; Ding, G.; Zhang, L. A hybrid genetic-particle swarm algorithm based on multilevel neighbourhood structure for flexible job shop scheduling problem. *Comput. Oper. Res.* **2021**, *135*, 105431. [CrossRef]
31. Wu, S.J.; You, Y.P.; Luo, F.Y. Genetic-Variable Neighborhood Search Algorithm with Elite Protection Strategy for Flexible Job Shop Scheduling Problem. *Comput. Eng. Appl.* **2020**, *56*, 236–243.
32. Zhou, P.; Li, X.; Zhang, H. An ant colony algorithm for job shop scheduling problem. In Proceedings of the Fifth World Congress on Intelligent Control and Automation, Hangzhou, China, 15–19 June 2004; Volume 4, pp. 2899–2903.

Article

Prediction Model of the Remaining Useful Life of the Drill Bit during Micro-Drilling of the Packaging Substrate

Xianwen Liu [1,2], Sha Tao [3], Tao Zhu [1], Zhaoguo Wang [1] and Hongyan Shi [1,2,*]

1   State Key Laboratory of Radio Frequency Heterogeneous Integration, Shenzhen University, Shenzhen 518060, China; liuxw04@163.com (X.L.); 2060291016@email.szu.edu.cn (T.Z.); w15604324793@163.com (Z.W.)
2   Institute of Nanosurface Science and Engineering, Shenzhen University, Shenzhen 518060, China
3   Shennan Circuits Co., Ltd., Shenzhen 518117, China; taosha_no_1@163.com
*   Correspondence: shy-no.1@163.com

Abstract: The packaging substrate plays a significant role in electrical connection, heat dissipation, and protection for the chips. With the characteristics of high hardness and the complex material composition of packaging substrates, drill bit failure is an austere challenge in micro-drilling procedures. In order to monitor the health state of the drill bit and predict its remaining useful life (RUL) in micro-drilling of packaging substrate, an improved RUL prediction model is established based on the similarity principle, degradation rate, and offset coefficient. And then, a micro-drilling experiment on packaging substrate is carried out to collect the axial drilling force through the precision drilling force measurement platform. Axial drilling force signals, which are processed via the Wiener filtering method, are used to analyze the effectiveness of the improved RUL prediction model. The experiment results indicate that, compared to the curves of the traditional RUL prediction model, the curves of the improved RUL prediction model present a higher fitting degree with the actual RUL curves. The average relative errors of the improved RUL prediction model are small and stable in all groups; all of the values are less than 15%, while the fluctuation of the average relative errors of the traditional model is greatly large, and the maximum value even reaches 74.43%. Therefore, taking the degradation rate and offset coefficient into account is a proper method to enhance the accuracy of the RUL prediction model. Furthermore, the improved RUL prediction model is a reliable theoretical support for the health state monitoring of drill bits during the micro-drilling of packaging substrates, which also acts as a potential method to improve micro hole processing efficiency for packaging substrates.

Keywords: remaining useful life; similarity principle; degradation rate; offset coefficient; micro-drilling; packaging substrate

Citation: Liu, X.; Tao, S.; Zhu, T.; Wang, Z.; Shi, H. Prediction Model of the Remaining Useful Life of the Drill Bit during Micro-Drilling of the Packaging Substrate. *Processes* **2023**, *11*, 2653. https://doi.org/10.3390/pr11092653

Academic Editor: Carlos Sierra Fernández

Received: 22 July 2023
Revised: 22 August 2023
Accepted: 1 September 2023
Published: 5 September 2023

1. Introduction

Micro hole drilling is the critical procedure in printed circuit board (PCB) fabrication, and the micro hole is essential to the electrical connection of PCBs [1,2]. Considering the requirements of heat dissipation and high-quality signal transmission, packaging substrates, as one kind of PCB, generally consist of low-outline copper foil, modified resin, modified glass fiber, and hard fillers. Additionally, the diameter of the micro hole in the packaging substrate is smaller than that of conventional PCBs. Therefore, the issue of drill bit life has drawn extensive attention in the micro-drilling of packaging substrates.

When it comes to the tool life and health state, many studies focused on the change in tool wear [3–8], and researchers believed that the range of oscillation [9] and vibration signal [10] were practical to explore the service state of tools. While considering the influence of the wear status of cutting tools on axial drilling force, Tansel et al. [11] presented a relationship model of axial force and tool wear in micro milling based on a BP neural network. The research result indicated that it was better to predict the tool status in cutting soft metal compared to the processing of hard metal. Malekian et al. [12] proposed a tool wear monitoring method utilizing various sensors, such as accelerometers, force, and acoustic emission sensors, in micro-milling. Comparison between the actual tool wear and the simulated results exhibited good agreement, particularly when the force, acceleration, and acoustic emission signals were fused together. In addition, this method could be extended to predict high-frequency bandwidth cutting forces. More importantly, Shi et al. [13] confirmed that drilling force was the vital reference to the failure of the micro drill bit, which could be the failure signal to the drill bit in micro-drilling of PCBs. Actually, the change in drill bit wear would finally reflect on the variation of axial drilling force, just as the aggravation of drill bit wear would generally lead to an increment of axial drilling force. Thereupon, the variation of axial drilling force is an intuitionistic reflection of the health state of the drill bit.

In research regarding health state and life prediction, RUL prediction is an effective methodology to ensure equipment safety, enhance the utilization rate, and reduce the maintenance cost [14,15]. In the area of batteries, the data-driven method for RUL prediction for lithium-ion batteries through an improved autoregressive model by particle swarm optimization was presented by Long et al. [16], and it was concluded that this prognostic approach could predict the RUL of batteries with a small error and was appropriate for on-board applications. However, Chen et al. investigated the RUL of lithium-ion batteries based on the hybrid deep learning model, and a new hybrid lithium-ion battery RUL prediction model that considered the channel attention mechanism and long short-term memory networks was proposed. Verification indicated that the prediction performance of the model was stable and was less influenced by different prediction starting points [17]. The research on the hybrid model on RUL of proton exchange membrane fuel cells showed that the maximum error in RUL prediction was 2.01% for the hybrid model, which provided credible RUL predictions with the highest accuracy [18].

Practical engineering applications face numerous challenges because of the complicated and unknown failure mechanisms under harsh working conditions. However, an effective method for enhancing the RUL prediction accuracy of machinery was provided by Zeng et al. [19], particularly in a noisy environment. In view of the fact that the RUL-prediction data of the planetary gearbox will result in statistical differences in the data distribution under the different service conditions and that the prediction accuracy will be greatly influenced, Liu et al. [20] proposed a domain-adaptive LSTM-DNN (a long, short-term memory network deep neural networks)-based method for remaining useful life prediction of planetary transmission. The experimental result exhibited that the proposed model was able to effectively extract degradation features from condition monitoring data under various operating conditions, and the generalization capability of the data-driven RUL prediction model was improved. Furthermore, it could adapt to the RUL prediction tasks under different operating conditions to a certain extent, which made up for the limitations of the traditional data-driven model. Eker et al. [21] presented a modification of a pure data-driven similarity-based prognostic approach. Their experimental results indicated that the modifications lessened the root mean squared error of the RUL estimations in two out of three datasets. It was also suggested that future studies would focus on the integration of a physics-based model with the modified similarity-based approach to achieve improved prediction of RUL. Xiong et al. [22] proposed a data analysis method to predict the RUL of aircraft engines based on running data. The experimental result verified that this approach showed excellent prognostic ability in predicting remaining life, so it was an effective reference for subsequent engine maintenance decisions. In the study on RUL of rolling bearings, a novelty RUL prediction model based on a bi-channel hierarchical vision transformer was established for the sake of improving prediction accuracy. Two different validation experiments were carried out by Hao et al., and it was found that the prediction accuracy of the novelty RUL prediction model could be increased by up to 9.43% and 43.10%, respectively, when compared with the current standard method [23].

Li et al. [24] utilized a support vector machine approach to accurately identify the tool wear state, followed by a Bayesian framework to deduce the RUL of the tool. Combining the preprocessing techniques, Bayesian hyperparameter optimization, and forward feature selection, Zegarra et al. [25] also provided a methodology to accurately predict tool wear at a lower computational cost. Experiments concerning the degradation modeling and RUL estimation approach using available degradation data for a deteriorating system indicated that the proposed approach could effectively model the degradation process for the individual system and acquire better results for RUL estimation [26].

Studies concerning RUL prediction, which takes the variation of axial drilling force as an object, are seldom reported, and the degradation rate and offset coefficient are of significance in the RUL prediction model. In this article, therefore, we focus on the establishment of the improved RUL prediction model of the drill bit during micro-drilling of packaging substrate, which is based on the similarity principle, degradation rate, and offset coefficient. Following that, a micro-drilling experiment on packaging substrate is carried out to validate the effectiveness of the improved RUL prediction model. The results indicate that the improved RUL prediction model is effective for the RUL prediction of drill bits during the micro-drilling of packaging substrates.

2. Establishment of the RUL Prediction Model

2.1. Traditional RUL Prediction Model

The prediction method based on similarity theory is a new direction in the research field of equipment remaining useful in recent years [27]. The basic idea based on the similarity principle is that the performance degradation curve of the current tool is matched with the historical failure data set by curve similarity, and the RUL of the current tool is predicted by weighting the RUL of each historical data set according to the similarity.

In this research, Euclidean distance is used for the quantification of similarity matching (Equation (1)):

$$L^i(\tau_{te}, \tau_i) = \sqrt{\sum_{j=1}^n \beta^{n-j}(Z_{te,\tau_{te}-n+j} - Z_{i,\tau_i-n+j})^2} \tag{1}$$

where β^{n-j} is the weighting function of distance, τ_{te} is the current recording spot of the tool, $\tau_i = (\tau_i^n, \tau_i^{n+1}, \cdots, \tau_i^F)$ is the collection point of the tool historical data of the current calculation similarity matching, and τ_i^F is the failure point of the tool history data. The similarity matching of each collection point is traversed to obtain the distance of the best matching between the current machining tool and the health state of each tool in the historical failure data set (Equation (2)):

$$L^i(\tau_{te}, \tau_i^*) = \min_{n \leq \tau_i \leq \tau_i^F} L^i(\tau_{te}, \tau_i) \tag{2}$$

where τ_i^* is the most matchable collection point in health state from the tool's historical data. Furthermore, the RUL of the most matchable data collection point of each tool in the tool historical failure data set can be expressed by Equation (3), and the weighting function is shown as Equation (4).

$$r^i(\tau_i^*) = (\tau_i^F - \tau_i^*)\Delta t \tag{3}$$

$$q_i = \frac{1/L^i(\tau_{te}, \tau_i^*)}{\sum_{j=1}^N 1/L^j(\tau_{te}, \tau_j^*)} \tag{4}$$

Δt is the time interval between two adjacent collection points. Finally, the RUL of current tool can be calculated by the following Equation (5):

$$R^{te}(\tau_{te}) = \sum_{i=1}^N q_i r^i(\tau_i^*) \tag{5}$$

2.2. Improved RUL Prediction Model

Under the condition of a limited size of the tool historical failure data set, the influence of the difference in performance degradation rate of similar performance degradation curves is necessary to be considered in the RUL prediction model, which is based on the similarity principle. However, in order to quantify the influence of the difference in performance degradation rate on the prediction model, the ratio of degradation rate of tool performance $v^{te}(\tau_{te})/v^i(\tau_1^*)$ and the function of variation on the ratio of tool degradation rate $m^i(\tau_i^*)$ need to be calculated.

Since the time-varying property of the performance degradation rate function, Peng et al. [20] suggested that the performance degradation process of a tool could be expressed by $X(\tau) \sim IG(\xi(\tau), \lambda\Lambda\xi^2(\tau))$, where, $\xi(\tau)$ is the mean value function of the performance degradation curve, which can be calculated by the following Equation (6):

$$\xi(\tau) = \int v(\tau)d\tau \tag{6}$$

Additionally, $\xi(\tau)$ can be expressed precisely by Equation (7) when the degradation curve has the properties of uncertainty and nonlinearity. Meanwhile, the average performance degradation rate over a longer time interval is selected to estimate the performance degradation rate of the tool, as Equation (8) shows.

$$v(\tau) = \partial\xi(\tau)/\partial(\tau) \tag{7}$$

$$v(\tau) = \frac{X_\tau - X_0}{\tau} \tag{8}$$

where $\tau = 1,2,\ldots,\tau^F$ is the current processing parameter collection point, X_0 is the initial value of the tool health index.

The Equation (5) would be described as the Equation (9) shown, while function $\varepsilon(*)$ is used to quantify the influence of the tool's performance degradation rate on the prediction results.

$$R^{te}(\tau_{te}) = \sum_{i=1}^{N} q_i\varepsilon(*)r^i(\tau_i^*) \tag{9}$$

Moreover, due to the performance degradation rates from the performance degradation curve of the current processing tool and the historical tool $v_{fu}^{te}(\tau_{te})$, $v_{fu}^i(\tau_i^*)$ are main factors in the prediction result, Equation (9), therefore, can be expressed as Equation (10).

$$R^{te}(\tau_{te}) = \sum_{i=1}^{N} q_i\varepsilon(v_{fu}^{te}(\tau_{te}), v_{fu}^i(\tau_i^*))r^i(\tau_i^*) \tag{10}$$

We assume that the future performance degradation rate of the current tool is the relative function of the tool performance degradation rate at the current processing parameter collection point, expressed by $v_{fu} = \vartheta(\tau)v_\tau$. Here, v_{fu} is the future performance degradation rate of the tool, v_τ is the tool performance degradation rate of the current processing parameter collection point, and $\vartheta(*)$ is a sequenced variate. Thus, $\varepsilon(*)$ is described as the Equation (11) shown.

$$\varepsilon(*) = \varepsilon(v_{fu}^{te}(\tau_{te}), v_{fu}^i(\tau_i^*)) = \varepsilon(v^{te}(\tau_{te}), v^i(\tau_i^*), \vartheta^{te}(\tau_{te}), \vartheta^i(\tau_i^*)) \tag{11}$$

Combining the Equations (10) and (11), the Equation (9) can be expressed by Equation (12) as follows:

$$R^{te}(\tau_{te}) = \sum_{i=1}^{N} q_i\varepsilon(v^{te}(\tau_{te}), v^i(\tau_i^*), \vartheta^{te}(\tau_{te}), \vartheta^i(\tau_i^*))r^i(\tau_i^*) \tag{12}$$

Because the RUL of tool and performance degradation rate is negatively correlated, so Equation (12) can be transferred into Equation (13):

$$R^{te}(\tau_{te}) = \sum_{i=1}^{N} q_i r^i(\tau_i^*) \left(\frac{\vartheta^{te}(\tau_{te})}{\vartheta^i(\tau_i^*)} \right) v^{te}(\tau_{te})/v^i(\tau_i^*) \tag{13}$$

Meanwhile, if $m^i(\tau_i^*) = \left(\frac{\vartheta^{te}(\tau_{te})}{\vartheta^i(\tau_i^*)} \right)$, so Equation (13) can be described as the following:

$$R^{te}(\tau_{te}) = \sum_{i=1}^{N} q_i r^i(\tau_i^*) m^i(\tau_i^*) v^{te}(\tau_{te})/v^i(\tau_i^*) \tag{14}$$

The RUL prediction value of the tool would skew to the intermediate value because of the limitation of historical data. To solve this issue, the offset coefficient of the tool RUL is proposed:

$$\delta(\tau_{te}) = 1 + \frac{R^{te}(\tau_{te}) - R_{mid}(\tau_{te})}{2(R_{\max}(\tau_{te}) - R_{mid}(\tau_{te}))} \tag{15}$$

where $R_{mid}(\tau_{te})$ is the mean value of the RUL collection point in tool historical data, $R_{\max}(\tau_{te})$ is the maximum value of the RUL collection point in tool historical data. Therefore, the improved RUL prediction model based on the similarity principle, the degradation rate, and the offset coefficient can be described as:

$$R*(\tau_{te}) = \delta(\tau_{te})R^{te}(\tau_{te}) = \delta(\tau_{te}) \sum_{i=1}^{N} q_i r^i(\tau_i^*) m^i(\tau_i^*) v^{te}(\tau_{te})/v^i(\tau_i^*) \tag{16}$$

3. Validation on the Improved RUL Prediction Model

Variation of axial drilling force is largely caused by drill bit wear, which also acts as a significant feature of the health state of the drill bit in micro-drilling of packaging substrate. Consequently, axial drilling force is used as the supporting data in validation of the RUL prediction model.

3.1. Experiment on Collection of Axial Drilling Force

3.1.1. Micro-Drilling Experiment

In this micro-drilling experiment, a drill bit diameter of 0.11 mm (Table 1) and packaging substrate with a specification of HL832NSF are selected. The properties of HL832NSF are shown in Table 2. In terms of processing parameters of the machine tool (HANS-F6MH, Shenzhen Han's CNC Technology Co., Ltd., Shenzhen, China), the spindle speed is 170 krpm, the feed rate is 45 mm/s (in order to acquire more breakage data of drill bits in a relative short time, the feed rate would be larger than the general value), and the main property parameters of the HANS-F6MH are as shown in Table 3.

Table 1. Drill bit structure parameters.

Drill Diameter	Point Angle	Helix Angle	Flute Land Ratio	Overall Length
0.11 mm	120°	40°	1:2	38.15 mm

Table 2. Properties of the packaging substrate (HL832NSF).

Dielectric Constant	Dissipation Factor	Young's Modulus	Density	Bending Strength
4.4	0.008	32 Gpa	2.0 g/cm^3	510 Mpa

Table 3. The main property parameters of HANS-F6MH.

Spindle Speed	Repeatability	Movement Speed	Range of Drill Diameter
200 krpm	5 μm	85 m/min	0.10–2.0 mm

In micro-drilling of packaging substrate, obtaining the original signals of drilling force precisely is the most critical process for the study of axial drilling force since those signals are extremely weak and the signal-to-noise ratio (SNR) is very low. Therefore, a high-precision dynameter system, manufactured by Kistler Company (Winterthur, Switzerland), is used to measure the axial drilling force; the corresponding product models are shown in Table 4. In consequence, the schematic diagram of axial drilling force measurement is shown in Figure 1.

Table 4. Equipment and software of the high precision dynameter system.

Equipment	Dynameter	Amplifier	A/D Card	Software
Product model	Kistler 9256 CQ01	5080A108004	2855A5	DynoWare 2825A-02-2

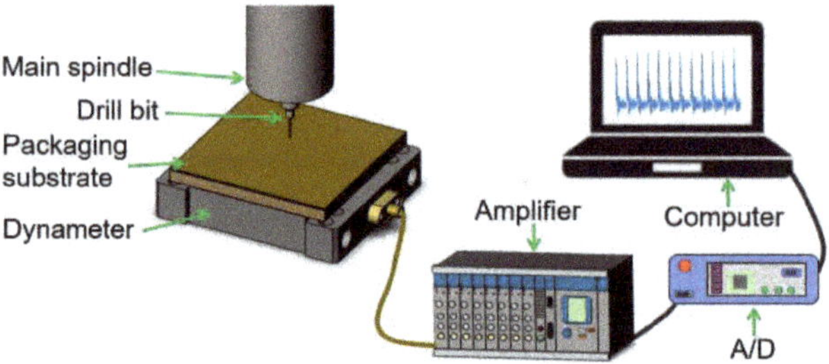

Figure 1. Schematic diagram of the axial drilling force measurement.

In this experiment, 25 drill bits are used for drilling holes in sequence, but if the drill bit breakage does not occur until the finished 1800 holes, drilling processes will be stopped, and this drill bit will be deemed to have reached a failure state. That is because the drill bit wear became very serious while finishing 1800 holes, and the micro hole quality could not satisfy the requirement. Since each drill bit will fail in different drilling stages, all drill bits are going to be divided into different groups via the drilling hole amount before failure appearance for the data analysis. Because the cyclic autocorrelation Wiener filter with time-domain accumulation was validated as a suitable method for weak and complex signals by the previous investigation of our team [28], it would be used for the processing of axial drilling force signals in this study. Moreover, the average force of 15 data points from 10 drilled holes during hole processing at different drilling stages was extracted for the analysis of RUL.

3.1.2. Results of Drill Bit Failure and Signal Processing of Axial Drilling Force Signal

The failure situation of 25 drill bits is shown in Table 5. It is clear that 19 drill bits fail before finishing 1800 holes, but failure does not occur in 6 drill bits while finishing 1800 holes. Furthermore, there are five groups for the 25 drill bits: the A group belongs to the failure appearance of about 650 holes; similarly, the B group is about 800 holes; the C group is in range of 1100–1450 holes; the D group is in range of 1600–1800 holes; and the E group belongs to the failure that does not appear while finishing 1800 holes.

Table 5. Failure situation of the drill bits.

Drill Bit Number	Failure (Hole Amount)	Drill Bit Number	Failure (Hole Amount)
A-1	650	C-4	1123
A-2	652	C-5	1440
A-3	639	D-1	1608
A-4	642	D-2	1795
A-5	643	D-3	1692
B-1	799	D-4	1769
B-2	808	E-1	1800
B-3	804	E-2	1800
B-4	808	E-3	1800
B-5	803	E-4	1800
C-1	1154	E-5	1800
C-2	1436	E-6	1800
C-3	1389		

Packaging substrate is a kind of composite material, and the noise source of drilling force signals in processing is complex, resulting in weak axial drilling force signals and quite low SNR. Simple filtering methods generally cannot effectively improve the SNR of axial drilling force signals. The Wiener filter principle holds excellent noise reduction performance, but before using of the Wiener filter to filter the axial drilling force signal, obtaining the desired signal of the original axial drilling force is needed first. Since the desired signals of the original axial drilling force cannot be acquired directly via drilling experiments, and the noise reduction process of the original axial drilling force signals cannot be completed by a single Wiener filter. Therefore, the original axial drilling force signal processed by a cyclic autocorrelated Wiener filter is needed, and the process flow is as follows:

(1) Calculating the periods of the axial drilling force signals through the cyclic stationary theory;
(2) Based on accumulation theory in the time domain, if the axial drilling force signals are sliced according to the calculated periods and the average value of each signal segment is acquired after time domain superposition, the preliminary filtered signal would be obtained;
(3) The signal accumulated in the time domain is used as the desired signal, and the original axial drilling force signals are filtered by the Wiener filter to obtain the final filtered signals.

The merit of this filtering method is that, while a single time-domain accumulation principle is used for filtering, although the SNR of the output signal is raised after filtering, the waveform of the output signal is difficult to match with the different original axial drilling force signal segments. When the average signal filtered by the time-domain accumulation principle is taken as the expected signal of the original axial drilling force signal and the original axial drilling force signal is filtered via the Wiener filter. Therefore, this filtering method improved the SNR of the output signal by 11.76 dB, and the consistency with the original axial force signal waveform would be enhanced [28].

In the collection of drilling force, the original signal of axial drilling force is as shown in Figure 2, and some of the Wiener-filtered signals of axial drilling force in the A group are as shown in Figure 3. It can be found that the Wiener-filtered signal is in good agreement with the original signal. Meanwhile, the Wiener-filtered signal also retains the advantage of a high signal-to-noise ratio. These results propose reliable support for the verification of the RUL prediction model.

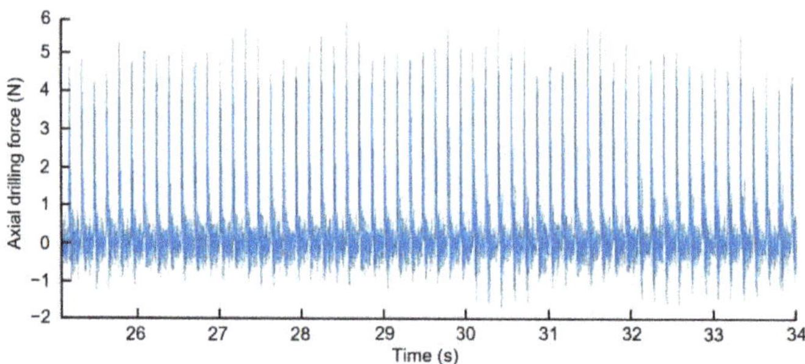

Figure 2. Original signal of the axial drilling force at different drilling stages.

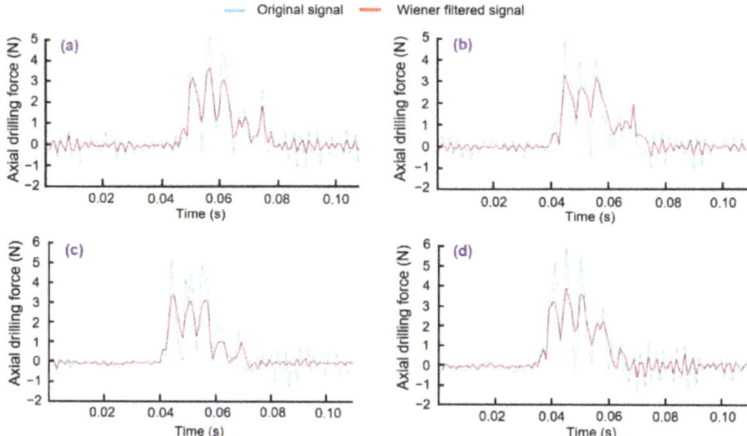

Figure 3. Wiener-filtered signals of axial drilling force in different drilling stages (drill bit number: A-1). (**a**) 80 holes; (**b**) 240 holes; (**c**) 480 holes; (**d**) 600 holes.

Extracting the axial drilling force in each drill bit when the interval of drilling hole amount is 80, and the first extraction of axial drilling force begins with drilling hole amount of 40. Thereupon, the variations of axial drilling force for each drill bit in different drilling stages are shown in Figures 4 and 5. In this study, testing failure data sets are selected at random for each group, and the remaining data sets are regarded as historical failure data sets. It means that the historical failure data sets are the reference library for the RUL prediction model of drill bits.

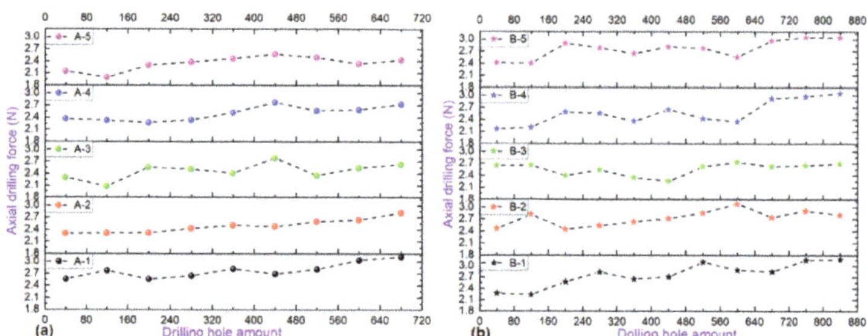

Figure 4. Axial drilling force at different drilling stages. (**a**) Drill bits in the A group; (**b**) drill bits in the B group.

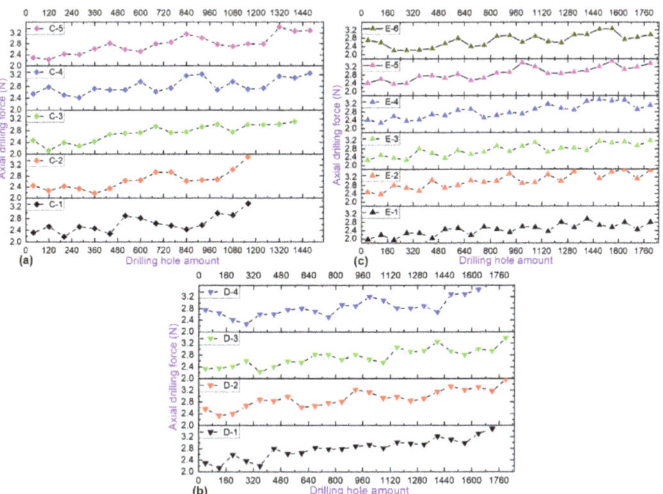

Figure 5. Axial drilling force at different drilling stages. (**a**) Drill bits in the C group; (**b**) drill bits in the D group; (**c**) drill bits in the E group.

3.2. Validation Results of the RUL Prediction Model

The RUL prediction method for cutting tools based on a degradation model was effective, but it took the roughness of the cutting surface as the failure criterion [29]. In this article, the testing failure data sets of all drill bits are put into the traditional RUL prediction model and the improved RUL prediction model, and their comparisons with the actual RUL are shown in Figures 6 and 7. From the changes in curves in each group, it can be seen that all curves of the improved RUL prediction model have higher fitting degrees with the actual RUL curve compared to the traditional RUL prediction model. Also, the RUL prediction based on the similarity principle was successfully used in cutting tools [30], aircraft engines [31], and many other machinery fields. Remarkably, it is found that the output results of the improved RUL prediction model generally have large deviations at the beginning stage. The main reasons for this phenomenon are that the increment of drill bit performance degradation rate in the same collection point interval within the initial range is less than that in the later period, which leads to a matching error when matching the drill bit performance degradation curve and then affects the calculation of the drill bit performance degradation rate function. So, under the combined effects of the two kinds of errors, the prediction results of the RUL of the drill bit have a large error at the beginning stage. But the higher fitting degree between the improved model curve and the actual RUL curve is going to be obtained with an increment in the drilling hole amount.

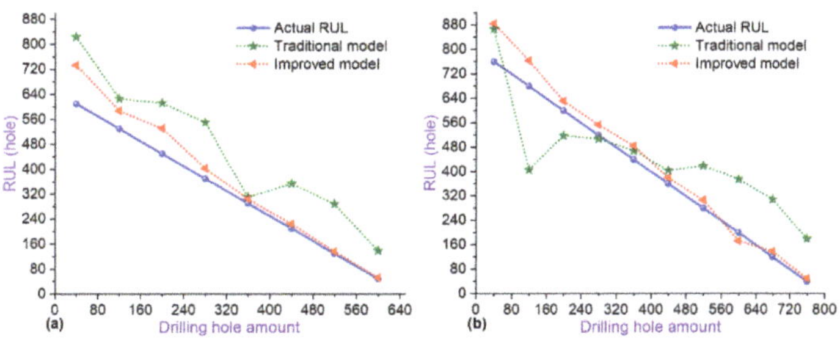

Figure 6. RUL prediction results of the two prediction models and actual RUL. (**a**) A group; (**b**) B group.

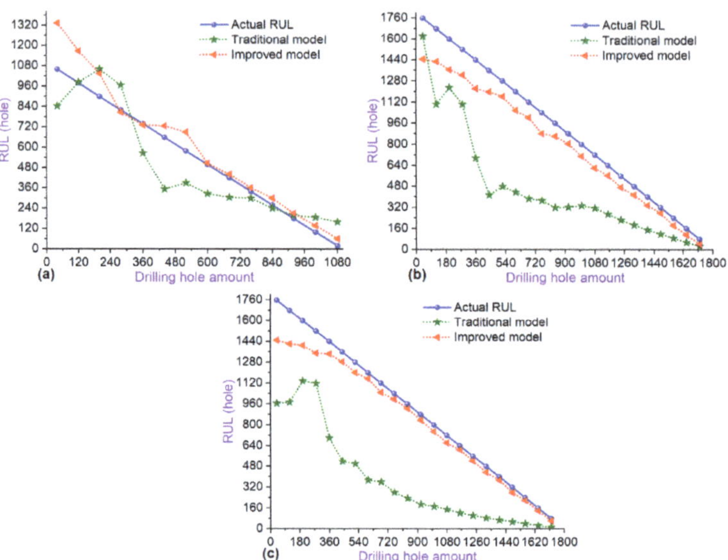

Figure 7. RUL prediction results of the two prediction models and the actual RUL. (**a**) C group; (**b**) D group; (**c**) E group.

The average relative errors of the two prediction models are calculated too, as shown in Figure 8. The average relative errors of the improved RUL prediction model are much smaller than those of the traditional RUL prediction model, which are also stable with a range of 8.97–14.18%. Conversely, the average relative errors of the traditional RUL prediction model for all groups hold a much larger fluctuation range, and the biggest one reaches 74.43%, while the smallest value is 31.81%.

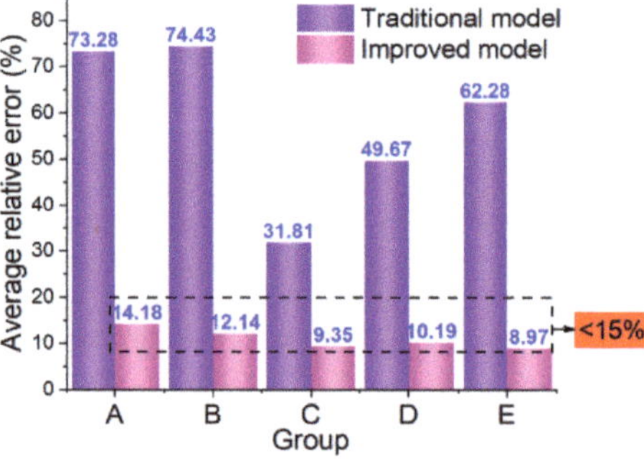

Figure 8. The average relative errors of the traditional prediction model and the improved prediction model in different groups.

Therefore, the output results of the improved RUL prediction model based on the similarity principle proposed in this paper are closer to the actual RUL curves, and its average relative error is less than 15%, which proves that considering the degradation rate difference and offset coefficient can significantly improve the performance of the RUL prediction model based on the similarity principle. However, in the case of a limited number of historical failure data samples, the traditional prediction model based on the similarity principle does not take into account the impact of sample performance degradation speed differences and data migration, so the accuracy of the prediction results is much lower than the method proposed in this paper. The RUL curves predicted by the improved model based on the similarity principle are reliable.

4. Conclusions

Based on the similarity principle and taking the degradation rate and offset coefficient into account, an improved RUL prediction model is established. And then, in order to verify the effectiveness of that model, a micro-drilling experiment on packaging substrate is carried out, the axial drilling force is collected through the precision measuring platform, and a cyclic autocorrelation Wiener filter with time-domain accumulation, which increased the SNR of the processed axial drilling force signal by 11.76 dB, is used for the signal processing. The validation results show that, compared to the traditional RUL prediction model, the curves of the improved RUL prediction model exhibit a higher fitting degree with the actual RUL curves. The average relative errors of the traditional RUL prediction model for all groups hold a much larger fluctuation range, and the biggest one reaches 74.43%, while the smallest value is 31.81%. But the average relative errors of the improved RUL prediction model are small and stable in all groups; all of the values are less than 15%. Therefore, the improved RUL prediction model is a reliable theoretical support for the health state monitoring of drill bits during the micro-drilling of packaging substrates, and it is also a potential method to improve micro hole processing efficiency for packaging substrates.

However, since the axial drilling force is classified linearly during the selection of the testing failure data set, it is untoward to the engineering application of this improved model. Thus, the extraction of testing failure data in a more reasonable way is expected to be investigated. Meanwhile, reducing the matching error of the drill bit performance degradation curve may be a feasible method to improve the prediction accuracy of the beginning stage.

Author Contributions: Conceptualization, H.S. and X.L.; methodology, X.L. and S.T.; validation, X.L. and S.T.; Investigation, S.T. and T.Z.; resources, H.S.; data curation, T.Z. and Z.W.; writing—original draft preparation, X.L.; writing—review and editing, H.S.; visualization, H.S.; supervision, H.S.; funding acquisition, H.S. All authors have read and agreed to the published version of the manuscript.

Funding: The research proposed in this paper was supported by the National Key Research and Development Program of China (No. 2020YFB1708500) and the National Natural Science Foundation of China (No. 52075344).

Data Availability Statement: Not applicable.

Conflicts of Interest: The authors declare no conflict of interest.

References

1. Kenny, S.; Baron, D.; Roelfs, B. A comprehensive packaging solution for next generation IC substrates. *Microelectron. Packag. Conf.* **2011**, *9*, 1–4.
2. Chang, J.Y.; Chang, T.H.; Chang, T.C. An Investigation into the package and printed circuit board assembly solutions of an ultrathin coreless flip-chip substrate. *J. Electron. Mater.* **2015**, *44*, 3855–3862. [CrossRef]
3. Iliescu, D.; Gehin, D.; Gutierrez, M.E.; Girot, F. Modeling and tool wear in drilling of CFRP. *Int. J. Mach. Tools Manuf.* **2010**, *50*, 204–213. [CrossRef]
4. Liu, X.W.; Shi, H.Y.; Huang, G.; Tao, S.; Gao, Z.S. Investigation on Drill Wear and Micro Hole Quality in High Speed Drilling of High Frequency Printed Circuit Board. *IOP Conf. Ser. Mater. Sci. Eng.* **2020**, *825*, 012034. [CrossRef]
5. Zheng, L.J.; Wang, C.Y.; Fu, L.Y.; Yang, L.P.; Qu, Y.P.; Song, Y.X. Wear mechanisms of micro-drills during dry high speed drilling of PCB. *J. Mater. Process. Technol.* **2012**, *212*, 1989–1997. [CrossRef]
6. Shi, H.Y.; Liu, X.W.; Gao, Z.S.; Huang, G.; Tao, S.; Fu, L.Y.; Liang, X. Micro hole drilling of high frequency printed circuit board containing hard fillers. *J. Micromech. Microeng.* **2021**, *31*, 055005. [CrossRef]
7. Imran, M.; Mativenga, P.T.; Withers, P.J. Assessment of machining performance using the wear map approach in micro-drilling. *Int. J. Adv. Manuf. Technol.* **2012**, *59*, 119–126. [CrossRef]
8. Li, J.; Lu, J.; Chen, C.; Ma, J.; Liao, X. Tool wear state prediction based on feature-based transfer learning. *Int. J. Adv. Manuf. Technol.* **2021**, *113*, 3283–3301. [CrossRef]
9. Shi, H.Y.; Lin, X.K.; Wang, Y. Characterization of drill bit breakage in pcb drilling process based on high speed video analysis. *Circuit World* **2017**, *43*, 89–96. [CrossRef]

10. Ren, Z.H.; Zheng, X.H.; An, Q.L.; Wang, C.Y.; Chen, M. Tool breakage feature extraction in pcb micro-hole drilling using vibration signals. *Adv. Mater. Res.* **2012**, *497*, 126–131. [CrossRef]
11. Tansel, I.; Arkan, T.; Bao, W. Tool wear estimation in micro-machining: Part I: Tool usage–cutting force relationship. *Int. J. Mach. Tools Manuf.* **2000**, *40*, 599–620. [CrossRef]
12. Malekian, M.; Park, S.; Jun, M. Tool wear monitoring of micro-milling operation. *J. Mater. Process. Tech.* **2009**, *209*, 4903–4914. [CrossRef]
13. Shi, H.Y.; Song, F.M.; Fu, L.Y. Experimental study on drilling force in printed circuit board micro drilling process. *Circuit World* **2011**, *37*, 24–29. [CrossRef]
14. Xue, B.; Xu, H.; Huang, X.; Zhu, K.; Xu, Z.; Pei, H. Similarity-based prediction method for machinery remaining useful life: A review. *Int. J. Adv. Manuf. Technol.* **2022**, *121*, 1501–1531. [CrossRef]
15. Lei, J.; Zhang, W.; Jiang, Z.; Gao, Z. A Review: Prediction method for the remaining useful life of the mechanicalsystem. *J. Fail. Anal. Prev.* **2022**, *22*, 2119–2137. [CrossRef]
16. Long, B.; Xian, W.M.; Jiang, L.; Liu, Z. An improved autoregressive model by particle swarm optimization for prognostics of lithium-ion batteries. *Microelectron. Reliab.* **2013**, *53*, 159–168. [CrossRef]
17. Chen, C.; Wei, J.; Li, Z.H. Remaining useful life prediction for lithium-ion batteries based on a hybrid deep learning model. *Processes* **2023**, *11*, 2333. [CrossRef]
18. Du, Q.; Zhan, Z.G.; Wen, X.F.; Zhang, H.; Tan, Y.; Li, S.; Pan, M. A hybrid model to assess the remaining useful life of proton exchange membrane fuel cells. *Processes* **2023**, *11*, 1583. [CrossRef]
19. Zeng, F.; Li, Y.; Jiang, Y.; Song, G. A deep attention residual neural network-based remaining useful life prediction of machinery. *Measurement* **2021**, *181*, 109642. [CrossRef]
20. Liu, Z.X.; Tan, C.B.; Liu, Y.X.; Li, H.; Cui, B.; Zhang, X. A study of a domain-adaptive LSTM-DNN-based method for remaining useful life prediction of planetary gearbox. *Processes* **2023**, *11*, 2002. [CrossRef]
21. Eker, O.F.; Camci, F.; Jennions, I.K. A Similarity-based prognostics approach for remaining useful life prediction. *PHM Soc. Eur. Conf.* **2014**, *2*. [CrossRef]
22. Xiong, X.; Yang, H.; Cheng, N.; Li, Q. Remaining useful life prognostics of aircraft engines based on damage propagation modeling and data analysis. In Proceedings of the 2015 8th International Symposium on Computational Intelligence and Design (ISCID), Hangzhou, China, 12–13 December 2015; pp. 143–147.
23. Hao, W.; Li, Z.X.; Qin, G.H.; Ding, K.; Lai, X.; Zhang, K. A novel prediction method based on bi-channel hierarchical vision transformer for rolling bearings' remaining useful life. *Processes* **2023**, *11*, 1153. [CrossRef]
24. Li, Y.; Xiang, Y.; Pan, B.; Shi, L. A hybrid remaining useful life prediction method for cutting tool considering the wear state. *Int. J. Adv. Manuf. Technol.* **2022**, *121*, 3583–3596. [CrossRef]
25. Zegarra, F.C.; Vargas-Machuca, J.; Coronado, A.M. Tool wear and remaining useful life (RUL) prediction based on reduced feature set and Bayesian hyperparameter optimization. *Prod. Eng. Res. Devel.* **2022**, *16*, 465–480. [CrossRef]
26. Pan, D.; Liu, J.; Cao, J. Remaining useful life estimation using an inverse Gaussian degradation model. *Neurocomputing* **2016**, *8*, 185–194. [CrossRef]
27. Zhong, X.H. Research on Equipment Remaining Useful Life Prediction Methods Based on Limited Failure Historical Data. Master's Thesis, Heifei University of Technology, Heifei, China, 2020.
28. Tao, S.; Gao, Z.; Shi, H. Characterization of printed circuit board micro-holes drilling process by accurate analysis of drilling force signal. *Int. J. Precis. Eng. Manuf.* **2022**, *23*, 131–138. [CrossRef]
29. Huang, Y.; Lu, Z.; Dai, W.; Zhang, W.; Wang, B. Remaining useful life prediction of cutting tools using an inverse Gaussian process model. *Processes* **2021**, *11*, 5011. [CrossRef]
30. Liu, Y.C.; Hu, X.F.; Zhang, W.J. Remaining useful life prediction based on health index similarity. *Reliab. Eng. Sys. Saf.* **2019**, *185*, 502–510. [CrossRef]
31. Chen, Z.Z.; Cao, S.C.; Mao, Z.J. Remaining useful life estimation of aircraft engines using a modified similarity and supporting vector machine (SVM) approach. *Energies* **2018**, *11*, 28. [CrossRef]

Article

Topology Optimization of Compliant Mechanisms Considering Manufacturing Uncertainty, Fatigue, and Static Failure Constraints

Dongpo Zhao [1] and Haitao Wang [2,*]

[1] School of Mechanical Engineering, Nanjing University of Science and Technology, Nanjing 210094, China; dpzhao1@163.com
[2] School of Mechanical and Electrical Engineering, Shenzhen Polytechnic University, Shenzhen 518055, China
* Correspondence: wanghaitao@szptedu.cn

Abstract: This study presents a new robust formulation for the topology optimization of compliant mechanisms, addressing the design challenges while considering manufacturability, static strength, and fatigue failure. A three-field density projection is implemented to control the minimum size of both real-phase and null-phase material structures to meet the manufacturing process requirements. The static strength is evaluated via the sum of the amplitude and the mean absolute value of the signed von Mises stress. The fatigue failure is solved via the modified Goodman criterion. The real output displacement is optimized by adding artificial springs to the prescribed value. This approach is implemented based on an improved solid isotropic material with penalization (SIMP) interpolation method to describe and solve the optimization model and derive the shape sensitivity of the optimization problem. Finally, two numerical examples are applied to illustrate the effectiveness of the presented method.

Keywords: compliant mechanism; topology optimization; fatigue failure; static strength; manufacturability

Citation: Zhao, D.; Wang, H. Topology Optimization of Compliant Mechanisms Considering Manufacturing Uncertainty, Fatigue, and Static Failure Constraints. *Processes* 2023, 11, 2914. https://doi.org/10.3390/pr11102914

Academic Editors: Maximilian Lackner and Antonino Recca

Received: 6 September 2023
Revised: 20 September 2023
Accepted: 2 October 2023
Published: 4 October 2023

1. Introduction

With the revolutionary changes in many fields, such as manufacturing, materials, information, biology, medicine, and national defense, caused by the wave of micro/nanotechnologies, compliant mechanisms have been widely used in the fields of micro/nano operations [1,2], nano-positioning stages [3,4], fast tool servos (FTSs) [5–7], and micro-electro-mechanical systems (MEMSs) [8,9]. Therefore, it is important to advance the development of the fundamental theory of compliant mechanisms and explore efficient design methods.

It is acknowledged in the literature that several approaches have been developed for designing compliant mechanisms, such as pseudo-rigid replacements [10] and constraint-based designs [5,11–16], as shown in Figure 1. However, these approaches have limitations as they do not allow for changes at the topological level, which can restrict performance improvements. To address the limitations of traditional design approaches to compliant mechanisms, a topology optimization approach is introduced to enhance the high-standard performance of compliant mechanisms [17–21]. Unlike other methods, a topology optimization approach is a systematic conceptual design approach that combines topological synthesis and scale synthesis, in which the geometric qualities and topological information of the structure are unknown, and the optimal material layout method is discovered [22–24]. Topology optimization can be classified into two types based on the representation of the structural morphology, namely, discrete [25] and continuous [17,26–32] optimization, as shown in Figure 1. Compared with discrete topology optimization, the advantage of continuous topology optimization lies in its ability to employ continuous mathematical optimization methods, which are typically more efficient and accurate [19]. Moreover, continuous topology optimization can consider more detailed structural features, such as curvature and edges.

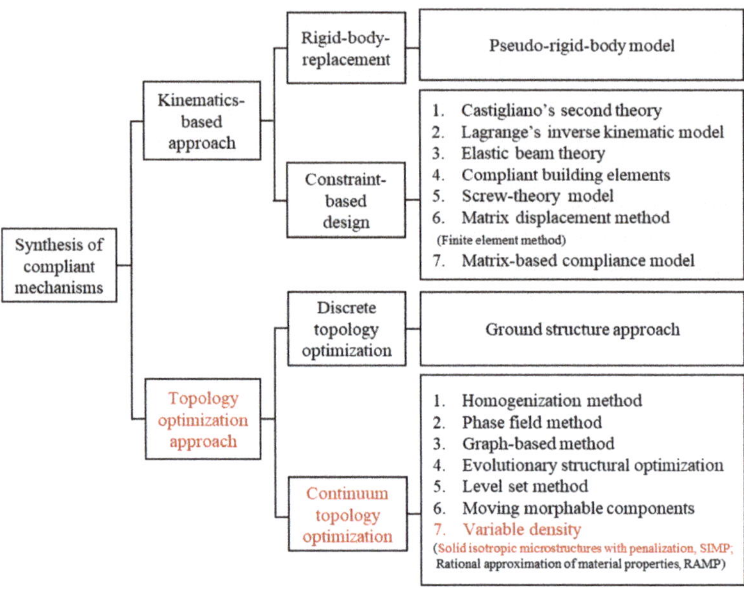

Figure 1. Synthesis of compliant mechanisms.

As the design theory of the continuous topology optimization of compliant mechanisms continues to improve, it becomes essential to consider the strength and fatigue life of these mechanisms during the design process. However, the current literature surveys reveal that the joint effect of static strength and fatigue failure is seldom taken into account in the design of compliant mechanisms. In the optimization process of compliant mechanisms, neglecting stress and fatigue failure can lead to suboptimal results for the stiffness and other overall responses, such as frequency. Thus, it is imperative to consider both static strength and fatigue failure when designing compliant mechanisms to ensure that their design meets the expected service life and performance requirements.

Since the pioneering work of Duysinx and Bendsøe [33], numerous techniques have been proposed to address static failure problems in continuum topology optimization. (see, e.g., [34–39]). Among them, compliant mechanism design problems have received increased attention in recent years [35–37] due to their potential for producing innovative solutions to engineering challenges. However, there are still several key challenges related to static strength, such as highly nonlinear behavior [33,40] and a large number of local stress constraints [41], as well as singularity phenomena [40,41]. To overcome these challenges, various techniques have been developed in continuum topology optimization. For example, static strength constraint relaxation techniques can be used to address the singularity phenomena and highly nonlinear behavior [33,40]. Additionally, aggregation approaches can be used to handle a large number of local stress constraints [41]. Zhu et al. [17] developed a multi-degree-of-freedom compliant mechanism that attained fully decoupled motion and exclusively addressed input–output coupling issues, disregarding strength considerations. Nevertheless, strength plays a pivotal role in ensuring the proper functioning of the mechanism. Liu et al. [23] developed new flexure hinges using topology optimization considering the static failure constraint. However, in practical applications, compliant mechanisms are subject to various reciprocating motions, which can induce alternating stresses and lead to fatigue damage and eventual failure of the mechanism [42,43]. However, their usability becomes relatively intricate when confronted with intricate problems encompassing multiple strength constraints. Therefore, it is essential to consider the fatigue performance of compliant mechanisms in the design optimization process so that the compliant mechanism can be satisfied with the required fatigue strength and life, thus ensuring its safe and reliable operation over the whole life cycle.

Addressing fatigue failure constraints is a complex and ongoing research topic and has received significant attention in the literature [43–45]. In recent years, several techniques have been proposed to tackle topology optimization problems subject to fatigue failure constraints [42,43,45–47]. Oest et al. [46] developed a static analysis method for the optimal design of continuum structures subject to fatigue failure constraints, which are formulated using the Palmgren–Miner's linear damage hypothesis, S-N curves, and Sines fatigue criterion. Nabaki et al. [43] employed a bi-directional evolutionary structural optimization (BESO) approach, incorporating a modified Goodman fatigue failure criterion into the analysis. However, it should be noted that the design reported in their study did not fully consider all intervals of the Goodman safe region; therefore, the design may not be entirely optimal.

Motivated by the above-mentioned challenges, this study proposes and examines a methodology for analyzing the static strength and fatigue failure constraints. Specifically, the modified Goodman failure criterion is applied directly in the sensitivity analysis to address these constraints. The high-cycle fatigue (HCF) approach is utilized for fatigue analysis under proportional loadings with constant amplitude. The static strength and fatigue failure constraints are converted into different stress constraint methods. The signed von Mises and modified Goodman criteria are then used to solve the problems of static strength and fatigue failure multi-performance constraints, respectively. Next, the study also investigates failure under the safe region condition of the modified Goodman criterion to ensure the reliability of compliant mechanisms. Then, the relaxation techniques for dealing with the singularity phenomenon and highly nonlinear behavior [33,40], and the P-norm approach is applied to different stress constraints for handling a large number of local stress constraints [41]. Additionally, a differentiable approximation formula is introduced as an alternative to the failure formula regarding the non-differentiability of stress components and design variables. This addresses the non-differentiable kinks related to the modified Goodman criteria. Finally, the modified Goodman criterion and signed von Mises stress are utilized to evaluate fatigue and static failure, respectively. Meanwhile, since different pieces of machining equipment have varying levels of accuracy, it is necessary to control the feature size of topology optimization results to avoid unmanufacturable structures such as thin rods and holes. Thus, a three-field density topology optimization formulation with eroded, intermediate, and dilated projections is applied to address manufacturing uncertainty in the layout optimization (see, e.g., [48–55]).

Based on the aforementioned analysis, this approach is implemented by utilizing an enhanced solid isotropic material with a penalization (SIMP) interpolation model, which accurately characterizes the material distribution to avoid numerical non-convergence issues [17–56]. The design problem is solved via the global convergence moving asymptote method (GCMMA) algorithm, which is based on sensitivity analysis [57]. Two numerical examples are presented to demonstrate the efficacy of the proposed method. Additionally, three distinct combinations of alternating and mean stresses, namely, von Mises, sines theory, and signed von Mises, are evaluated to test layout optimization. Thus, this paper is organized as follows: In Section 2, we present the proposed maximum output displacement, as well as the static strength, fatigue failure, and manufacturability. The optimization problem statement is outlined in Section 3, while the design problem is solved via the GCMMA algorithm based on sensitivity analysis in Section 4. The numerical examples are presented in Section 5 to demonstrate the effectiveness of the proposed method. Finally, the conclusions of this study are provided in Section 6.

2. Problem Formulation

This study presents a novel methodology for the topology optimization of compliant mechanisms, which incorporates multiperformance coupled manufacturability analysis. The optimization objective is reformulated to maximize the output displacement while considering static strength, fatigue failure, and manufacturability. The proposed approach employs a density filter with threshold projection to address manufacturability concerns, which are integrated into both the objective function and constraint function. In this study, the material distribution is described via a modified SIMP interpolation model that establishes a relationship between the artificial material density and Young's modulus of the element as [7]:

$$E(\rho_i) = E_{min} + \rho_i{}^p (E_0 - E_{min}), \rho_i \in [0, 1] \tag{1}$$

where the i-th element density, $\rho_i \in [0, 1]$, is the design variable. $E(\rho_i)$ represents Young's modulus of i-th element; E_0 and $E_{min} > 0$ denote elastic modulus of solid and void parts, respectively; and $p > 1$ is the penalty to provide a better description of the materials in the void (0) or solid (1) state [17,56].

2.1. Methods of the Manufacturability

In the manufacturability method, a modified robust topology optimization formulation based on three-field density technique will be applied [49,58]; it uses one design variable, ρ_i; one filtered density, $\tilde{\rho}_i$; as well as three relative densities $\bar{\tilde{\rho}}_i^{(e)}$, $\bar{\tilde{\rho}}_i^{(i)}$, and $\bar{\tilde{\rho}}_i^{(d)}$. Among them, the three relative densities approach are real physical densities.

With threshold projection, the relative densities, $\bar{\tilde{\rho}}_i^{(e)}$, $\bar{\tilde{\rho}}_i^{(i)}$, and $\bar{\tilde{\rho}}_i^{(d)}$, are correlated with design variables, ρ_i, via filtered densities, $\tilde{\rho}_i$ [58]. Thus, three relative densities of an element, i, can be described as:

$$\bar{\tilde{\rho}}_i = \frac{\tanh(\beta\eta) + \tanh\left(\beta\left(\tilde{\rho}_i - \eta\right)\right)}{\tanh(\beta\eta) + \tanh\left(\beta(1 - \eta)\right)} \tag{2}$$

where β denotes a projection parameter, which is used to control a steepness of an approximated Heaviside function. η and $\tilde{\rho}_i$ represent a projection level of a smoothed Heaviside approximation threshold and the filtered relative density of element i, respectively. $\tilde{\rho}_i$ can be obtained via linear projection, which can be written as:

$$\tilde{\rho}_i = \frac{\sum_{i=N_i} \omega_{ij} v_j \rho_j}{\sum_{j=N_i} \omega_{ij} v_j} \tag{3}$$

where v_j denotes volume of element j. $N_i = \left\{ j : \text{dist}(i, j) \leq r_f \right\}$ denotes the set of neighbours lying within the radius R of the filter of i-th element, where r_f represents the size of the neighbourhood or filter. $\text{dist}(i, j)$ and ω_{ij} are the distance and a function of the distance between neighboring elements, respectively. ω_{ij} is described as:

$$\omega_{ij} = r_f - \text{dist}(i, j), j \in N_i \tag{4}$$

The relative density, $\bar{\tilde{\rho}}_i$, is incorporated into layout optimization, and the regularized SIMP interpolation model can be written as:

$$E\left(\bar{\tilde{\rho}}_i\right) = E_{min} + \bar{\tilde{\rho}}_i{}^p (E_0 - E_{min}), \bar{\tilde{\rho}}_i \in [0, 1] \tag{5}$$

In addition, the derivative of a filtered density, $\tilde{\rho}_i$, with respect to the design variables, ρ_j, will be determined as:

$$\frac{\partial \tilde{\rho}_i}{\partial \rho_j} = \frac{\omega_{ij} v_j}{\sum_{j=N_i} \omega_{ij} v_j} \tag{6}$$

Since the 0/1 projection, $\overline{\tilde{\rho}}_i$, is a function of the filtered density, $\tilde{\rho}_i\left(\tilde{\rho}_i\right)$, the sensitivities of the objective and constraint function can be obtained via a chain rule:

$$\frac{\partial \psi}{\partial \rho_j} = \sum_{i=N_i} \frac{\partial \psi}{\partial \overline{\tilde{\rho}}_i} \frac{\partial \overline{\tilde{\rho}}_i}{\partial \tilde{\rho}_i} \frac{\partial \tilde{\rho}_i}{\partial \rho_j} \tag{7}$$

2.2. Methods of the Objective Function

The optimization problem of maximizing output port displacement is a well-known benchmark in layout optimization, and it represents an extension of the standard min–max method [58]. When utilizing the GCMMA algorithm to solve the min–max problem, it is imperative to ensure that the objective value in the resulting robust topology optimization does not become negative. To address this issue, a constant term of 100 is introduced into the topology optimization formulation. This approach does not alter the optimized design outcomes, failure rates, or corresponding amplification factors. Therefore, the optimization problem for the objective function can be described as follows:

$$\min_{\rho} : f_{obj} = max\left\{ u_{out} = \mathbf{L}_{out}^T \mathbf{U}_{out} + 100 \right\} \tag{8}$$

where $\mathbf{L}_{out}^T$ represents a unit length vector with zeros at all degrees of freedom except at the output point, where it takes the value of one. The displacement vector is denoted by $\mathbf{U}_{out}$. Furthermore, u_{out} corresponds to the three relative densities, $\overline{\tilde{\rho}}_i^{(e)}$, $\overline{\tilde{\rho}}_i^{(i)}$, and $\overline{\tilde{\rho}}_i^{(d)}$, which represent three different designs. Note that $\mathbf{L}_{out}^T$ and $\mathbf{U}_{out}$ correspond to different densities, and they are represented by different values.

2.3. Methods of the Static Strength and Fatigue Failure

To ensure effective prevention of structural and mechanical failures and to address issues related to strength and reliability, fatigue analysis utilizes the stress–life approach under constant and proportional loading conditions. As shown in Figure 2a, the HCF approach with proportional loadings and constant amplitude is employed to test for fatigue failure within the linear elastic range of the structure. A sinusoidal load can be applied to the structure to generate a stress state history, as depicted in Figure 2b. The vibration amplitude, σ_a, and mean stress, σ_m, can be determined from the maximum stress, σ_{max}, and minimum stress, σ_{min}, which can be calculated as:

$$\sigma_a = \frac{\sigma_{max} - \sigma_{min}}{2} ; \sigma_m = \frac{\sigma_{max} + \sigma_{min}}{2} \tag{9}$$

Mean stress values and vibration amplitudes can be determined using equivalent static finite element analysis. To implement equivalent static analysis, the following equation can be used:

$$\mathbf{KU} = \mathbf{F}_{max} \tag{10}$$

where $\mathbf{K}$ and $\mathbf{U}$ denote a global stiffness matrix and displacement vector, respectively. $\mathbf{F}_{max}$ represents an array for maximum force, F_{max}, which is employed to calculate the amplitude stress scaling factor, c_a, and the mean stress scaling factor, c_m, respectively:

$$c_a = \frac{1 - \left(\frac{F_{min}}{F_{max}}\right)}{2} ; c_m = \frac{1 + \left(\frac{F_{min}}{F_{max}}\right)}{2} \tag{11}$$

where F_{min} represents the minimum force.

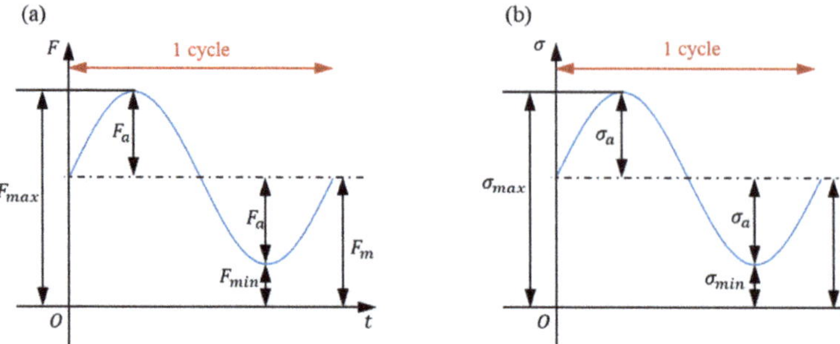

Figure 2. One cycle of the history in high cycle fatigue approach: (**a**) one-cycle constant and proportional loading history, and (**b**) one-cycle stress history.

In order to calculate the von Mises stress, the element centroid is selected as the stress evaluation point for each element, and the solid material stress vector at the stress evaluation point i can be expressed in Voigt notation as $\boldsymbol{\sigma}_i = \left[\sigma_{i,x}, \sigma_{i,y}, \tau_{i,xy}\right]^T$. Finally, the element displacement vector is calculated using finite element analysis to obtain the corresponding alternating and mean stresses at the stress evaluation point [41,43].

$$\underbrace{\boldsymbol{\sigma}_{a_i}}_{3\times1} = c_a \underbrace{\boldsymbol{\sigma}_i}_{3\times1} = \underbrace{\mathbf{D}}_{3\times3} \underbrace{\mathbf{B}}_{3\times8} \underbrace{\mathbf{u}_i}_{8\times1} \; ; \; \underbrace{\boldsymbol{\sigma}_{m_i}}_{3\times1} = c_m \underbrace{\boldsymbol{\sigma}_i}_{3\times1} = c_m \underbrace{\mathbf{D}}_{3\times3} \underbrace{\mathbf{B}}_{3\times8} \underbrace{\mathbf{u}_i}_{8\times1} \tag{12}$$

where $\mathbf{B}$ denotes the strain-displacement matrix of the element centroid. $\mathbf{u}_i$ represents the nodal displacement vector for the i-th element. $\mathbf{D}$ denotes the constitutive matrix of a fully solid material, which can be defined in a plane stress problem as:

$$\mathbf{D} = \frac{E_0}{1-\nu^2} \begin{bmatrix} 1 & \nu & 0 \\ \nu & 1 & 0 \\ 0 & 0 & \frac{1-\nu}{2} \end{bmatrix} \tag{13}$$

The von Mises stress evaluation can be employed to calculate an elemental alternating and mean stress [43]:

$$\sigma_{a_i,vm} = \left(\boldsymbol{\sigma}_{a_i}^T \cdot \mathbf{V} \cdot \boldsymbol{\sigma}_{a_i}\right)^{1/2} = \sqrt{\sigma_{a_{i,x}}^2 + \sigma_{a_{i,y}}^2 - \sigma_{a_{i,x}} \cdot \sigma_{a_{i,y}} + 3 \cdot \tau_{a_{i,xy}}^2}$$

$$\sigma_{m_i,vm} = \left(\boldsymbol{\sigma}_{m_i}^T \cdot \mathbf{V} \cdot \boldsymbol{\sigma}_{m_i}\right)^{1/2} = \begin{cases} \sqrt{\sigma_{m_{i,x}}^2 + \sigma_{m_{i,y}}^2 - \sigma_{m_{i,x}} \cdot \sigma_{m_{i,y}} + 3 \cdot \tau_{m_{i,xy}}^2}, & \sigma_{m_{i,x}} \geq 0 \\ -\sqrt{\sigma_{m_{i,x}}^2 + \sigma_{m_{i,y}}^2 - \sigma_{m_{i,x}} \cdot \sigma_{m_{i,y}} + 3 \cdot \tau_{m_{i,xy}}^2}, & \sigma_{m_{i,x}} < 0 \end{cases} \tag{14}$$

To avoid the nonlinear behavior and singularity phenomenon of an optimization problem, the q-p relaxation approach is employed to deal with the element stress [40,41]:

$$\begin{cases} \hat{\sigma}_{a_i,vm} = \varphi\left(\tilde{\tilde{\rho}}_i\right) \sigma_{a_i,vm} \\ \hat{\sigma}_{m_i,vm} = \varphi\left(\tilde{\tilde{\rho}}_i\right) \sigma_{m_i,vm} \end{cases}, \varphi\left(\tilde{\tilde{\rho}}_i\right) = \tilde{\tilde{\rho}}_i^q \tag{15}$$

where q is the stress relaxation coefficient [41].

The fatigue failure criterion can be evaluated using a modified Goodman diagram, as shown in Figure 3a. The alternating stress, σ_a, is constrained by the fatigue stress, σ_{N_f}, for infinite life cycles. σ_m, σ_y, and σ_{ut} indicate the mean stress, the yielding stress, and the ultimate stress, respectively.

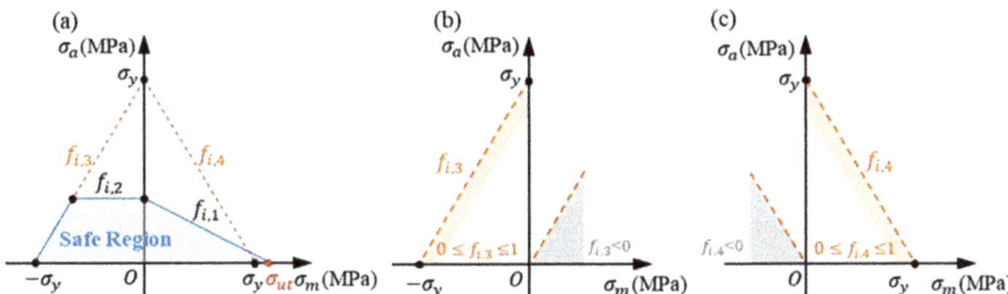

Figure 3. Modified Goodman diagram fatigue criterion: (**a**) modified Goodman diagram; the static failure envelopes of the present optimization problem (**b**) $f_{i,3}$ and (**c**) $f_{i,4}$.

The fatigue failure constraints of the compliant mechanism are transformed into two stress restrictions for analyzing fatigue failure [42]:

$$f_{i,1} = \frac{\hat{\sigma}_{a_i,vm}}{\sigma_{N_f}} + \frac{\max[\hat{\sigma}_{m_i,vm}, 0]}{\sigma_{ut}} \leq 1, \hat{\sigma}_{m_i,vm} \geq 0, \; f_{i,2} = \frac{\hat{\sigma}_{a_i,vm}}{\sigma_{N_f}} \leq 1, \hat{\sigma}_{m_i,vm} < 0 \qquad (16)$$

where σ_{N_f} denotes an association with a fatigue limit of the elements for allowable life cycles, which can be defined based on Basquin's equation as:

$$\sigma_{N_f} = \sigma'_f \left(2N_f\right)^{b_f} \qquad (17)$$

where σ'_f represents the fatigue strength coefficient, b_f represents the fatigue strength exponent (readers can refer to reference [43]).

As shown in Figure 3b,c, to prevent static failure, the maximum absolute value of the sum of the alternating stress and the mean stress should be less than the yield strength at the third constraint, $L_{i,3}$, and fourth constraint, $L_{i,4}$. Thus, $L_{i,3}$ and $L_{i,4}$ can be expressed as:

$$f_{i,3} = \frac{\hat{\sigma}_{a_i,vm} - \hat{\sigma}_{m_i,vm}}{\sigma_y} \leq 1; \; f_{i,4} = \frac{\hat{\sigma}_{a_i,vm} + \hat{\sigma}_{m_i,vm}}{\sigma_y} \leq 1 \qquad (18)$$

Determining whether the stress state is in compression or tension using mean stress measures, like von Mises equivalent stress, is challenging because it requires this measurement to always be positive. However, the static failure criteria of $L_{i,3}$ and $L_{i,4}$ are not always positive. Thus, this study employs a maximum operator with zero value and element-level static failure to address the above issues:

$$f_{i,3} = \max\left[\frac{\hat{\sigma}_{a_i,vm} - \hat{\sigma}_{m_i,vm}}{\sigma_y}, 0\right]; f_{i,4} = \max\left[\frac{\hat{\sigma}_{a_i,vm} + \hat{\sigma}_{m_i,vm}}{\sigma_y}, 0\right] \qquad (19)$$

where $L_{i,3}$ and $L_{i,4}$ will be positive and can be employed in a P-norm approach. This research introduces maximum operators [42], which non-differentiable operators approach employ in static failure criteria. Thus, a and b may be expressed as:

$$\Psi_{\max}(a,b) = \frac{a+b}{2} + \frac{\sqrt{(a-b)^2 + \varepsilon}}{2} \approx \begin{cases} a, a \geq b \\ b, a < b \end{cases} \qquad (20)$$

where ε is a small positive value [42].

In order to avoid the massive computation for the stress constraint for each element, we employed the P-norm approach to measure global stress [38,41,42]:

$$\hat{f}_{pn_{i,j}} = \left(\sum_{i=1}^{N} f_{i,j}^{P}\right)^{1/P}; j = 1, 2, \cdots, 4 \qquad (21)$$

where P is the P-norm aggregation parameter.

In order to more closely relate the P-norm stress to the practical stress, this study introduces a normalized adaptive constraint scaling method to modify the p-norm stress:

$$f_{i,j,max} \approx f_{pn_{i,j}} = c_k \hat{f}_{pn_{i,j}} \tag{22}$$

where c_k is the standardized adaptive constraint scaling coefficient computed at each optimization iteration, $k \geq 1$. The iteration steps, $k \geq 1$, c_k, can be expressed as:

$$c_k = \alpha_c \frac{f_{i,j,max}^{(k-1)}}{\hat{f}_{pn_{i,j}}^{(k-1)}} + (1 - \alpha_c)c_{k-1}, \kappa \geq 1 \tag{23}$$

where $\alpha_k \in (0, 1]$ is a control parameter [41].

3. The Optimization Problem Statement

This study proposes a novel topology optimization framework utilizing modified Goodman's fatigue and static failure criteria, along with a three-field density approach to account for manufacturing uncertainty. Consequently, an enhanced and robust topology optimization strategy for compliant mechanism modeling, grounded in both fatigue and static considerations, can be expressed as follows:

$$\min_{\rho} : f_{obj}\left(\widetilde{\rho}_i^{(e)}, \widetilde{\rho}_i^{(i)}, \widetilde{\rho}_i^{(d)}\right) = max\left\{ u_{out}^{(\widetilde{\rho}_i^{(e)}, \widetilde{\rho}_i^{(i)}, \widetilde{\rho}_i^{(d)})} = \mathbf{L}_{out}^T\left(\widetilde{\rho}_i^{(e)}, \widetilde{\rho}_i^{(i)}, \widetilde{\rho}_i^{(d)}\right)\mathbf{U}_{out}\left(\widetilde{\rho}_i^{(e)}, \widetilde{\rho}_i^{(i)}, \widetilde{\rho}_i^{(d)}\right) + 100 \right\}$$

$$s.t. : \frac{V}{V^*} = \frac{\Sigma_{i=N_i}\widetilde{\rho}_i^{(e)} v_i + \Sigma_{i=N_i}\widetilde{\rho}_i^{(i)} v_i + \Sigma_{i=N_i}\widetilde{\rho}_i^{(d)} v_i}{3V^*} \leq \delta$$

$$f_{i,1}^{(\widetilde{\rho}_i^{(e)}, \widetilde{\rho}_i^{(i)}, \widetilde{\rho}_i^{(d)})} = \frac{\hat{\sigma}_{a_{i,vm}}}{\sigma_{N_f}} + \frac{max\left[\hat{\sigma}_{m_{i,vm}}, 0\right]}{\sigma_{ut}} \leq 1, \hat{\sigma}_{m_{i,vm}} \geq 0$$

$$f_{i,2}^{(\widetilde{\rho}_i^{(e)}, \widetilde{\rho}_i^{(i)}, \widetilde{\rho}_i^{(d)})} = \frac{\hat{\sigma}_{a_{i,vm}}}{\sigma_{N_f}} \leq 1, \hat{\sigma}_{m_{i,vm}} < 0 \tag{24}$$

$$f_{i,3}^{(\widetilde{\rho}_i^{(e)}, \widetilde{\rho}_i^{(i)}, \widetilde{\rho}_i^{(d)})} = \frac{\hat{\sigma}_{a_{i,vm}} - \hat{\sigma}_{m_{i,vm}}}{\sigma_y} \leq 1$$

$$f_{i,4}^{(\widetilde{\rho}_i^{(e)}, \widetilde{\rho}_i^{(i)}, \widetilde{\rho}_i^{(d)})} = \frac{\hat{\sigma}_{a_{i,vm}} + \hat{\sigma}_{m_{i,vm}}}{\sigma_y} \leq 1$$

$$\mathbf{K}\left(\widetilde{\rho}_i^{(e)}\right)\mathbf{U}\left(\widetilde{\rho}_i^{(e)}\right) = \mathbf{F}; \mathbf{K}\left(\widetilde{\rho}_i^{(i)}\right)\mathbf{U}\left(\widetilde{\rho}_i^{(i)}\right) = \mathbf{F}; \mathbf{K}\left(\widetilde{\rho}_i^{(d)}\right)\mathbf{U}\left(\widetilde{\rho}_i^{(d)}\right) = \mathbf{F}$$

$$0 \leq \rho_{min} \leq \rho_i \leq 1; i = 1, 2, \cdots, N$$

where the practical and allowable volumes are denoted by V and V^*, respectively. N represents the total number of elements. The density fields, $\widetilde{\rho}_i^{(e)}$, $\widetilde{\rho}_i^{(i)}$, and $\widetilde{\rho}_i^{(d)}$, are obtained using different threshold values, η_e, η_i, and η_d, as prescribed by Equation (2). Specifically, the intermediate density, eroded density, and dilated density fields are denoted by $\widetilde{\rho}_i^{(e)}$, $\widetilde{\rho}_i^{(i)}$, and $\widetilde{\rho}_i^{(d)}$, respectively, with threshold values of $\eta_e = 1 - \eta$, $\eta_i = 0.5$, and $\eta_d = \eta$, respectively, where $\eta < 0.5 < 1 - \eta$.

With an improved robust topology optimization for compliant mechanism formulation, the relationship between the a manufacturing error b and the threshold parameter η_e for a given filter radius R can be obtained as [58,59]:

$$\eta_e = \begin{cases} \frac{1}{4}\left(\frac{b}{R}\right)^2 + \frac{1}{2}\frac{b}{R}, \frac{b}{R} \in [0, 1] \\ -\frac{1}{4}\left(\frac{b}{R}\right)^2 + \frac{b}{R}, \frac{b}{R} \in [1, 2] \\ 1, \frac{b}{R} \in [2, \infty) \end{cases} \tag{25}$$

Moreover, the normalized length scale on the intermediate design as η function of the threshold a for the robust formulation is shown in Figure 4.

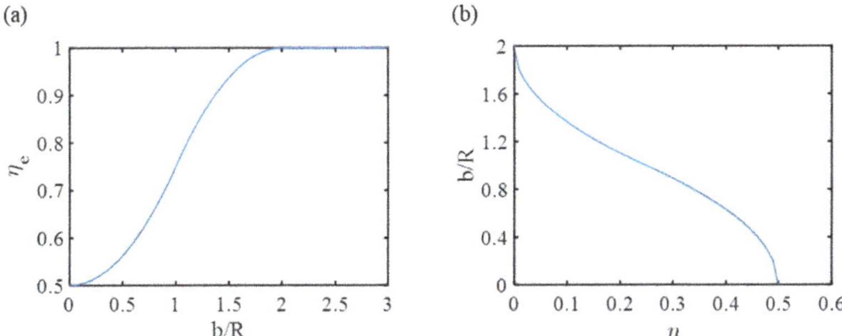

Figure 4. Normalized length scale on the intermediate design as η function of the threshold a for the robust formulation: (**a**) η_e versus b/R curve useful for design, and (**b**) the η versus b/R curve useful for design.

4. Sensitivity Analysis

The optimization problem with multiple constraintsis considered. To efficiently solve the optimization problem using gradient-based topology optimization algorithms, the first-order derivatives of the design objectives and all constraints with respect to the design variables, $\widetilde{\rho}_i^{(e)}$, $\widetilde{\rho}_i^{(i)}$, and $\widetilde{\rho}_i^{(d)}$.

4.1. Sensitivity of the Optimization Objective

To obtain the sensitivity of the objective function, we can rewrite the function by adding a zero function based on the adjoint method, which can be expressed as:

$$u_{out} = \mathbf{L}_{out}^T \mathbf{U}_{out} - \lambda^T (\mathbf{K}\mathbf{U}_{out} - \mathbf{F}_{in}) \tag{26}$$

Among them, λ is the arbitrary adjoint vector; thus, the output displacement, u_{out}, may be described using the element variable:

$$\frac{\partial u_{out}}{\partial \widetilde{\rho}_i} = \mathbf{L}_{out}^T \frac{\partial \mathbf{U}_{out}}{\partial \widetilde{\rho}_i} - \frac{\partial \lambda^T}{\partial \widetilde{\rho}_i} \mathbf{K}\mathbf{U}_{out} + \frac{\partial \lambda^T}{\partial \widetilde{\rho}_i} \mathbf{F}_{in} - \lambda^T \frac{\partial \mathbf{K}}{\partial \widetilde{\rho}_i} \mathbf{U}_{out} - \lambda^T \mathbf{K} \frac{\partial \mathbf{U}_{out}}{\partial \widetilde{\rho}_i} = \left(\mathbf{L}_{out}^T - \lambda^T \mathbf{K}\right) \frac{\partial \mathbf{U}_{out}}{\partial \widetilde{\rho}_i} - \lambda^T \frac{\partial \mathbf{K}}{\partial \widetilde{\rho}_i} \mathbf{U}_{out} -$$
$$\frac{\partial \lambda^T}{\partial \widetilde{\rho}_i}(\mathbf{K}\mathbf{U}_{out} - \mathbf{F}_{in}) = \left(\mathbf{L}_{out}^T - \lambda^T \mathbf{K}\right) \frac{\partial \mathbf{U}_{out}}{\partial \widetilde{\rho}_i} - \lambda^T \frac{\partial \mathbf{K}}{\partial \widetilde{\rho}_i} \mathbf{U}_{out} \tag{27}$$

Since $\mathbf{L}_{out}^T$ is an arbitrary vector, we can set $\mathbf{L}_{out}^T - \lambda^T \mathbf{K}$; thus, we can obtain the sensitivity of the output displacement, u_{out}, which can be written as:

$$\frac{\partial u_{out}}{\partial \widetilde{\rho}_i} = -\lambda^T \frac{\partial \mathbf{K}}{\partial \widetilde{\rho}_i} \mathbf{U}_{out} \tag{28}$$

4.2. Sensitivity of the Static Strength and Fatigue Failure

According to Equation (7), the sensitivity of the fatigue and static failure with modified P-norm fatigue criteria in Equation (22) can be derived:

$$\frac{\partial f_{pn,i,j}}{\partial \widetilde{\rho}_i} = \sum_{i=1}^{N} \frac{\partial f_{pn,i,j}}{\partial f_{i,j}} \frac{\partial f_{i,j}}{\partial \widetilde{\rho}_i}, j = 1, 2 \ldots 4 \tag{29}$$

The derivatives of $\frac{\partial f_{pn,i}}{\partial L_{i,j}}$ may be written as:

$$\frac{\partial f_{pn,i,j}}{\partial f_{i,j}} = \frac{c_{k,j}\left[\sum_{i=1}^{N} f_{i,j}^{P}\right]^{\frac{1}{P}-1}}{p}(f_{i,j})^{P-1}, j = 1,2\ldots4 \tag{30}$$

Among them, the derivative of the term $\frac{\partial f_{i,j}}{\partial \widetilde{\rho}_i}$ can be written as:

$$\begin{aligned}
\frac{\partial f_{i,1}}{\partial \widetilde{\rho}_i} &= \frac{\partial f_{i,1}}{\partial \hat{\sigma}_{a_i,vm}}\frac{\partial \hat{\sigma}_{a_i,vm}}{\partial \widetilde{\rho}_i} + \frac{\partial f_{i,1}}{\partial \hat{\sigma}_{m_i,vm}}\frac{\partial \hat{\sigma}_{m_i,vm}}{\partial \widetilde{\rho}_i} \\[4pt]
\frac{\partial f_{i,2}}{\partial \widetilde{\rho}_i} &= \frac{\partial f_{i,2}}{\partial \hat{\sigma}_{a_i,vm}}\frac{\partial \hat{\sigma}_{a_i,vm}}{\partial \widetilde{\rho}_i} \\[4pt]
\frac{\partial f_{i,3}}{\partial \widetilde{\rho}_i} &= \frac{\partial f_{i,3}}{\partial \hat{\sigma}_{a_i,vm}}\frac{\partial \hat{\sigma}_{a_i,vm}}{\partial \widetilde{\rho}_i} - \frac{\partial f_{i,3}}{\partial \hat{\sigma}_{m_i,vm}}\frac{\partial \hat{\sigma}_{m_i,vm}}{\partial \widetilde{\rho}_i} \\[4pt]
\frac{\partial f_{i,4}}{\partial \widetilde{\rho}_i} &= \frac{\partial f_{i,4}}{\partial \hat{\sigma}_{a_i,vm}}\frac{\partial \hat{\sigma}_{a_i,vm}}{\partial \widetilde{\rho}_i} + \frac{\partial f_{i,4}}{\partial \hat{\sigma}_{m_i,vm}}\frac{\partial \hat{\sigma}_{m_i,vm}}{\partial \widetilde{\rho}_i}
\end{aligned} \tag{31}$$

Meanwhile, the derivative of the term $\frac{\partial f_{i,j}}{\partial \hat{\sigma}_{a_i,vm}}$ and $\frac{\partial f_{i,j}}{\partial \hat{\sigma}_{m_i,vm}}, j \neq 2$, can be expressed as:

$$\frac{\partial f_{i,1}}{\partial \hat{\sigma}_{a_i,vm}} = \frac{1}{\sigma_{N_f}}; \frac{\partial f_{i,1}}{\partial \hat{\sigma}_{m_i,vm}} = \frac{1}{\sigma_{ut}}$$

$$\frac{\partial f_{i,2}}{\partial \hat{\sigma}_{a_i,vm}} = \frac{1}{\sigma_{N_f}}$$

$$\frac{\partial f_{i,3}}{\partial \hat{\sigma}_{a_i,vm}} = \frac{1}{2\sigma_y} + \frac{1}{4\sigma_y}\left[\frac{\frac{\hat{\sigma}_{a_i,vm}-\hat{\sigma}_{m_i,vm}}{\sigma_y}}{\sqrt{\left(\frac{\hat{\sigma}_{a_i,vm}-\hat{\sigma}_{m_i,vm}}{\sigma_y}\right)^2+\varepsilon}}\right]; \frac{\partial f_{i,3}}{\partial \hat{\sigma}_{m_i,vm}} = -\frac{1}{2\sigma_y} - \frac{1}{4\sigma_y}\left[\frac{\frac{\hat{\sigma}_{a_i,vm}-\hat{\sigma}_{m_i,vm}}{\sigma_y}}{\sqrt{\left(\frac{\hat{\sigma}_{a_i,vm}-\hat{\sigma}_{m_i,vm}}{\sigma_y}\right)^2+\varepsilon}}\right]$$

$$\frac{\partial f_{i,4}}{\partial \hat{\sigma}_{a_i,vm}} = \frac{1}{2\sigma_y} + \frac{1}{4\sigma_y}\left[\frac{\frac{\hat{\sigma}_{a_i,vm}+\hat{\sigma}_{m_i,vm}}{\sigma_y}}{\sqrt{\left(\frac{\hat{\sigma}_{a_i,vm}+\hat{\sigma}_{m_i,vm}}{\sigma_y}\right)^2+\varepsilon}}\right]; \frac{\partial f_{i,4}}{\partial \hat{\sigma}_{m_i,vm}} = \frac{1}{2\sigma_y} + \frac{1}{4\sigma_y}\left[\frac{\frac{\hat{\sigma}_{a_i,vm}+\hat{\sigma}_{m_i,vm}}{\sigma_y}}{\sqrt{\left(\frac{\hat{\sigma}_{a_i,vm}+\hat{\sigma}_{m_i,vm}}{\sigma_y}\right)^2+\varepsilon}}\right] \tag{32}$$

and

$$\frac{\partial \hat{\sigma}_{a_i,vm}}{\partial \widetilde{\rho}_i} = \frac{\partial \hat{\sigma}_{a_i,vm}}{\partial \sigma_{a_i,vm}}\left(\frac{\sigma_{a_i,vm}}{\partial \sigma_{a_i}}\right)^T\frac{\partial \sigma_{a_i}}{\partial \widetilde{\rho}_i} + \frac{\hat{\sigma}_{a_i,vm}}{\partial \varphi}\frac{\partial \varphi}{\partial \widetilde{\rho}_i}; \frac{\partial \hat{\sigma}_{m_i,vm}}{\partial \widetilde{\rho}_i} = \frac{\partial \hat{\sigma}_{m_i,vm}}{\partial \sigma_{m_i,vm}}\left(\frac{\sigma_{m_i,vm}}{\partial \sigma_{m_i}}\right)^T\frac{\partial \sigma_{m_i}}{\partial \widetilde{\rho}_i} + \frac{\hat{\sigma}_{m_i,vm}}{\partial \varphi}\frac{\partial \varphi}{\partial \widetilde{\rho}_i} \tag{33}$$

The derivatives of von Mises alternating and mean stresses in Equation (34) for the two-dimensional plane stress situation can be described as:

$$\frac{\partial \sigma_{a_i,vm}}{\partial \sigma_{a_i}} = \begin{bmatrix}\frac{\partial \sigma_{a_i,vm}}{\partial \sigma_{a_i,x}} \\ \frac{\partial \sigma_{a_i,vm}}{\partial \sigma_{a_i,y}} \\ \frac{\partial \sigma_{a_i,vm}}{\partial \tau_{a_i,xy}}\end{bmatrix} = \begin{bmatrix}\frac{1}{2\sigma_{a_i,vm}}\left(2\sigma_{a_i,x} - \sigma_{a_i,y}\right) \\ \frac{1}{2\sigma_{a_i,vm}}\left(2\sigma_{a_i,y} - \sigma_{a_i,x}\right) \\ \frac{3}{\sigma_{a_i,vm}}\left(\tau_{a_i,xy}\right)\end{bmatrix}; \frac{\partial \sigma_{m_i,vm}}{\partial \sigma_{m_i}} = \begin{bmatrix}\frac{\partial \sigma_{m_i,vm}}{\partial \sigma_{m_i,x}} \\ \frac{\partial \sigma_{m_i,vm}}{\partial \sigma_{m_i,y}} \\ \frac{\partial \sigma_{m_i,vm}}{\partial \tau_{m_i,xy}}\end{bmatrix} = \begin{bmatrix}\frac{1}{2\sigma_{m_i,vm}}\left(2\sigma_{m_i,x} - \sigma_{m_i,y}\right) \\ \frac{1}{2\sigma_{m_i,vm}}\left(2\sigma_{m_i,y} - \sigma_{m_i,x}\right) \\ \frac{3}{\sigma_{m_i,vm}}\left(\tau_{m_i,xy}\right)\end{bmatrix} \tag{34}$$

The derivation of stress vector, $\hat{\sigma}_i$, with respect to the design variables, $\widetilde{\rho}_i$, can be written as:

$$\frac{\partial \sigma_{a_i}}{\partial \widetilde{\rho}_i} = c_a \cdot \mathbf{D} \cdot \mathbf{B} \cdot \mathbf{L}_i \cdot \frac{\partial \mathbf{U}_i}{\partial \widetilde{\rho}_i}; \frac{\partial \sigma_{m_i}}{\partial \widetilde{\rho}_i} = c_m \cdot \mathbf{D} \cdot \mathbf{B} \cdot \mathbf{L}_i \cdot \frac{\partial \mathbf{U}_i}{\partial \widetilde{\rho}_i}, \frac{\partial \mathbf{u}_i}{\partial \widetilde{\rho}_i} = \mathbf{L}_i \cdot \frac{\partial \mathbf{U}_i}{\partial \widetilde{\rho}_i} \tag{35}$$

where $\mathbf{L}_i$ represents the global virtual unit load vector, denoted by $[0,0,\cdots,1,\cdots,0]^T$, the i-th component value is 1 at the corresponding input point, O_i, and all other component values are 0. The concomitant variable can expressed as $\mathbf{u}_i = \mathbf{L}_i\mathbf{U}$.

Meanwhile, the adjoint method is employed to solve $\frac{\partial \mathbf{u}_i}{\partial \tilde{\rho}_i}$, which can be calculated using $\frac{\partial \mathbf{K}}{\partial \tilde{\rho}_i}\mathbf{U} + \mathbf{K}\frac{\partial \mathbf{U}}{\partial \tilde{\rho}_i} = 0$; thus, the following can be expressed:

$$
\begin{aligned}
\frac{\partial f_{pn,i,1}}{\partial \tilde{\rho}_i} &= \frac{\partial f_{pn,i,1}}{\partial f_{i,1}} \left\{ \begin{array}{l} \frac{\partial f_{i,1}}{\partial \tilde{\sigma}_{a_i,vm}} \left[-\tilde{\rho}_i^{\,q} \left(\frac{\sigma_{a_i,vm}}{\partial \sigma_{a_i}} \right)^T c_a \cdot \mathbf{D} \cdot \mathbf{B} \cdot \mathbf{L}_i \cdot \mathbf{K}^{-1} \frac{\partial \mathbf{K}}{\partial \tilde{\rho}_i} \mathbf{U}_i + \sigma_{a_i,vm} q \tilde{\rho}_i^{\,q-1} \right] \\[2ex] + \frac{\partial f_{i,1}}{\partial \tilde{\sigma}_{m_i,vm}} \left[-\tilde{\rho}_i^{\,q} \left(\frac{\sigma_{m_i,vm}}{\partial \sigma_{m_i}} \right)^T c_m \cdot \mathbf{D} \cdot \mathbf{B} \cdot \mathbf{L}_i \cdot \mathbf{K}^{-1} \frac{\partial \mathbf{K}}{\partial \tilde{\rho}_i} \mathbf{U}_i + \sigma_{m_i,vm} q \tilde{\rho}_i^{\,q-1} \right] \end{array} \right\} \\[3ex]
\frac{\partial f_{pn,i,2}}{\partial \tilde{\rho}_i} &= \frac{\partial f_{pn,i,2}}{\partial f_{i,2}} \left\{ \frac{\partial f_{i,2}}{\partial \tilde{\sigma}_{a_i,vm}} \left[-\tilde{\rho}_i^{\,q} \left(\frac{\sigma_{a_i,vm}}{\partial \sigma_{a_i}} \right)^T c_a \cdot \mathbf{D} \cdot \mathbf{B} \cdot \mathbf{L}_i \cdot \mathbf{K}^{-1} \frac{\partial \mathbf{K}}{\partial \tilde{\rho}_i} \mathbf{U}_i + \sigma_{a_i,vm} q \tilde{\rho}_i^{\,q-1} \right] \right\} \\[3ex]
\frac{\partial f_{pn,i,3}}{\partial \tilde{\rho}_i} &= \frac{\partial f_{pn,i,3}}{\partial f_{i,3}} \left\{ \begin{array}{l} \frac{\partial f_{i,3}}{\partial \tilde{\sigma}_{a_i,vm}} \left[-\tilde{\rho}_i^{\,q} \left(\frac{\sigma_{a_i,vm}}{\partial \sigma_{a_i}} \right)^T c_a \cdot \mathbf{D} \cdot \mathbf{B} \cdot \mathbf{L}_i \cdot \mathbf{K}^{-1} \frac{\partial \mathbf{K}}{\partial \tilde{\rho}_i} \mathbf{U}_i + \sigma_{a_i,vm} q \tilde{\rho}_i^{\,q-1} \right] \\[2ex] - \frac{\partial f_{i,3}}{\partial \tilde{\sigma}_{m_i,vm}} \left[-\tilde{\rho}_i^{\,q} \left(\frac{\sigma_{m_i,vm}}{\partial \sigma_{m_i}} \right)^T c_m \cdot \mathbf{D} \cdot \mathbf{B} \cdot \mathbf{L}_i \cdot \mathbf{K}^{-1} \frac{\partial \mathbf{K}}{\partial \tilde{\rho}_i} \mathbf{U}_i + \sigma_{m_i,vm} q \tilde{\rho}_i^{\,q-1} \right] \end{array} \right\} \\[3ex]
\frac{\partial f_{pn,i,4}}{\partial \tilde{\rho}_i} &= \frac{\partial f_{pn,i,4}}{\partial f_{i,4}} \left\{ \begin{array}{l} \frac{\partial f_{i,4}}{\partial \tilde{\sigma}_{a_i,vm}} \left[-\tilde{\rho}_i^{\,q} \left(\frac{\sigma_{a_i,vm}}{\partial \sigma_{a_i}} \right)^T c_a \cdot \mathbf{D} \cdot \mathbf{B} \cdot \mathbf{L}_i \cdot \mathbf{K}^{-1} \frac{\partial \mathbf{K}}{\partial \tilde{\rho}_i} \mathbf{U}_i + \sigma_{a_i,vm} q \tilde{\rho}_i^{\,q-1} \right] \\[2ex] + \frac{\partial f_{i,4}}{\partial \tilde{\sigma}_{m_i,vm}} \left[-\tilde{\rho}_i^{\,q} \left(\frac{\sigma_{m_i,vm}}{\partial \sigma_{m_i}} \right)^T c_m \cdot \mathbf{D} \cdot \mathbf{B} \cdot \mathbf{L}_i \cdot \mathbf{K}^{-1} \frac{\partial \mathbf{K}}{\partial \tilde{\rho}_i} \mathbf{U}_i + \sigma_{m_i,vm} q \tilde{\rho}_i^{\,q-1} \right] \end{array} \right\}
\end{aligned} \tag{36}
$$

Equation (37) can be rewritten in the following form:

$$
\begin{aligned}
\frac{\partial f_{pn,i,1}}{\partial \tilde{\rho}_i} &= \frac{\partial f_{pn,i,1}}{\partial f_{i,1}} \left[\begin{array}{l} \left(Z_{a,1} + \frac{\partial f_{i,1}}{\partial \tilde{\sigma}_{a_i,vm}} \sigma_{a_i,vm} q \tilde{\rho}_i^{\,q-1} \right) \\[2ex] + \left(Z_{m,1} + \frac{\partial f_{i,1}}{\partial \tilde{\sigma}_{m_i,vm}} \sigma_{m_i,vm} q \tilde{\rho}_i^{\,q-1} \right) \end{array} \right] \\[3ex]
\frac{\partial f_{pn,i,2}}{\partial \tilde{\rho}_i} &= \frac{\partial f_{pn,i,2}}{\partial f_{i,2}} \left(Z_{a,2} + \frac{\partial f_{i,2}}{\partial \tilde{\sigma}_{a_i,vm}} \sigma_{a_i,vm} q \tilde{\rho}_i^{\,q-1} \right) \\[3ex]
\frac{\partial f_{pn,i,3}}{\partial \tilde{\rho}_i} &= \frac{\partial f_{pn,i,3}}{\partial f_{i,3}} \left[\begin{array}{l} \left(Z_{a,3} + \frac{\partial f_{i,3}}{\partial \tilde{\sigma}_{a_i,vm}} \sigma_{a_i,vm} q \tilde{\rho}_i^{\,q-1} \right) \\[2ex] - \left(Z_{m,3} + \frac{\partial f_{i,3}}{\partial \tilde{\sigma}_{m_i,vm}} \sigma_{m_i,vm} q \tilde{\rho}_i^{\,q-1} \right) \end{array} \right] \\[3ex]
\frac{\partial f_{pn,i,4}}{\partial \tilde{\rho}_i} &= \frac{\partial f_{pn,i,4}}{\partial f_{i,4}} \left[\begin{array}{l} \left(Z_{a,4} + \frac{\partial f_{i,4}}{\partial \tilde{\sigma}_{a_i,vm}} \sigma_{a_i,vm} q \tilde{\rho}_i^{\,q-1} \right) \\[2ex] + \left(Z_{m,4} + \frac{\partial f_{i,4}}{\partial \tilde{\sigma}_{m_i,vm}} \sigma_{m_i,vm} q \tilde{\rho}_i^{\,q-1} \right) \end{array} \right]
\end{aligned} \tag{37}
$$

In order to avoid the calculation of $\mathbf{K}^{-1}$, we defined the adjoint variables, $\mathcal{T}_a$ and $\mathcal{T}_m$, as follows:

$$
\begin{aligned}
\mathcal{T}_{a,j}^T &= \tilde{\rho}_i^{\,q} \frac{\partial f_{i,j}}{\partial \tilde{\sigma}_{a_i,vm}} \left(\frac{\sigma_{a_i,vm}}{\partial \sigma_{a_i}} \right)^T c_a \cdot \mathbf{D} \cdot \mathbf{B} \cdot \mathbf{L}_i \cdot \mathbf{K}^{-1}, j = 1,2,3,4 \\[2ex]
\mathcal{T}_{m,j\neq 2}^T &= \tilde{\rho}_i^{\,q} \frac{\partial f_{i,j}}{\partial \tilde{\sigma}_{m_i,vm}} \left(\frac{\sigma_{m_i,vm}}{\partial \sigma_{m_i}} \right)^T c_m \cdot \mathbf{D} \cdot \mathbf{B} \cdot \mathbf{L}_i \cdot \mathbf{K}^{-1}, j = 1,3,4
\end{aligned} \tag{38}
$$

where $\mathbf{K}\mathcal{T}_{a,j}$ and $\mathbf{K}\mathcal{T}_{m,j\neq 2}$ can be calculated as:

$$
\begin{aligned}
\mathbf{K}\mathcal{T}_{a,j} &= \tilde{\rho}_i^{\,q} \frac{\partial f_{i,j}}{\partial \tilde{\sigma}_{a_i,vm}} c_a \cdot (\mathbf{D} \cdot \mathbf{B} \cdot \mathbf{L}_i \cdot)^T \frac{\sigma_{a_i,vm}}{\partial \sigma_{a_i}} \\[2ex]
\mathbf{K}\mathcal{T}_{m,j\neq 2} &= \tilde{\rho}_i^{\,q} \frac{\partial f_{i,j}}{\partial \tilde{\sigma}_{m_i,vm}} c_m \cdot (\mathbf{D} \cdot \mathbf{B} \cdot \mathbf{L}_i \cdot)^T \frac{\sigma_{m_i,vm}}{\partial \sigma_{m_i}}
\end{aligned} \tag{39}
$$

Therefore, $Z_{a,j}$ and $Z_{m,j\neq 2}$ can be further simplified as:

$$Z_{a,j} = -\mathcal{T}_{a,j}^T \frac{\partial \mathbf{K}}{\partial \widetilde{\widetilde{\rho}}_i} \mathbf{U}_i; Z_{m,j\neq 2} = -\mathcal{T}_{m,j\neq 2}^T \frac{\partial \mathbf{K}}{\partial \widetilde{\widetilde{\rho}}_i} \mathbf{U}_i \qquad (40)$$

4.3. Sensitivity of the Volume

The derivative of volume constraint with respect to element design variables can be obtained:

$$\frac{\partial V}{\partial \widetilde{\widetilde{\rho}}_i} = \frac{\partial \left(\sum_{i\in N} \widetilde{\widetilde{\rho}}_i^{(e)} v_i + \sum_{i\in N} \widetilde{\widetilde{\rho}}_i^{(i)} v_i + \sum_{i\in N} \widetilde{\widetilde{\rho}}_i^{(d)} v_i \right)}{3\partial \widetilde{\widetilde{\rho}}_i} = v_i \qquad (41)$$

where the volume constraint is set by the mean of three designs.

5. Numerical Implementation

Based on the above analysis, an improved robust topology optimization formulation will be employed using eroded, intermediate, and dilated projections to optimize three different designs. The three designs will be developed using the threshold projection filter, which is briefly explained in Section 2.1. When the initial value of η is 0.5, densities below the threshold are projected to 0, while those above are projected to 1, utilizing the full potential of the threshold projection filter. By varying the threshold parameter η, the occurrence of manufacturing errors can be simulated. Increasing the value of η results in an eroded design, as more densities are projected to 0, while a decrease represents a dilated design.

The proposed topology optimization methodology for the compliant mechanism is illustrated in Figure 5 using MATLAB R2020a software. The optimization process starts by defining the initial design domain, boundary conditions, material parameters, algorithm parameters, and pre-FEA operations. Next, the objective and constraint functions are established, and a density filter is employed to avoid the occurrence of checkerboard patterns and mesh dependency issues. A robust topology optimization formulation, incorporating erosion, intermediate, and dilation projections, is then implemented to ensure that minimum size [49]. In addition, the objective and constraint functions are proposed in this research. The GCMMA method is employed to update design variables until convergence [57]. The optimization process terminates when either of the following two criteria is met: (1) the variation in design variables is less than 0.001 or (2) the current number of cycles reaches the predefined maximum number of steps, typically 350. The numerical parameters utilized in the topology optimization of the compliant mechanism are summarized in Table 1.

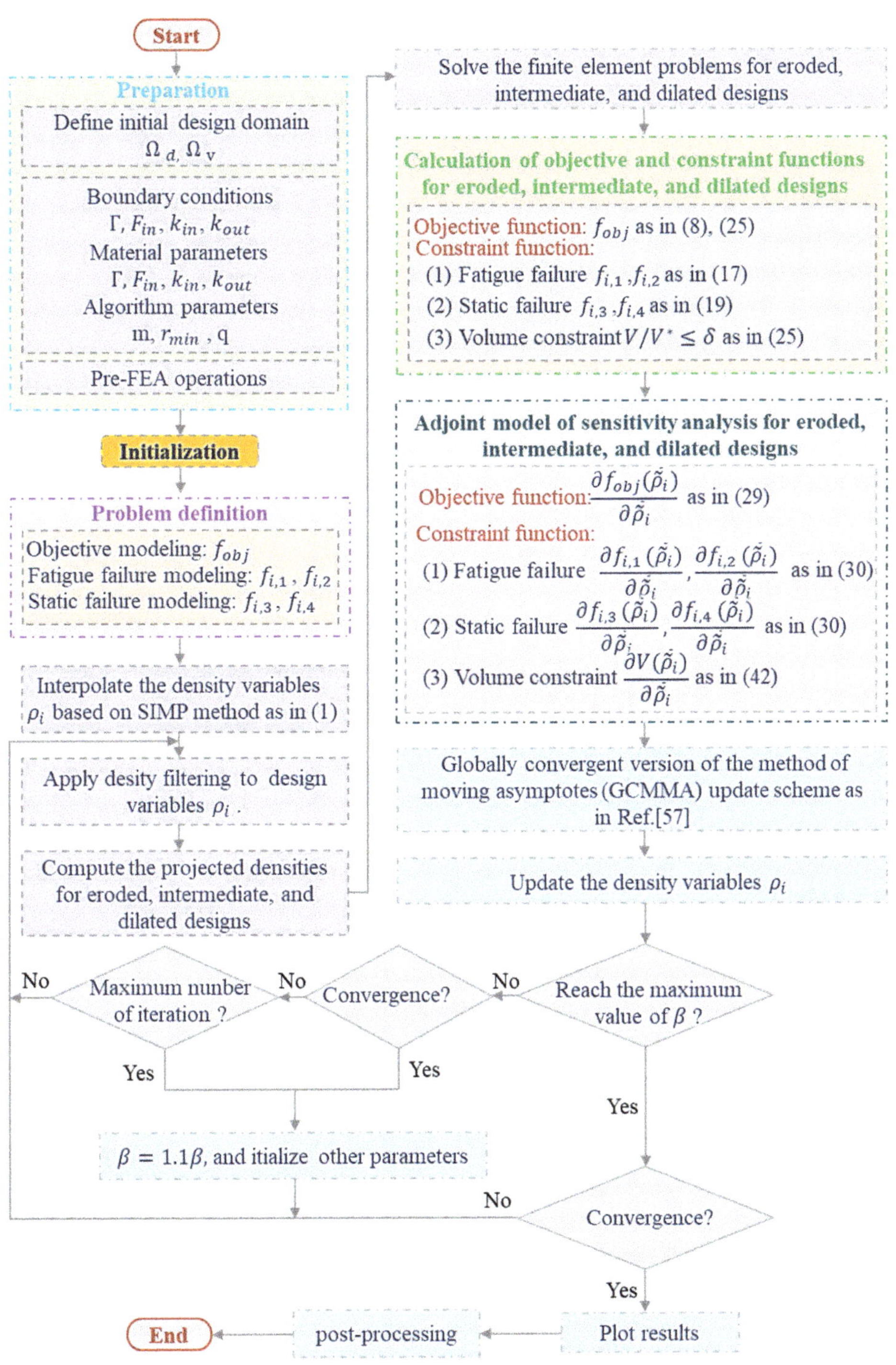

Figure 5. A flowchart of the topology optimization of compliant mechanism optimization algorithm.

Table 1. Parameters for topology optimization of compliant mechanisms.

Parameter	Symbol	Value	Parameter	Symbol	Value
Elastic modulus for solid element	E_0	68.9 GPa	Fatigue limit of the elements	σ_{N_f}	96.2 MPa
Elastic modulus for void element	E_{min}	10^{-9} MPa	Yielding stress	σ_y	276 MPa
Poisson's ratio	μ	0.33	Ultimate stress	σ_{ut}	310 MPa
Penalty parameter	p	3	Fatigue strength coefficient	σ_f'	658.75 MPa
Material density	ρ_0	2770 kg/m^3	Fatigue strength exponents	b_f	-0.1326
Volume fraction	δ	0.3	Allowable life cycles	N_f	10^6
Stress relaxation coefficient	q	1	Filter radius	R	$0.03 \times a$
Initial scaling coefficient	c_0	1	Small positive value	ε	10^{-5}
P-norm aggregation parameter	P	12	Maximum force	F_{max}	800 N
Control parameter	α_k	0.5	Minimum force	F_{min}	-200 N

6. Numerical Examples

In this section, the proposed method is tested using the force inverter and gripper problems, with a focus on maximizing the output displacement through optimization formulation. The material properties tested in these experiments are outlined in Table 1. For the force inverter problem, the input and output spring stiffness values are set to $k_{in} = 4$ N/μm and $k_{out} = 1$ N/μm, respectively. Meanwhile, for the gripper problem, the input and output spring stiffness values are set to $k_{in} = 5$ N/μm and $k_{out} = 2$ N/μm, respectively.

In order to ensure an optimal design without any risk of failure, it is necessary for the element's alternating and mean stresses to be located within the safe region of the modified Goodman diagram, as illustrated in Figure 2a. To test the layout optimization, three different combinations of alternating and mean stresses are evaluated:

$$
\begin{aligned}
&\text{von Mises :}
\begin{cases}
\hat{\sigma}_{a_{i,vm}} = \sqrt{\hat{\sigma}_{a_{i,x}}^2 + \hat{\sigma}_{a_{i,y}}^2 - \hat{\sigma}_{a_{i,x}}^2 \hat{\sigma}_{a_{i,y}}^2} \\
\hat{\sigma}_{m_{i,vm}} = \sqrt{\hat{\sigma}_{m_{i,x}}^2 + \hat{\sigma}_{m_{i,y}}^2 - \hat{\sigma}_{m_{i,x}}^2 \hat{\sigma}_{m_{i,y}}^2}
\end{cases} \\[6pt]
&\text{Sines theory :}
\begin{cases}
\hat{\sigma}_{a_{i,vm}} = \sqrt{\hat{\sigma}_{a_{i,x}}^2 + \hat{\sigma}_{a_{i,y}}^2 - \hat{\sigma}_{a_{i,x}} \hat{\sigma}_{a_{i,y}} + 3\hat{\tau}_{a_{i,xy}}^2} \\
\hat{\sigma}_{m_{i,vm}} = \hat{\sigma}_{m_{i,x}} + \hat{\sigma}_{m_{i,y}}
\end{cases} \\[6pt]
&\text{Signed von Mises :}
\begin{cases}
\hat{\sigma}_{a_{i,vm}} = \sqrt{\hat{\sigma}_{a_{i,x}}^2 + \hat{\sigma}_{a_{i,y}}^2 - \hat{\sigma}_{a_{i,x}} \hat{\sigma}_{a_{i,y}} + 3\hat{\tau}_{a_{i,xy}}^2} \\
\hat{\sigma}_{m_{i,vm}} =
\begin{cases}
\sqrt{\hat{\sigma}_{m_{i,x}}^2 + \hat{\sigma}_{m_{i,y}}^2 - \hat{\sigma}_{m_{i,x}} \hat{\sigma}_{m_{i,y}} + 3\hat{\tau}_{m_{i,xy}}^2}, & \hat{\sigma}_{m_{i,x}} \geq 0 \\
-\sqrt{\hat{\sigma}_{m_{i,x}}^2 + \hat{\sigma}_{m_{i,y}}^2 - \hat{\sigma}_{m_{i,x}} \hat{\sigma}_{m_{i,y}} + 3\hat{\tau}_{m_{i,xy}}^2}, & \hat{\sigma}_{m_{i,x}} < 0
\end{cases}
\end{cases}
\end{aligned}
\tag{42}
$$

6.1. Numerical Examples of the Inverter

The first example is the force inverter, which is depicted in Figure 6a. The design domain, Ω_d, has dimensions of $a \times a = 100 \times 100$ mm^2, and the filter radius is $R = 3$ mm. The smoothness parameter, β, is determined through a trial-and-error process and is set to 25. The eroded, dilated, and intermediate (real) designs have threshold values of $\eta_e = 0.7$, $\eta_d = 0.3$, and $\eta_i = 0.5$, respectively.

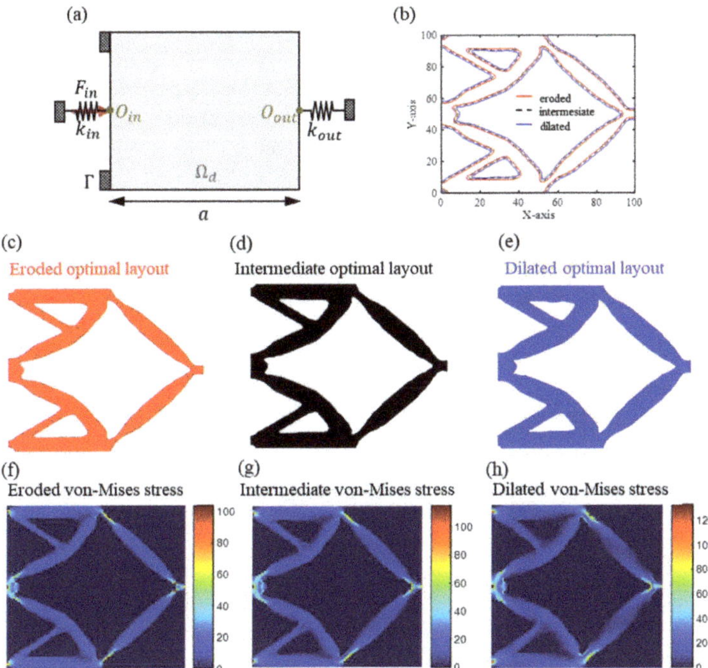

Figure 6. The force inverter. (**a**) The design domain; (**b**) the contours of different designs: eroded $\eta_e = 0.7$, intermediate $\eta_i = 0.5$, and dilated $\eta_d = 0.3$; optimized topologies for three levels of fatigue constraints (**c**) eroded designs, (**d**) intermediate designs, and (**e**) dilated designs; the von Mises stress distribution of (**f**) eroded, (**g**) intermediate, and (**h**) dilated designs.

Figure 6b displays the different thresholds for the contours of the different designs, and Figure 6c–e show the contours of different designs optimized topologies for three levels of fatigue and static failure constraints. The corresponding von Mises stress distributions of the eroded, intermediate, and dilated designs are shown in Figure 6f–h, respectively. The von Mises stress is estimated to be approximately 100 MPa. To prevent static failure, which is also referred to as one-time loading failure, the maximum absolute value of the sum of the alternating stress and the mean stress should be less than the yield strength at the second constraint. To ensure that the force inverter satisfies this constraint, the modified Goodman fatigue criteria, $L_{i,1}$ and $L_{i,2}$, and the static failure criteria, $L_{i,3}$ and $L_{i,4}$, are employed for the three-field density projection with eroded, intermediate, and dilated projections, as shown in Table 2.

Table 2. The resultant values of the force inverter.

Parameter	Symbol	Eroded	Intermediate	Dilated
Output displacement	u_{out}	99.8876	99.8850	99.8909
Amplification ratio	A_r	0.69	0.733	0.727
Volume fraction	V	0.252	0.302	0.346
Fatigue failure 1	$L_{i,1}$	0.84	0.89	1.26
Fatigue failure 2	$L_{i,2}$	0.678	0.75	0.865
Static failure 1	$L_{i,3}$	0.378	0.418	0.482
Static failure 2	$L_{i,4}$	0.378	0.418	0.482
Gray level indicator	Mnd	2.5%	1.5%	1.8%

To verify the results obtained from the MATLAB implementation, three different combinations were employed to evaluate the fatigue and static damage resistance of the optimal design. As illustrated in Figure 7a–c, the eroded, intermediate, and dilated designs of the Goodman fatigue criteria were tested under different conditions. In this Figure 7a–c, the color pink is utilized to represent the von Mises theory, blue is employed to signify the Sines theory, and green is used to denote the Sines von Mises theory. The outcomes demonstrate that the force inverter is capable of satisfying both the fatigue and static failure constraints, and all test conditions fall within the modified Goodman safety zone. Furthermore, it is evident that the fatigue life of the inverter is infinite when subjected to the designated load conditions during the design phase. It should be noted that the output displacement of the compliant mechanism may be reduced due to the limitations imposed via the partial constraints on fatigue and static failure for maximum output displacement. Meanwhile, the different designs for the different thresholds are displayed in Figure 8a–c.

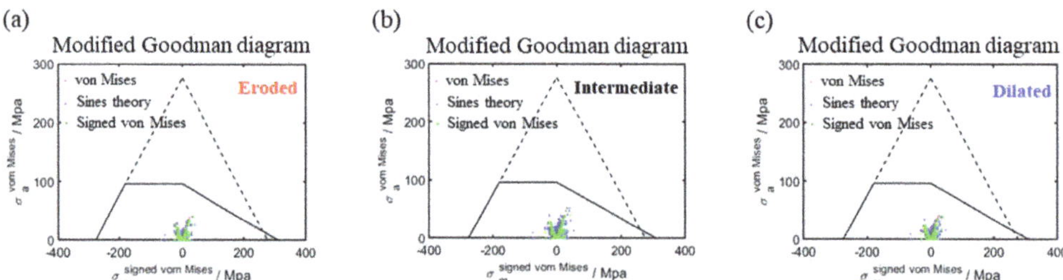

Figure 7. The eroded, intermediate, and dilated designs of the Goodman fatigue criteria of force inverter: (**a**) the eroded $\eta_e = 0.7$, (**b**) intermediate $\eta_i = 0.5$, and (**c**) dilated $\eta_d = 0.3$ designs of the threshold projection filters.

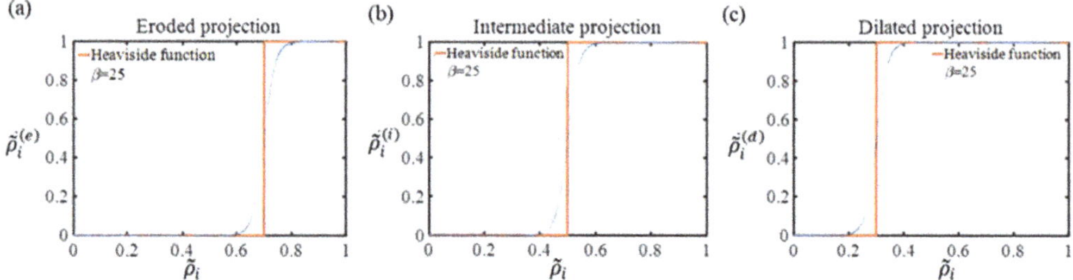

Figure 8. The threshold projection filters of force inverter: (**a**) the eroded $\eta_e = 0.7$, (**b**) intermediate $\eta_i = 0.5$, and (**c**) dilated $\eta_d = 0.3$ designs of the threshold projection filters.

To check the maximum output displacement of the optimal design, the deformation configurations of the obtained inverter are shown in Figure 9a–c, corresponding to the eroded, intermediate, and dilated deformation configurations of the inverter, respectively. We also presented the corresponding data for the inverter with three threshold projections, as shown in Table 2. The maximum output displacements of three threshold projection are 99.8876, 99.8850, and 99.8909, and the corresponding amplification ratios are 0.69, 0.733, and 0.727, respectively. It can be observed that the dilated value, $L_{i,1}$, is greater than 1 and is close to 1, which may be caused via numerical instability and may not affect our final design results. Meanwhile, the discreteness of the obtained design adds a gray level indicator, Mnd, measurer for gray level measurements. When every element exhibits intermediate density, or $\tilde{\tilde{\rho}}_i = 0.5$, the indicator's value is Mnd = 100%. The indicator's value is Mnd = 0% if all elements have densities of 0 or 1.

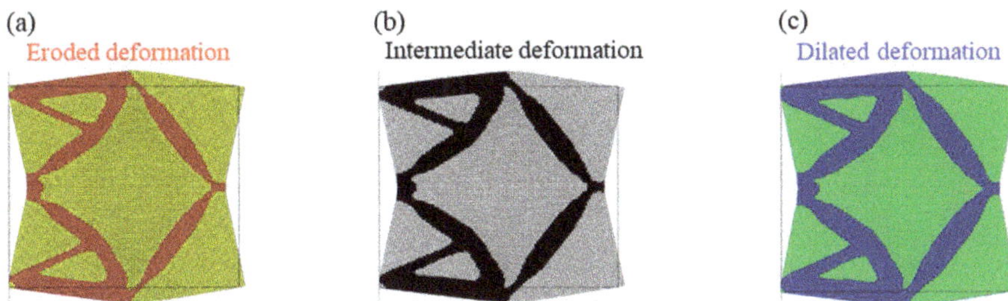

Figure 9. Deformation configuration of force inverter, (**a**) the eroded $\eta_e = 0.7$, (**b**) intermediate $\eta_i = 0.5$, and (**c**) dilated $\eta_d = 0.3$.

Finally, the convergence history of the intermediate threshold projections of the compliant inverter are depicted in Figure 10, which are the objective function and volume constraint. Note: since the displacement directions of the output end and the input end are opposite, the optimization result converges downward, and an additional 100 mm displacement is added to the objective function, which is consistent with our design philosophy.

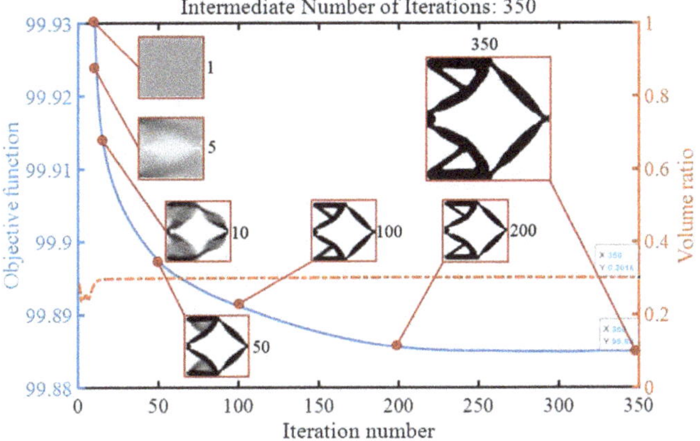

Figure 10. Optimization history for the force inverter.

6.2. Numerical Examples of the Gripper

The gripper presented in Figure 11a serves as the second example, with its design domain dimensions specified as a $\times$ b $= 160 \times 100$ mm^2. The analysis domain comprises a rectangular design domain denoted by Ω_d, and a rectangular void non-design domain labeled as Ω_v. The rectangular void non-design domain, Ω_v, has an area of b$/2 \times$ b$/2 = 50 \times 50$ mm^2. An input force of $F_{in,1}$ is applied at the input port, $O_{in,1}$, and the area, Γ, is fixed. The filter radius is set to $R = 4.8$ mm, while the smoothness parameter is specified as $\beta = 30$. The dilated design is set to $\eta_d = 0.3$, whereas the eroded design is established at $\eta_e = 0.7$, and the intermediate design is assigned a value of $\eta_i = 0.5$.

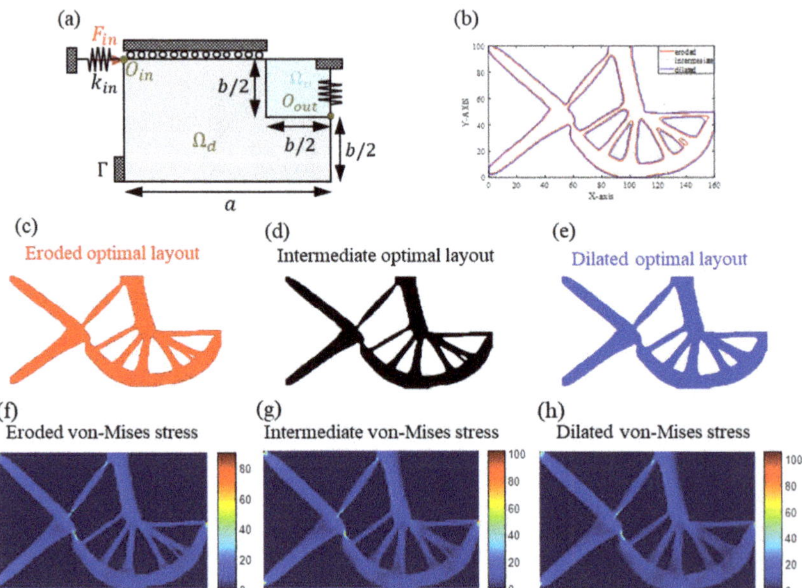

Figure 11. The compliant gripper. (**a**) The design domain; (**b**) the contours of different designs: eroded $\eta_e = 0.7$, intermediate $\eta_i = 0.5$, and dilated $\eta_d = 0.3$; optimized topologies for three levels of fatigue constraints: (**c**) eroded designs, (**d**) intermediate designs, and (**e**) dilated designs; the von Mises stress distribution of (**f**) eroded, (**g**) intermediate, and (**h**) dilated designs.

The optimized topologies for different designs with respect to three levels of fatigue and static failure constraints are displayed in Figure 11c–e in terms of their corresponding contours. Moreover, Figure 11f–h depict the corresponding von Mises stress distribution of the eroded, intermediate, and dilated designs, which were found to be approximately 100 MPa. Furthermore, the modified Goodman fatigue criteria, $L_{i,1}$ and $L_{i,2}$, as well as the static failure criteria, $L_{i,3}$ and $L_{i,4}$, were employed in the three-field density projection, utilizing the eroded, intermediate, and dilated projections, as depicted in Table 3. The three threshold projection maximum output displacements are 99.9698, 99.9665, and 99.9657, with corresponding amplification ratios of 0.204, 0.231, and 0.238, respectively. Both the fatigue and static failure constraints of the three-threshold projection meet the strength design requirements, ensuring an infinite lifespan of the mechanism under a given load. Furthermore, the design incorporates a gray level indicator, Mnd, for measuring gray levels discretely. The maximum gray level is 2.8%, indicating that the design achieves high precision in its gray level measurements. Furthermore, the eroded, intermediate, and dilated designs of the Goodman fatigue criteria were tested in three different combinations, as shown in Figure 12a–c.

Table 3. The resultant values of the compliant gripper.

Parameter	Symbol	Eroded	Intermediate	Dilated
Output displacement	u_{out}	99.9698	99.9665	99.9657
Amplification ratio	A_r	0.204	0.231	0.238
Volume fraction	V	0.266	0.307	0.327
Fatigue failure 1	$L_{i,1}$	0.697	0.786	0.817
Fatigue failure 2	$L_{i,2}$	0.587	0.663	0.689
Static failure 1	$L_{i,3}$	0.327	0.370	0.384

Table 3. *Cont.*

Parameter	Symbol	Eroded	Intermediate	Dilated
Static failure 2	$L_{i,4}$	0.327	0.370	0.384
Gray level indicator	Mnd	2.8%	0.9%	1.0%

Note: It should be noted that the lower output displacement performance is justifiable due to the robust strength-constrained approach, which faces an additional challenge in terms of strength feasibility. This challenge must be ensured for three distinct fields of relative densities, which further adds to the complexity of the problem.

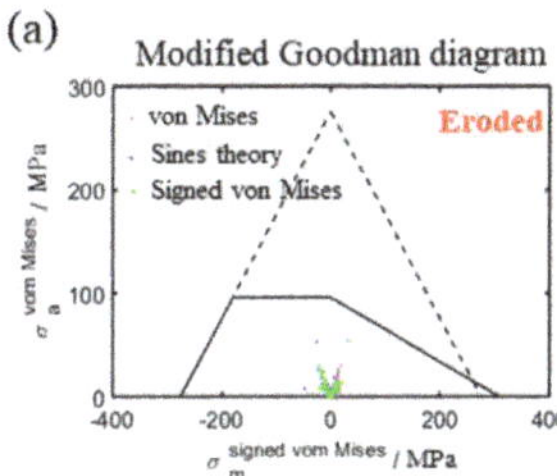
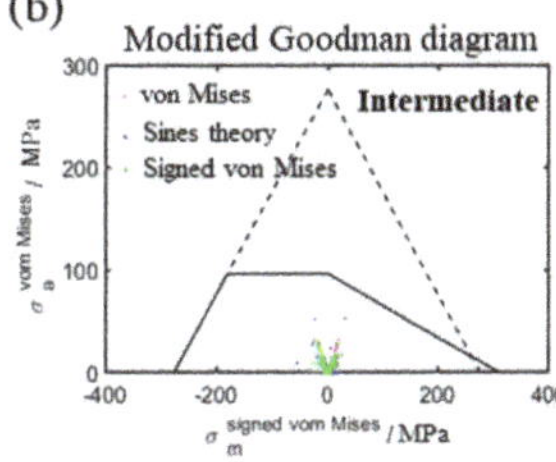
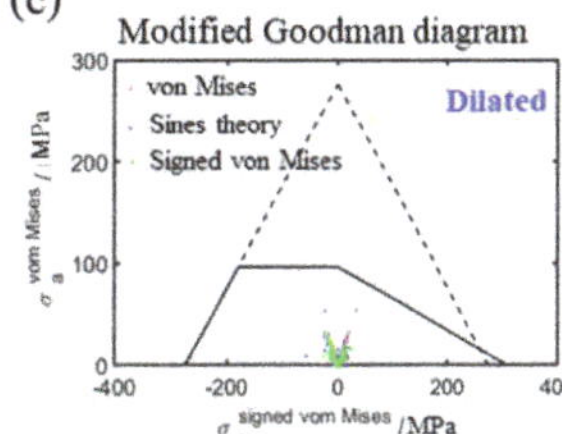

Figure 12. The eroded, intermediate, and dilated designs of the Goodman fatigue criteria of compliant gripper: (**a**) the eroded $\eta_e = 0.7$, (**b**) intermediate $\eta_i = 0.5$, and (**c**) dilated $\eta_d = 0.3$ designs of the threshold projection filters.

The deformation configurations of the compliant gripper are illustrated in Figure 13a–c, which correspond to the eroded, intermediate, and dilated configurations of the inverter, respectively. Moreover, the feature size that controls the topology optimization results effectively prevents the appearance of unmanufacturable thin levers, holes, and other structures. Finally, Figure 14 depicts the convergence history of the intermediate threshold projections of the compliant gripper, which includes the objective function and volume constraint.

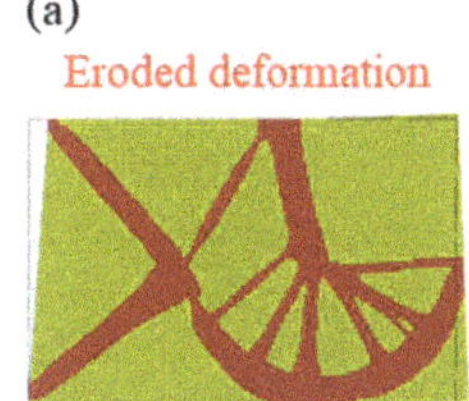

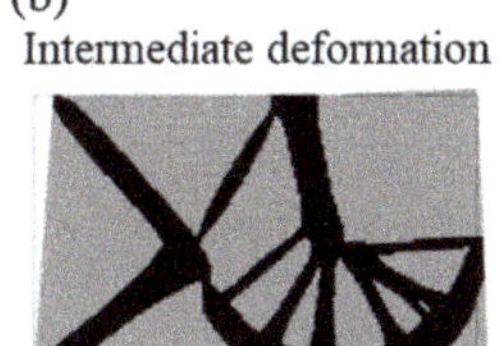

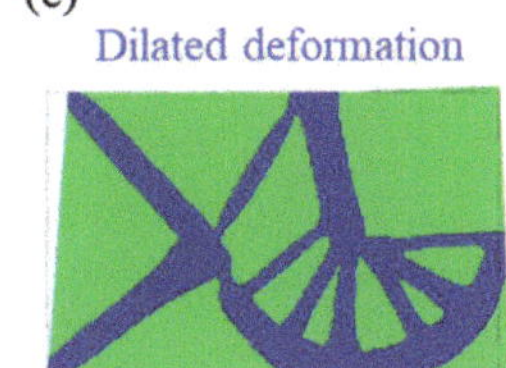

Figure 13. Deformation configurations of compliant gripper: (**a**) eroded $\eta_e = 0.7$, (**b**) intermediate $\eta_i = 0.5$, and (**c**) dilated $\eta_d = 0.3$.

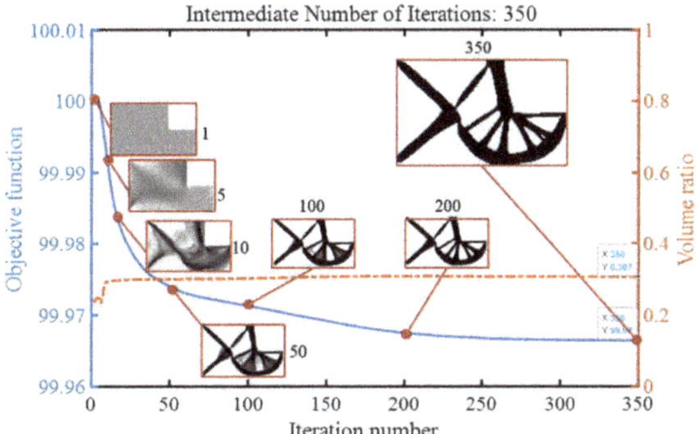

Figure 14. Optimization history for the compliant gripper.

7. Conclusions

This study presents a novel approach for topology optimization of compliant mechanisms that addresses issues related to static strength, fatigue failure, and manufacturability. The proposed method involves converting the static strength and fatigue failure constraints into different stress constraints, which are then addressed using stress relaxation techniques and p-norm aggregation approaches in the context of continuum topology optimization. Furthermore, the introduced maximum operator approach is utilized to address non-differentiable kink issues. Considering that different pieces of machining equipment have different machining accuracies, a three-field density projection approach is employed to ensure manufacturability of the optimized design. The proposed method is implemented using an improved SIMP interpolation model, and the design problem is solved using the GCMMA algorithm based on sensitivity analysis. Finally, the effectiveness of the proposed method is demonstrated through two numerical examples:

1. The von Mises stresses in the force inverter and compliant gripper were found to be approximately 120 MPa and 100 MPa, respectively. These stresses were below the material's strength limit of 275 MPa.
2. Compared with the previous topology optimization without fatigue constraints, the fatigue-constrained topology optimization can more effectively suppress the one-node hinge connection problems and avoid the phenomenon of stress concentration. Moreover, the maximum stress value of the compliant mechanism obtained using the fatigue-constrained topology optimization was lower, and the stress distribution was more uniform.
3. The three-field density projection approach was successfully employed to control the minimum size in the layout optimization, thereby meeting the manufacturing process requirements. In addition, a gray level indicator, Mnd, was utilized to measure the gray level, and the maximum gray level of the real design was found to be less than 1.5%. The effectiveness of the proposed method was effectively demonstrated through two numerical examples.

Author Contributions: Methodology, H.W.; Formal analysis, D.Z.; Investigation, D.Z.; Supervision, H.W.; Funding acquisition, H.W. All authors have read and agreed to the published version of the manuscript.

Funding: This work was supported by the Shenzhen Natural Science Foundation University Stability Support Project (grant no.: 20200821110721002), the Shenzhen Polytechnic University High-level Talent Start-up Project Funds (6021310028K), and the National Natural Science Foundation of China (U1913213).

Data Availability Statement: Not applicable.

Conflicts of Interest: The authors declare that they have no known competing financial interests or personal relationships that could have appeared to influence the work reported in this study.

References

1. Wang, R.; Zhang, X. Optimal design of a planar parallel 3-dof nanopositioner with multi-objective. *Mech. Mach. Theory* **2017**, *112*, 61–83. [CrossRef]
2. Alberola, J.A.M.; Fassi, I. Cyber-physical systems for micro-/nano-assembly operations: A survey. *Curr. Robot. Rep.* **2021**, *2*, 33–41. [CrossRef]
3. Wang, X.; Meng, Y.; Huang, W.-W.; Li, L.; Zhu, Z.; Zhu, L. Design, modeling, and test of a normal-stressed electromagnetic actuated compliant nano-positioning stage. *Mech. Syst. Signal Process.* **2023**, *185*, 109753. [CrossRef]
4. Gu, G.-Y.; Zhu, L.-M.; Su, C.-Y.; Ding, H.; Fatikow, S. Modeling and control of piezo-actuated nanopositioning stages: A survey. *IEEE Trans. Autom. Sci. Eng.* **2014**, *13*, 313–332. [CrossRef]
5. Zhou, R.; Zhu, Z.-H.; Kong, L.; Yang, X.; Zhu, L.; Zhu, Z. Development of a high-performance force sensing fast tool servo. *IEEE Trans. Ind. Inform.* **2021**, *18*, 35–45. [CrossRef]
6. Zhao, D.; Zhu, Z.; Huang, P.; Guo, P.; Zhu, L.; Zhu, Z. Development of a piezoelectrically actuated dual-stage fast tool servo. *Mech. Syst. Signal Process.* **2020**, *144*, 106873. [CrossRef]

7. Zhao, D.; Du, H.; Wang, H.; Zhu, Z. Development of a novel fast tool servo using topology optimization. *Int. J. Mech. Sci.* **2023**, *250*, 108283. [CrossRef]
8. Sano, P.; Verotti, M.; Bosetti, P.; Belfiore, N.P. Kinematic synthesis of a d-drive mems device with rigid-body replacement method. *J. Mech. Des.* **2018**, *140*, 075001. [CrossRef]
9. Yeon, A.; Yeo, H.G.; Roh, Y.; Kim, K.; Seo, H.-S.; Choi, H. A piezoelectric micro-electro-mechanical system vector sensor with a mushroom-shaped proof mass for a dipole beam pattern. *Sens. Actuators A Phys.* **2021**, *332*, 113129. [CrossRef]
10. Salinić, S.; Nikolić, A. A new pseudo-rigid-body model approach for modeling the quasi-static response of planar flexure-hinge mechanisms. *Mech. Mach. Theory* **2018**, *124*, 150–161. [CrossRef]
11. Kenton, B.J.; Leang, K.K. Design and control of a three-axis serial-kinematic high-bandwidth nanopositioner. *IEEE/ASME Trans. Mechatron.* **2011**, *17*, 356–369. [CrossRef]
12. Ryu, J.W.; Lee, S.-Q.; Gweon, D.-G.; Moon, K.S. Inverse kinematic modeling of a coupled flexure hinge mechanism. *Mechatronics* **1999**, *9*, 657–674. [CrossRef]
13. Ling, M.; Cao, J.; Zeng, M.; Lin, J.; Inman, D.J. Enhanced mathematical modeling of the displacement amplification ratio for piezoelectric compliant mechanisms. *Smart Mater. Struct.* **2016**, *25*, 075022. [CrossRef]
14. Li, C.; Chen, S.-C. Design of compliant mechanisms based on compliant building elements. Part I: Principles. *Precis. Eng.* **2023**, *81*, 207–220. [CrossRef]
15. Chen, F.; Dong, W.; Yang, M.; Sun, L.; Du, Z. A pzt actuated 6-dof positioning system for space optics alignment. *IEEE/ASME Trans. Mechatron.* **2019**, *24*, 2827–2838. [CrossRef]
16. Zhu, Z.; Zhou, X.; Liu, Z.; Wang, R.; Zhu, L. Development of a piezoelectrically actuated two-degree-of-freedom fast tool servo with decoupled motions for micro-/nanomachining. *Precis. Eng.* **2014**, *38*, 809–820. [CrossRef]
17. Zhu, B.; Chen, Q.; Jin, M.; Zhang, X. Design of fully decoupled compliant mechanisms with multiple degrees of freedom using topology optimization. *Mech. Mach. Theory* **2018**, *126*, 413–428. [CrossRef]
18. Lum, G.Z.; Teo, T.J.; Yeo, S.H.; Yang, G.; Sitti, M. Structural optimization for flexure-based parallel mechanisms-towards achieving optimal dynamic and stiffness properties. *Precis. Eng.* **2015**, *42*, 195–207. [CrossRef]
19. Jin, M.; Zhang, X. A new topology optimization method for planar compliant parallel mechanisms. *Mech. Mach. Theory* **2016**, *95*, 42–58. [CrossRef]
20. Pinskier, J.; Shirinzadeh, B.; Ghafarian, M.; Das, T.K.; Al-Jodah, A.; Nowell, R. Topology optimization of stiffness constrained flexure-hinges for precision and range maximization. *Mech. Mach. Theory* **2020**, *150*, 103874. [CrossRef]
21. Pham, M.T.; Yeo, S.H.; Teo, T.J.; Wang, P.; Nai, M.L.S. A decoupled 6-dof compliant parallel mechanism with optimized dynamic characteristics using cellular structure. *Machines* **2021**, *9*, 5. [CrossRef]
22. Zhu, B.; Zhang, X.; Zhang, H.; Liang, J.; Zang, H.; Li, H.; Wang, R. Design of compliant mechanisms using continuum topology optimization: A review. *Mech. Mach. Theory* **2020**, *143*, 103622. [CrossRef]
23. Liu, M.; Zhan, J.; Zhu, B.; Zhang, X. Topology optimization of flexure hinges with a prescribed compliance matrix based on the adaptive spring model and stress constraint. *Precis. Eng.* **2021**, *72*, 397–408. [CrossRef]
24. Liu, C.-H.; Chen, Y.; Yang, S.-Y. Topology optimization and prototype of a multimaterial-like compliant finger by varying the infill density in 3d printing. *Soft Robot.* **2022**, *9*, 837–849. [CrossRef] [PubMed]
25. Dorn, W.S. Automatic design of optimal structures. *J. Mec.* **1964**, *3*, 25–52.
26. Nishiwaki, S.; Frecker, M.I.; Min, S.; Kikuchi, N. Topology optimization of compliant mechanisms using the homogenization method. *Int. J. Numer. Methods Eng.* **1998**, *42*, 535–559. [CrossRef]
27. Takezawa, A.; Nishiwaki, S.; Kitamura, M. Shape and topology optimization based on the phase field method and sensitivity analysis. *J. Comput. Phys.* **2010**, *229*, 2697–2718. [CrossRef]
28. Wang, N.F.; Zhang, X.M. Topology optimization of compliant mechanisms using pairs of curves. *Eng. Optim.* **2015**, *47*, 1497–1522. [CrossRef]
29. Ansola, R.; Veguería, E.; Canales, J.; T'arrago, J.A. A simple evolutionary topology optimization procedure for compliant mechanism design. *Finite Elem. Anal. Des.* **2007**, *44*, 53–62. [CrossRef]
30. Luo, Z.; Tong, L.; Wang, M.Y.; Wang, S. Shape and topology optimization of compliant mechanisms using a parameterization level set method. *J. Comput. Phys.* **2007**, *227*, 680–705. [CrossRef]
31. Wang, R.; Zhang, X.; Zhu, B. Imposing minimum length scale in moving morphable component (mmc)-based topology optimization using an effective connection status (ecs) control method. *Comput. Methods Appl. Mech. Eng.* **2019**, *351*, 667–693. [CrossRef]
32. Luo, Z.; Chen, L.; Yang, J.; Zhang, Y.; Abdel-Malek, K. Compliant mechanism design using multi-objective topology optimization scheme of continuum structures. *Struct. Multidiscip. Optim.* **2005**, *30*, 142–154. [CrossRef]
33. Duysinx, P.; Bendsøe, M.P. Topology optimization of continuum structures with local stress constraints. *Int. J. Numer. Methods Eng.* **1998**, *43*, 1453–1478. [CrossRef]
34. de Troya, M.A.S.; Tortorelli, D.A. Adaptive mesh refinement in stress-constrained topology optimization. *Struct. Multidiscip. Optim.* **2018**, *58*, 2369–2386. [CrossRef]
35. Chu, S.; Gao, L.; Xiao, M.; Luo, Z.; Li, H. Stress-based multi-material topology optimization of compliant mechanisms. *Int. J. Numer. Methods Eng.* **2018**, *113*, 1021–1044. [CrossRef]

36. de Assis Pereira, A.; Cardoso, E.L. On the influence of local and global stress constraint and filtering radius on the design of hinge-free compliant mechanisms. *Struct. Multidiscip. Optim.* **2018**, *58*, 641–655. [CrossRef]

37. Conlan-Smith, C.; James, K.A. A stress-based topology optimization method for heterogeneous structures. *Struct. Multidiscip. Optim.* **2019**, *60*, 167–183. [CrossRef]

38. Deng, H.; Vulimiri, P.S.; To, A.C. An efficient 146-line 3d sensitivity analysis code of stress-based topology optimization written in matlab. *Optim. Eng.* **2021**, *23*, 1733–1757. [CrossRef]

39. Roin, T.; Montemurro, M.; Pailh, J. Stress-based topology optimization through non-uniform rational basis spline hyper-surfaces. *Mech. Adv. Mater. Struct.* **2022**, *29*, 3387–3407. [CrossRef]

40. Bruggi, M. On an alternative approach to stress constraints relaxation in topology optimization. *Struct. Multidiscip. Optim.* **2008**, *36*, 125–141. [CrossRef]

41. Le, C.; Norato, J.; Bruns, T.; Ha, C.; Tortorelli, D. Stress-based topology optimization for continua. *Struct. Multidiscip. Optim.* **2010**, *41*, 605–620. [CrossRef]

42. Jeong, S.H.; Choi, D.-H.; Yoon, G.H. Fatigue and static failure considerations using a topology optimization method. *Appl. Math. Model.* **2015**, *39*, 1137–1162. [CrossRef]

43. Nabaki, K.; Shen, J.; Huang, X. Evolutionary topology optimization of continuum structures considering fatigue failure. *Mater. Des.* **2019**, *166*, 107586. [CrossRef]

44. Holmberg, E.; Torstenfelt, B.; Klarbring, A. Fatigue constrained topology optimization. *Struct. Multidiscip. Optim.* **2014**, *50*, 207–219. [CrossRef]

45. Collet, M.; Bruggi, M.; Duysinx, P. Topology optimization for minimum weight with compliance and simplified nominalstress constraints for fatigue resistance. *Struct. Multidiscip. Optim.* **2017**, *55*, 839–855. [CrossRef]

46. Oest, J.; Lund, E. Topology optimization with finite-life fatigue constraints. *Struct. Multidiscip. Optim.* **2017**, *56*, 1045–1059. [CrossRef]

47. Chen, Z.; Long, K.; Wen, P.; Nouman, S. Fatigue-resistance topology optimization of continuum structure by penalizing the cumulative fatigue damage. *Adv. Eng. Softw.* **2020**, *150*, 102924. [CrossRef]

48. Sigmund, O. Manufacturing tolerant topology optimization. *Acta Mech. Sin.* **2009**, *25*, 227–239. [CrossRef]

49. Wang, F.; Jensen, J.S.; Sigmund, O. Robust topology optimization of photonic crystal waveguides with tailored dispersion properties. *JOSA B* **2011**, *28*, 387–397. [CrossRef]

50. Schevenels, M.; Lazarov, B.S.; Sigmund, O. Robust topology optimization accounting for spatially varying manufacturing errors. *Comput. Methods Appl. Mech. Eng.* **2011**, *200*, 3613–3627. [CrossRef]

51. Lazarov, B.S.; Schevenels, M.; Sigmund, O. Robust design of large-displacement compliant mechanisms. *Mech. Sci.* **2011**, *2*, 175–182. [CrossRef]

52. Guo, X.; Zhang, W.; Zhang, L. Robust structural topology optimization considering boundary uncertainties. *Comput. Methods Appl. Mech. Eng.* **2013**, *253*, 356–368. [CrossRef]

53. Zhang, W.; Kang, Z. Robust shape and topology optimization considering geometric uncertainties with stochastic level set perturbation. *Int. J. Numer. Methods Eng.* **2017**, *110*, 31–56. [CrossRef]

54. da Silva, G.A.; Beck, A.T.; Sigmund, O. Topology optimization of compliant mechanisms with stress constraints and manufacturing error robustness. *Comput. Methods Appl. Mech. Eng.* **2019**, *354*, 397–421. [CrossRef]

55. da Silva, G.A.; Beck, A.T.; Sigmund, O. Topology optimization of compliant mechanisms considering stress constraints, manufacturing uncertainty and geometric nonlinearity. *Comput. Methods Appl. Mech. Eng.* **2020**, *365*, 112972. [CrossRef]

56. Liu, M.; Zhan, J.; Zhu, B.; Zhang, X. Topology optimization of compliant mechanism considering actual output displacement using adaptive output spring stiffness. *Mech. Mach. Theory* **2020**, *146*, 103728. [CrossRef]

57. Svanberg, K. *Mma and Gcmma-Two Methods for Nonlinear Optimization*; KTH: Stockholm, Sweden, 2007; Volume 1, pp. 1–15.

58. Wang, F.; Lazarov, B.S.; Sigmund, O. On projection methods, convergence and robust formulations in topology optimization. *Struct. Multidiscip. Optim.* **2011**, *43*, 767–784. [CrossRef]

59. Qian, X.; Sigmund, O. Topological design of electromechanical actuators with robustness toward over-and under-etching. *Comput. Methods Appl. Mech. Eng.* **2013**, *253*, 237–251. [CrossRef]

 *processes*

Article

Effect of Hot Filament Chemical Vapor Deposition Filament Distribution on Coated Tools Performance in Milling of Zirconia Ceramics

Louis Luo Fan [1], Wai Sze Yip [1,2], Zhanwan Sun [1,3], Baolong Zhang [1] and Suet To [1,2,*]

[1] State Key Laboratory of Ultra-Precision Machining Technology, Department of Industrial and Systems Engineering, The Hong Kong Polytechnic University, Hung Hom, Kowloon, Hong Kong SAR, China; louisluo.fan@connect.polyu.hk (L.L.F.); lenny.ws.yip@polyu.edu.hk (W.S.Y.); zw.sun@gdut.edu.cn (Z.S.); baolong-edison.zhang@connect.polyu.hk (B.Z.)

[2] The Hong Kong Polytechnic University Shenzhen Research Institute, Shenzhen 518063, China

[3] State Key Laboratory of Precision Electronic Manufacturing Technology and Equipment, Guangdong University of Technology, Guangzhou 510006, China

* Correspondence: sandy.to@polyu.edu.hk

Abstract: Zirconia ceramics (ZrO_2) have been used for a variety of applications due to their superior physical properties, including in machining tools and dentures. Nonetheless, due to its extreme hardness and brittleness in both sintered and half-sintered forms, zirconia is difficult to machine. In this study, half-sintered zirconia blocks are milled with tungsten carbide milling tools which are coated with diamond film using hot filament chemical vapor deposition (HFCVD) at various substrate-to-filament distances. The objective was to determine the effect of substrate-to-filament distances on the coating thickness, diamond purity, coating grain size, and ZrO_2 machining performance during HFCVD. The experimental results show that, in HFCVD, the grain size and coating thickness of the diamond film on milling tools tend to decrease when the substrate-to-filament distances decrease. Tool failure happened at a cutting time of 200 min for all coated tools. However, the machining quality in terms of surface topology, surface roughness, and tool condition is superior for diamond-coated milling tools with smaller grain sizes and thinner thicknesses. It can be concluded that diamond milling tools with a smaller grain size and lesser thickness produced under shorter substrate-to-filament distances have a superior machining performance and a longer tool life. This study could potentially be used for parameter optimization in the production of coated tools.

Keywords: hot filament chemical vapor deposition; diamond-coated tools; micro-milling; zirconia; surface integrity

Citation: Fan, L.L.; Yip, W.S.; Sun, Z.; Zhang, B.; To, S. Effect of Hot Filament Chemical Vapor Deposition Filament Distribution on Coated Tools Performance in Milling of Zirconia Ceramics. *Processes* **2023**, *11*, 2773. https://doi.org/10.3390/pr11092773

Academic Editor: Hideki KITA

Received: 31 May 2023
Revised: 24 July 2023
Accepted: 31 July 2023
Published: 16 September 2023

1. Introduction

Zirconia (ZrO_2) is an inorganic, non-metallic material with a low thermal conductivity and high chemical inertness; it is mechanical-wear resistant and electrically conductive [1]. Since the mid-1970s, a growing number of nations have increased their efforts and resource allocations into the development of zirconia series products, thereby expanding the functional materials application field of zirconia. As a result of this continuous investment and innovation into zirconia ceramics, it has become one of the highest-performance new materials favored by industrial policies in a number of countries and is widely used in our daily tasks and lives. Zirconia ceramics can be used to manufacture a variety of products for industrial applications, such as machining tools [2], dentures [3], and surface coatings [4], as well as the raw material for telecommunication components, etc.

Numerous technologies exist for machining zirconia ceramics, including laser-assisted machining [5] and abrasive machining [6], but the processing cost is relatively high, the processing efficiency is relatively low, and the precision level of the machined surface is remarkably low. Industrial zirconia ceramics are extremely hard and brittle [7], which makes large-scale machining challenging. Due to their typical physical and mechanical properties, zirconia ceramic materials can only be initially processed by grinding them under typical conditions. ZrO_2 is a typical zirconia ceramic material that is brittle, posing the greatest challenge to machining. Principally, there are three distinct types of ZrO_2: pre-sintered, half-sintered, and fully sintered. Due to its extreme hardness and brittleness, it is nearly impossible to machine fully sintered ZrO_2 using conventional methods [8]. Typically, grinding or additional sintering of pre-sintered or partially sintered ZrO_2 is required. Pre-sintered ZrO_2 is significantly softer than other forms of ZrO_2. In order to obtain observable and quantifiable results, lengthy experiments involving cutting tests are required [9]. This study selected half-sintered ZrO_2 as the cutting material used for the experiments. It is softer than fully sintered ZrO_2 and can still be machined using conventional methods such as milling, but its hardness level is significantly higher than pre-sintered materials. Diamond-coated cutting tools made by h filament chemical vapor deposition (HFCVD) have many excellent properties, including high hardness [9], exceptional thermal conductivity [10], and low friction [11], as well as a low expansion coefficient [12], and this coating technology is considerably less expensive and allows for greater flexibility in tool geometry. The diamond-coated tools for zirconia ceramic workpiece machining are anticipated to have a significantly longer tool life and produce a superior machining quality than that of non-coated tools.

Numerous studies have investigated the connection between various parameters and coating performance. Numerous experiments were conducted to determine the effects of the influential factors of HFCVD, such as pretreatment strategies, methane concentration, chamber pressure, and reaction temperature. Due to the vast difference between the filament temperature and the background temperature of the furnace, there is a sharp temperature gradient between the hot filament and the substrate during HFCVD. In addition, because the temperature range for the synthesis of sp^3 carbon bonding is very narrow, a slight change in the substrate-to-filament distance has a significant impact on the purity and surface morphology of the diamond coating. Nevertheless, a clear illustration of the effect of the substrate-to-filament distance on the milling performance of diamond-coated tools on semi-sintered zirconia ceramics is still lacking. Therefore, the purpose of this paper is to determine the relationship between the milling performance of zirconia with HFCVD-coated tools and the substrate-to-filament distance in the HFCVD process. The coating parameters, particularly the substrate-to-filament distance, of an HFCVD-coated tool for milling zirconia are chosen as the main variables within a series of experiments conducted with different cutting parameters. This helps in determining the optimal cutting parameter for various tolerance and application requirements in order to maximize tool life while maintaining the required energy efficiencies and surface finishes.

In this paper, the structure is organized as follows: Section 2 introduces the procedures and mechanisms of HFCVD diamond coating and ZrO_2. Section 3 illustrates the experimental setup of the HFCVD process and ZrO_2 cutting test. Section 4, explains the results and discussion including the effect that the distance between the substrate and filament had on the grain size, the coating thickness of HFCVD-coated milling tools, the purity of sp^3 carbon within diamond coated milling tools, tool wear, and surface quality. Lastly, Section 5 summarizes and concludes this study.

2. Theory

2.1. Decarburization and Cobalt Etching in HFCVD Diamond Coating

Cobalt granules are typically added to the powder metallurgy process in order to bind tungsten carbide rods and tools. This is due to the high solubility of carbon with transition elements such as cobalt during the HFCVD diamond synthesis process; cobalt induces graphitization during the formation of the diamond deposit, reducing the purity of the diamond coating and the adhesion strength between the diamond coating and the tungsten carbide substrate surface [13]. Gaseous cobalt diffuses into the diamond coating during the diamond deposition process in HFCVD, increasing the yield of sp^2

bonding (graphite phase) and decreasing the yield of sp^3 bonding. In contrast to deposition on the substrate surface, carbon would diffuse into the cobalt granules in the carbide substrate [14]. To prevent or reduce the negative effects of cobalt and carbon diffusion, a series of pretreatments involving Murakami etching and nitric acid etching must be performed prior to the HFCVD process to remove surface and subsurface cobalt.

2.2. The Effect of Substrate to Filament Distance in HFCVD Coating Process

HFCVD is widely utilized in industries for diamond synthesis applications. However, there are a number of deposition parameters in HFCVD that have a significant effect on the surface morphology and growth rate of the diamond film. In the HFCVD process, when the concentration of CH_4 near the filament rises due to the flow of methane, the concentration of the methyl radical (CH_3) rises, which diffuses onto the surface of the substrate. The CH_3 further reacts with hydrogen ions to produce methanide radicals (CH_2) and methylidyne radicals (CH), which ultimately reduce to carbon atoms and are deposited on the substrate to facilitate diamond nucleation and growth. In addition, the increase in CH_4 concentration encourages secondary nucleation, which decreases grain size [15]. On the surface of the crystals formed by the initial nucleation, a second nucleus forms, inhibiting their growth. Furthermore, the second nuclei smoothen the surface by filling any surface irregularities [16]. This indicates that as methane flow increases, growth rate will increase while grain size will decrease [15,16].

As for the effect of H_2 concentration in the HFCVD process, it is closely related to the concentration of hydrogen ions, which is necessary for receiving the hydrogen ions released during the previously mentioned methane fragmentation. In addition, it eliminates sp^2 carbon, which reduces the yield of graphite and other carbon forms that are not diamonds. When the concentration of H_2 reaches a critical level, the sp^3 carbon will also be etched away at a rate faster than the diamond nucleation and deposition rates, thereby inhibiting the nucleation and growth of diamond crystals [17].

The concentration gradient of CH_3, CH_2, and CH decreases with an increasing distance between the substrate and the filament. Therefore, as the substrate approaches the filament, it is exposed to a greater concentration of CH_3, CH_2, and CH radicals and a shorter diffusion distance, which increases surface diffusion on the substrate's surface. This improves the growth rate of the diamond film. However, the substrate's temperature increases when it is close to the filament. Since sp^3 carbon can only form within a relatively narrow temperature range, placing the substrate too close to the filament may impede diamond growth.

2.3. ZrO_2 Cutting Mechanism

Due to the high hardness of ZrO_2, the main material removal mechanism of sintered ZrO_2 is focused on its brittle or semi-brittle nature. In addition to removal via brittle fracture mode, ZrO_2 can be removed via plastic deformation under the cutting conditions of a low feed rate and a high spindle speed [18–20].

Under the case of micro milling with a ball end mill, the cutting speed, v_c, can be calculated using Equation (1) when the value of the inclination angle of the tool axis β is zero (vertical milling), as shown below,

$$v_c = 2\pi n \sqrt{a_p(D - a_p)}, \quad \text{where } \beta = 0 \tag{1}$$

where n is the spindle speed, D is the tool diameter and a_p is the axial the depth of cut [21] (Figure 1).

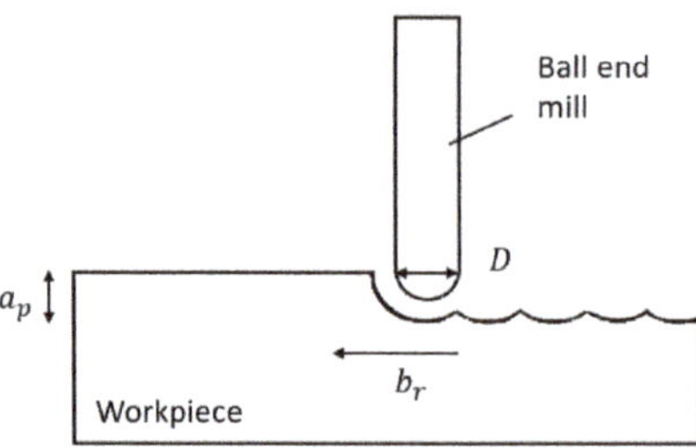

Figure 1. Kinematics of vertical micro-ball end milling.

The chip formation and tool wear during the zirconium micro-milling process are simplified and summarized for clarity in Figure 2. Due to the brittleness of zirconium, the cutting chips of it in micro-milling are discontinuous [22]. The tool wear that develops during a process can be divided into three stages, the first stage consists of abrasive and adhesive wear, in which coating dents caused by abrasive wear allow residue materials from the workpiece to adhere to the coating surface [23]. The second stage of tool wear is characterized by attrition wear and apparent delamination of coating [24]. The indentations formed during the first stage enhance the formation of cracks caused by debonding, resulting in coating delamination. In the third and final stage of tool wear, the coating has completely delaminated, exposing the substrate material and resulting in abrasive wear on the substrate [25] which causes tool failure [26].

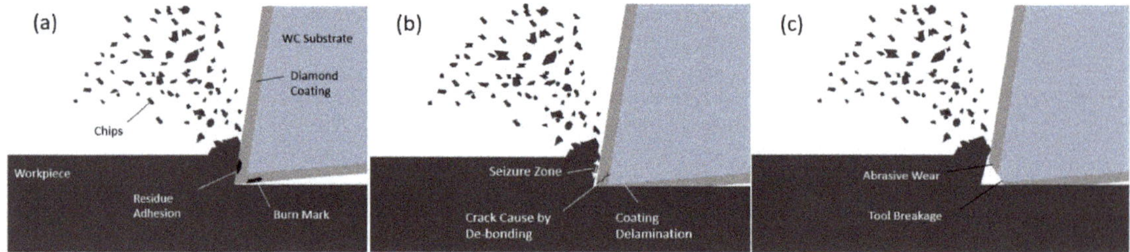

Figure 2. Three main stages of tool wear (**a**) stage 1: adhesive and abrasive wear, (**b**) attrition wear and delamination, (**c**) substrate material breakage.

3. Experimental Setup

Uncoated tungsten carbide ball end mills (diameter: 2 mm, flute length: 12 mm, overall length: 50 mm, outer diameter: 4 mm) were chemically pretreated before HFCVD diamond deposition. The tools were first treated with Murakami solution (85% KOH: 99.5% $K_3[Fe(CN)_6]$: H_2O, 1:1:10) to increase surface roughness and improve mechanical interlocking between the substrate and HFCVD-deposited diamond layer. Before the second pretreatment, the instruments were cleaned and dried. The instruments were ultrasonically etched in a 98% $HNO:H_2O$, 1:10 solution for cobalt removal.

Following the cobalt etching of the ball end mills, a seeding procedure was carried out to facilitate the nucleation of diamond during the initial phase of the diamond deposition. The instruments were treated in an ultrasonic bath for 30 min with a 3–6 micron diamond powder ethanol solution, followed by a 30 second treatment with 99 percent ethanol to remove excess diamond powder. The tools were then placed within the HFCVD furnace. The tools were divided into two sets, A and B, with substrate-to-filament distances 10 and 30, respectively. The parameter for HFCVD coating is shown in Table 1. After the HFCVD diamond deposition on the milling tools, the tools performed scanning electron microscopy (SEM) imaging and Raman analysis to determine their various coating properties, such as coating grain size, coating thickness, and the percentage of sp^3 bonds present in the coatings produced. Figure 3 shows the HFCVD machine and related facilities used in this study for the diamond coating of drilling tools.

Table 1. HFCVD Deposition Parameters.

Deposition Set	Cobalt Etching Time (min)	Seeding (min)	CH$_4$ Flow (sccm)	H$_2$ Flow (sccm)	Substrate Distance from Filament (mm)	Chamber Pressure (mbar)	Power (kW)
A	6	20	2000	30	10	10	4
B	6	20	2000	30	30	10	4

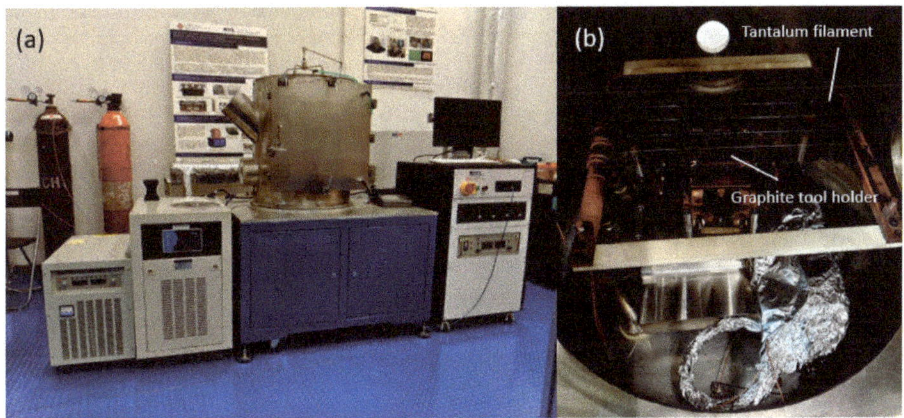

Figure 3. The HFCVD and the related facilities for diamond coating of drilling tools, (**a**) overall installation of HFCVD facilities, (**b**) inner view of HFCVD reaction chamber.

Twelve diamond-coated milling tools (six milling tools from set A and six milling tools from set B) were then used to conduct a series of ZrO$_2$ cutting tests with the following parameters: spindle speed of 22 kprm, feed rates of 100 mm/min and 200 mm/min, cutting times of 200 min, 400 min, and 800 min, and a depth of cut of 0.6 mm. The sample size was 30 mm × 20 mm. Table 2 includes the cutting parameters. After the cutting tests, the ZrO$_2$ workpiece surface was analyzed by an optical profiling system to determine surface roughness and surface profile in order to deduce the cutting quality of coated end mills manufactured under two different substrate-to-filament distances in relation to the different cutting parameters.

Table 2. ZrO$_2$ Milling Parameters.

Sample	Spindle Speed (kprm)	Feed Rate (mm/min)	Cutting Time (min)	Depth of Cut (mm)
1	22	100	200	0.6
2	22	100	400	0.6
3	22	100	800	0.6
4	22	200	200	0.6
5	22	200	400	0.6
6	22	200	800	0.6

4. Result and Discussion

4.1. The Effects of Distance between Substrate and Filament on Grain Size and Coating Thickness of HFCVD Coated Milling Tools

Figure 1a–d are SEM images of the coating produced using deposition on set A and set B tools, respectively. According to Figure 4, the average grain size of the coating produced using deposition for set B is 9.37 μm, which is considerably larger than the grains produced using deposition for set A, which are only 3.19 μm. In terms of coating thickness, the coating produced using deposition sor set B has an average grain size of 43.56 μm, which is approximately 5.77 μm thicker than the coating produced using deposition for set A. The SEM results indicate that the coating thickness increases as grain size increases. As a result of the growth rate during diamond deposition, the growth rate obtained using deposition for set B is considerably higher than for set A. When coated tools are produced under the coating parameters for set A at a shorter substrate-to-filament distance, the surface of the substrate is exposed to a higher concentration of CH_3, CH_2, and CH radicals, the carbon sources for diamond synthesis. This should theoretically increase the growth rate. However, the substrate's temperature increases when it is close to the filament. Due to the fact that sp^3 carbon can only form within a narrow temperature range [27], when the substrate is too close to the filament the substrate temperature would exceed the optimal temperature for diamond synthesis, reducing the growth rate and the coating purity [28], which causes the thinner coating thickness and smaller grain size for the milling tools in set A. In addition, the increased concentration of CH_3, CH_2, and CH radicals resulting from the coating parameters of set A promotes secondary nucleation, resulting in a smaller average grain size compared to set B. In conclusion, when the distance between the substrate and the filament decreases in HFCVD processes, the grain size and coating thickness of the diamond film on milling tools become generally smaller and thinner.

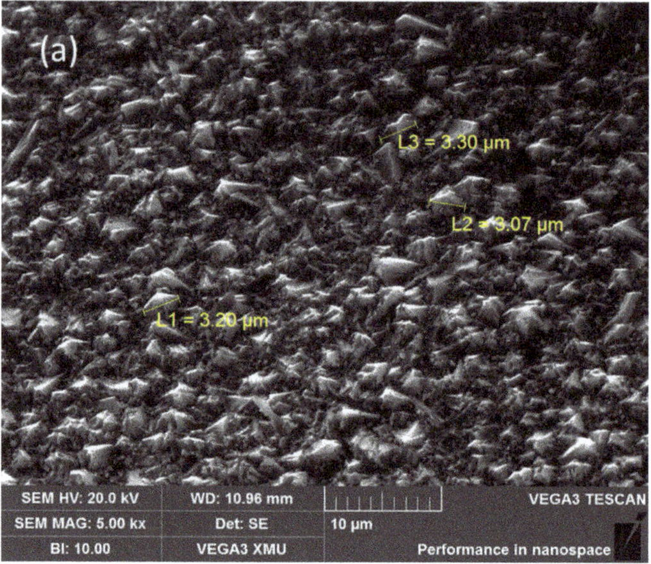

Figure 4. *Cont.*

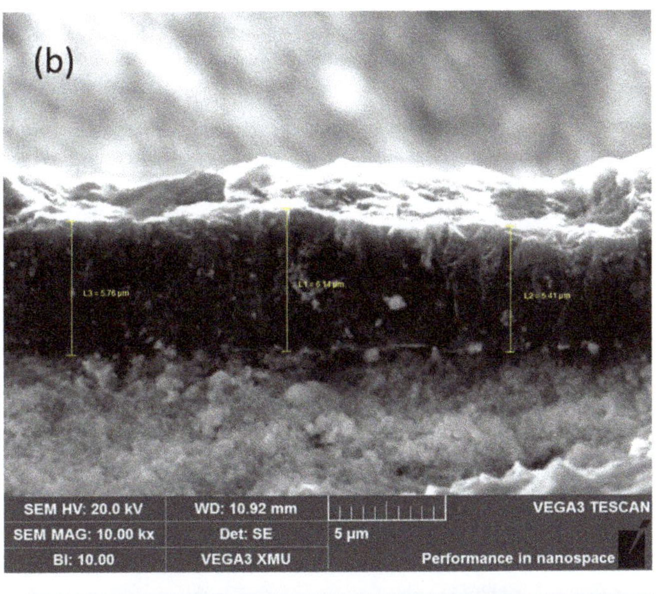

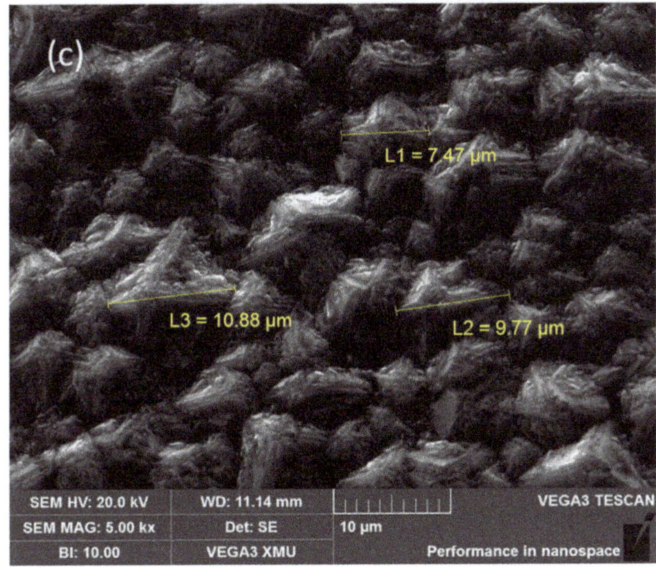

Figure 4. *Cont.*

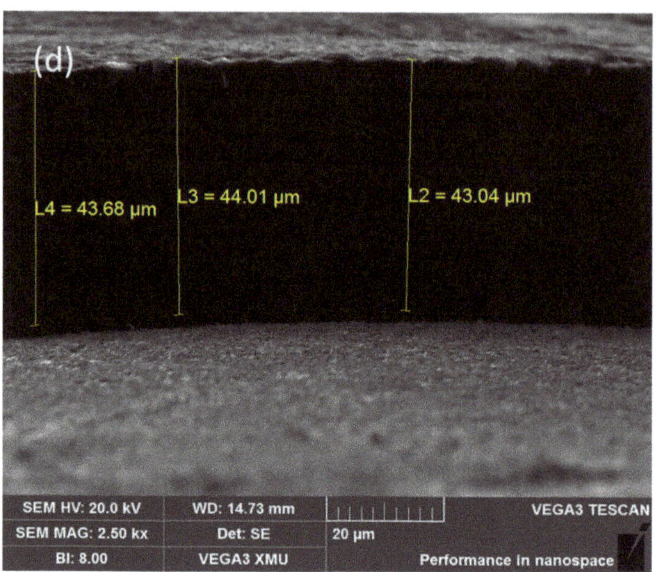

Figure 4. SEM images of diamond-coated tools produced by HFCVD under set A (**a**) grain size, (**b**) coating thickness, and set B (**c**) grain size, (**d**) coating thickness.

4.2. The Effects of Distance between Substrate and Filament on the Purity of sp^3 Carbon of Diamond-Coated Milling Tools

Raman spectroscopy is a nondestructive technique for analyzing materials based on the interaction between light and the material. Raman spectroscopy can provide detailed information about the chemical structure, phase and morphology, crystallinity, and molecular interactions of the coated milling tools. Typically, a Raman spectrum comprises a certain number of Raman peaks, each of which represents the wavelength position and intensity of the Raman scattered light corresponding to that peak. For each peak, a particular molecular bond vibration can be identified.

With the Renishaw Micro-Raman Spectroscopy System^{TM}, it is possible to obtain Raman spectrograms of the diamond coating for tools in sets A and B, displaying the intensity distributions of various characteristic peaks as illustrated in Figures 5 and 6, respectively. Multiple peak fitting based on the Gaussian model clearly demonstrates the composition of the coating, which consists of disordered sp^3 carbon, C-N vibrations, diamond, disordered graphite, distorted sp^3 carbon, graphite, etc. Once all the relevant characteristic peaks have been identified, as shown in Tables 3 and 4 which show the Raman spectra for the peak fitting of diamond-coated milling tools for both sets A and B, the area of integration can be calculated for quantitative analysis. The purity of sp^3 carbon in set A can be calculated to be approximately 98.47 percent when sp^2 and sp^3 carbon are added together. In contrast, the sp^3 purity of set B exceeds 99 percent. Comparing the sp^3 proportion results, one can conclude that for an increase in the distance between the substrate and the filament in a HFCVD process, the purity of the sp^3 carbon in the diamond-coated milling tools is improved.

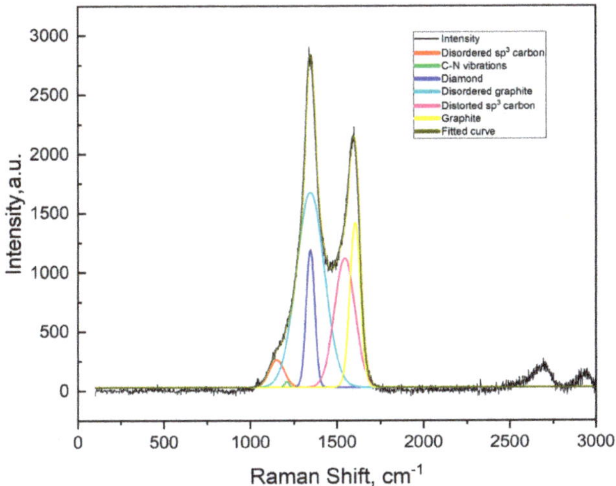

Figure 5. Raman spectrum for peak fitting of diamond-coated milling tools of set A.

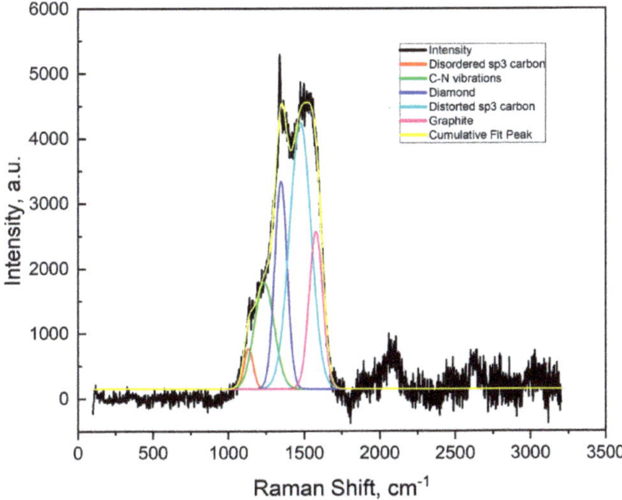

Figure 6. Raman spectrum peak fitting of diamond-coated milling tools of set B.

Table 3. Integration results for peak fitting of diamond-coated milling tools of set A.

Assignment	Area of Integration
Disordered sp^3 carbon	117,832
C-N vibrations	94,698
Diamond	164,886
Disordered graphite	427,524
Distorted sp^3 carbon	253,574
Graphite	197,225
Sp^3/sp^2	64.3809
Sp^3 purity	0.9847

Table 4. Integration result for peak fitting of diamond-coated milling tools of set B.

Assignment	Area of Integration
Disordered sp^3 carbon	523,629
C-N vibrations	738,764
Diamond	801,438
Distorted sp^3 carbon	1,203,299
Graphite	750,086
Sp^3/sp^2	252.8075
Sp^3 purity	0.9961

4.3. The Effects of Distance between Substrate and Filament on Tool Wear

The progression of tool wear for a diamond-coated tool during milling includes the adhesive wear, delamination of coated materials, tool breakage/loss of tools, and the geometry of the coated and substrate materials, according to the cutting mechanism of a diamond-coated tool [29]. High cutting temperature causes a continuous adhesion of workpiece materials, to the rake and flank faces of the cutting tool at first. As a result, the relief angle of the cutting edge decreases, and the cutting motion changes to plowing and rubbing. Abrasive wear increases as a result of the constant compression stress, friction heat, and high cutting temperature. During continuous cutting, delamination occurs along the cutting edge of a coated tool, exposing the sharp edge of the tool substrate. The wear process may include graphitization in the diamond at high cutting temperatures, as well as attrition in columnar diamond grains. As a result of the constant cutting, the exposed tungsten carbide edge quickly wore away. This results in a variation in the tool geometry, rounding of the cutting edge, and tool breakage.

A similar logic is applied to the ball end mill conditions in this study. Figure 7 illustrates SEM images of ball end mills from sets A and B, which the cutting time is 200 min. According to Figure 7, an end mill tool edge from set A shows a shiny surface on the drilling edges, indicating that the tungsten carbide substrate has already been exposed; however, it is not yet to the stage of tool breakage. For the end mill tool edge of set B, however, the tool breakage has already occurred at the same cutting time as the drilling tool of set A. As shown in Figure 7b, a variation in the tool geometry has been observed on the drilling edge. As a result, under the same cutting conditions, the drilling tool of set B has tool breakage, whereas the end mill tool of set A is still in the delamination stage, demonstrating that the tool wear level of set A is lower than that of set B. The aforementioned testing results demonstrated that the milling tools of set B have already failed at 200 min of cutting, but the milling tools of set A are still able to continue cutting.

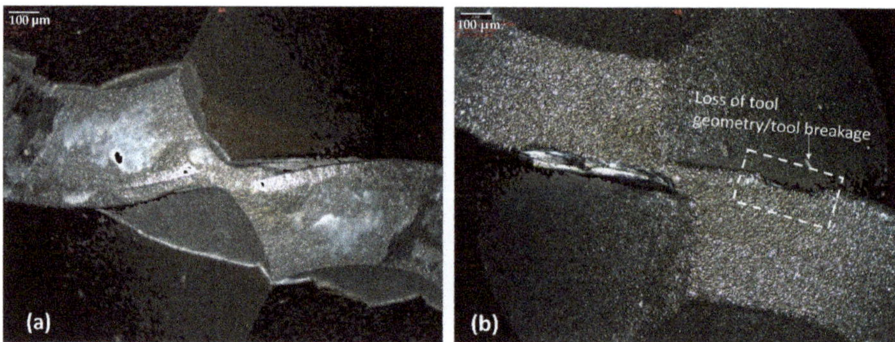

Figure 7. SEM images of drilling tools from (**a**) set A, and (**b**) set B, for cutting time 200 min.

4.4. Influence of Distance between Substrate and Filament on Surface Quality

Table 5 summarized the surface roughness of milled surfaces by diamond-coated milling tools for sets A and B. Figure 8 indicates that a higher feed rate (200 mm/min) in milling of zirconia ceramic using diamond-coated milling tools results in a lower average surface roughness value, particularly when the cutting time is at 200 min. This is because larger grain size experiences more severe friction, thus experiences more wear under high spindle speed [30]. Moreover, film cracks tend to propagate through the columnar particles, leading to a decrease in toughness and adhesion. This result is consistent with the tool condition described in Section 4.1, which indicates that set B milling tools failed at 200 min. On the other hand, the reason that set A and set B tools have comparable surface roughness at 800 min is because both sets have experienced significant tool wear as a result of prolonged machining time.

Table 5. Surface roughness of milled surfaces by diamond-coated milling tools for sets A and B.

Sample	Surface Roughness (nm)	
	Set A	Set B
1	470.667	620.667
2	405.667	949.333
3	478.333	235.333
4	395.333	618.667
5	159.667	487.666
6	423.667	360

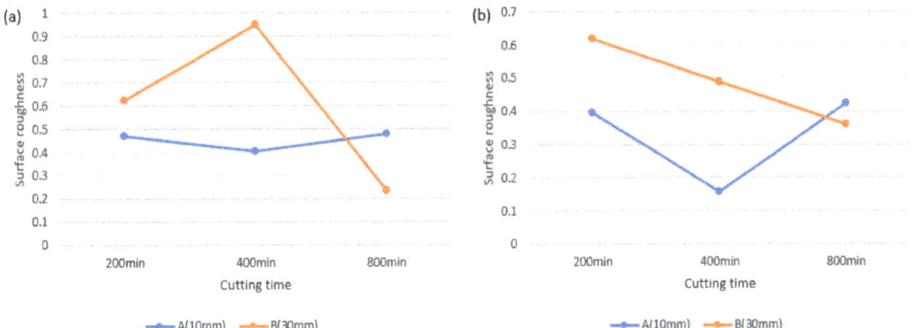

Figure 8. Surface roughness vs. cutting time for sets A and B diamond milling tools (**a**) at feed rate 100 mm/min and, (**b**) 200 mm/min.

To comprehend the effect of substrate-to-filament distance on the cutting quality of coated end mills under different cutting parameters, it is necessary to examine the average surface roughness. Figure 9 indicates that a higher feed rate (200 mm/min) results in a lower surface roughness value for both set A and B diamond-coated milling tools, particularly when the cutting time is 400 min. Comparing the surface produced by set A and set B, the surface roughness of the machined surface of a diamond-coated milling tool from set B is significantly higher than that of a tool from set A due to the larger grain size, which causes the individual grains and the coating to experience a higher level of stress during the milling process. As mentioned above, under the larger microcrystalline diamond (MCD) structure, film cracks propagate through the columnar particles; a thicker coating has longer columnar particles which further enhance crack propagation, reducing the adhesion and toughness of the coating [31]. However, for a smaller grain size coating such as the diamond coating produced using deposition for set A, the structure is less continuous, thus reducing crack propagation. In addition, tool wear increases with the grain size [29] and the substrate material, tungsten carbide, is brittle, resulting in the accumulation of internal stress along the coating, interlayer, and substrate subsurface. This causes tool delamination and even abrasive wear, resulting in severe tool wear or even the failure of the diamond-coated milling tools of set B. At 800 min, both set A and set B tools had a similar surface roughness distribution, as both sets had already experienced significant tool wear and failure.

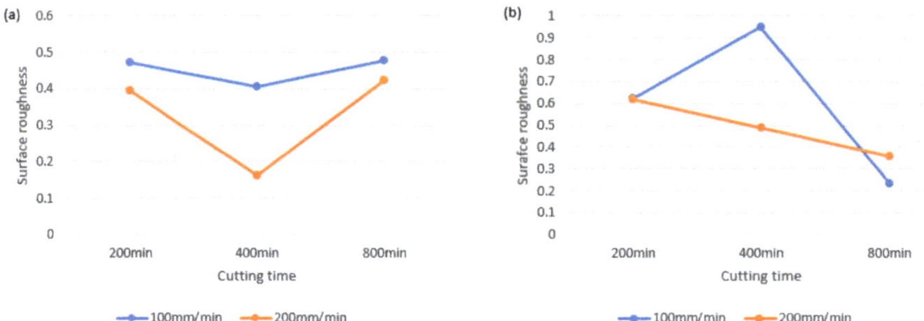

Figure 9. Surface roughness against cutting time at 100 mm/min and 200 mm/min feed rates (**a**) set A diamond milling tools and, (**b**) set B diamond milling tools.

Figure 10 shows the surface topology of the surface machined with set A and B diamond milling tools at a feed rate of 100 mm/min and 200 mm/min for 200 min, 400 min and 800 min, respectively. Due to peeling of the zirconia ceramic surface in milling, the machined surfaces produced by set B diamond-coated milling tools exhibit numerous micro-voids and cavities, which are evenly distributed on the machined surface. The micro-void on the machined surface is a typical surface defect that appeared during tool milling with a diamond-coated tool. These voids can serve as sites for crack initiation, leading to the collapse of near-surface cavities and the initiation and propagation of cracks along grain boundaries [32]. As a result, a network of various voids is found on the surface which causes stress concentration on the machined surfaces, which in turn affects the mechanical properties of components made with a diamond-coated milling zirconia surface. As mentioned above, the larger grain size and thicker coating of set B diamond-coated tools promote crack propagation and mean that they experience a larger internal stress which leads to tool failure and reduces the cutting quality. In conclusion, the diamond-coated milling tool from set A produces a surface with fewer surface defects. In conjunction with the results and justification of surface roughness in Sections 4.2 and 4.3, it can be concluded that a diamond-coated tool with an appropriately thin coating thickness and finer grain size produces a machined surface with a higher quality, and that a shorter substrate-to-filament distance will produce a surface with higher quality on zirconia ceramics.

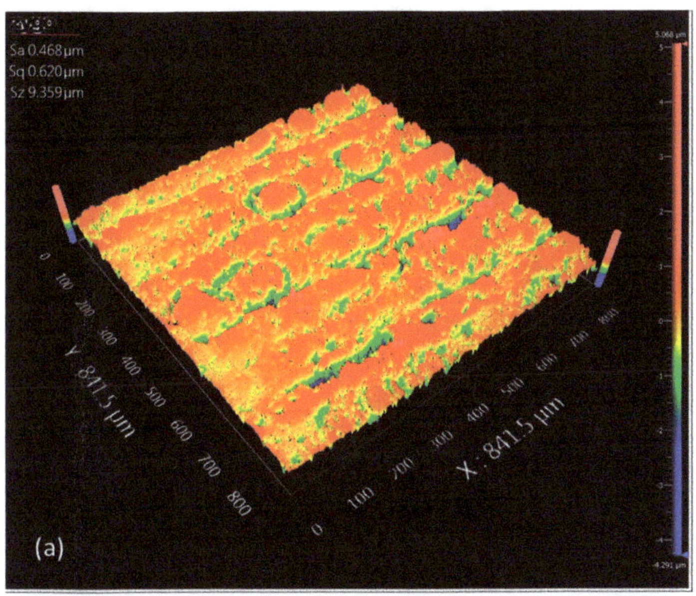

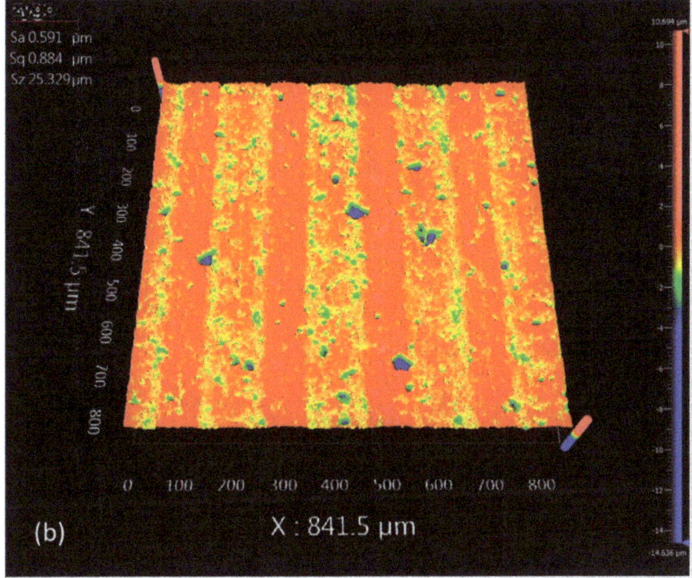

Figure 10. *Cont.*

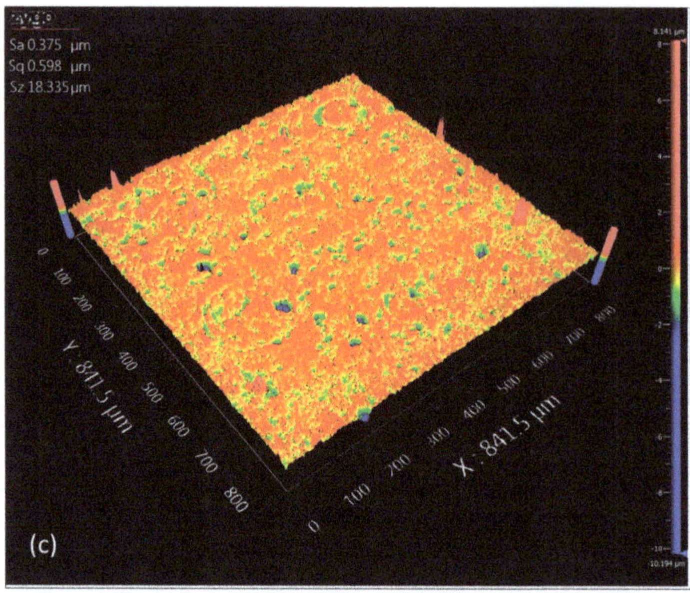

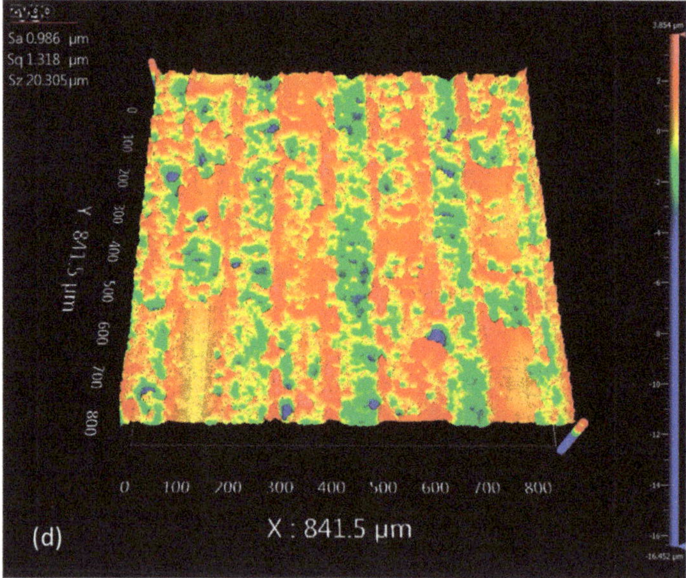

Figure 10. *Cont.*

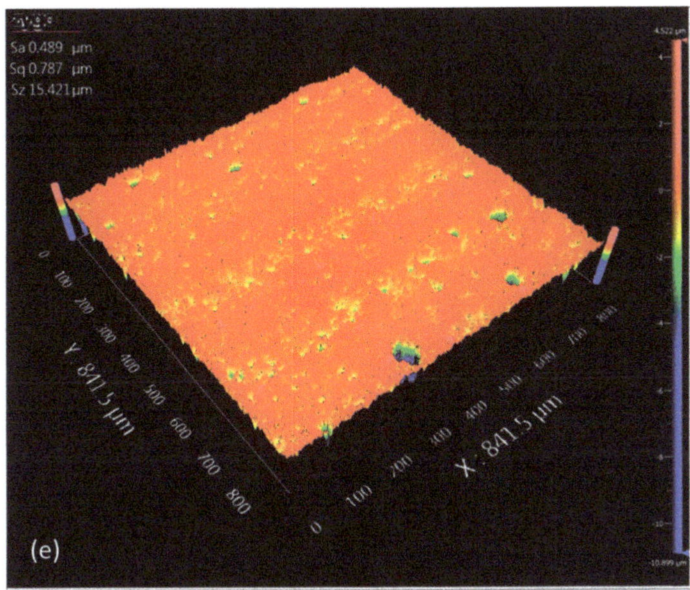

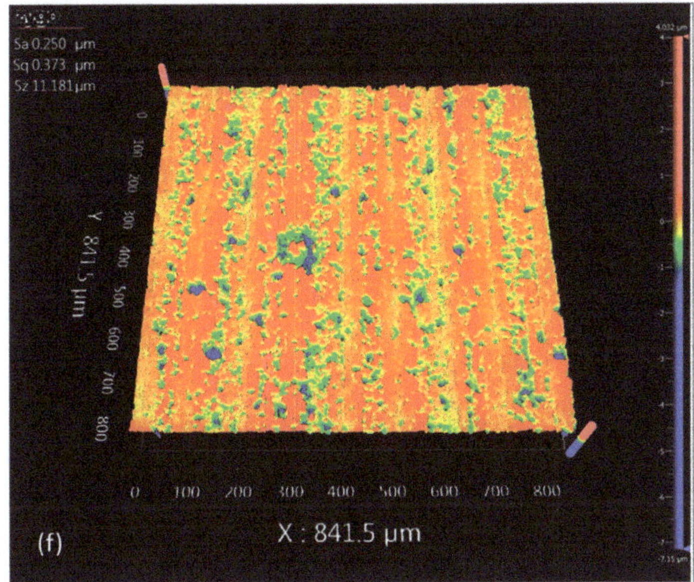

Figure 10. *Cont.*

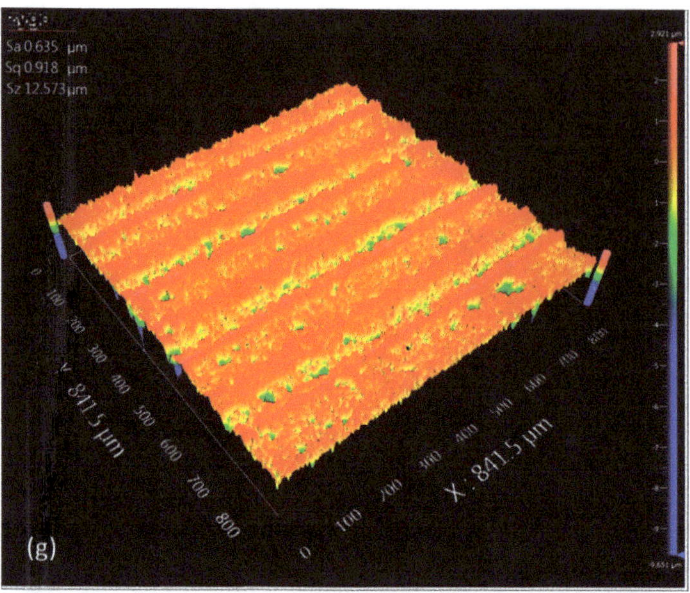

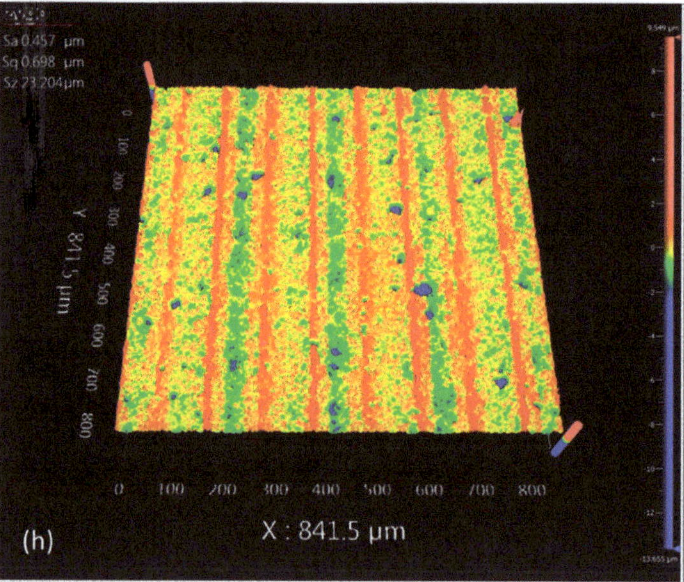

Figure 10. *Cont.*

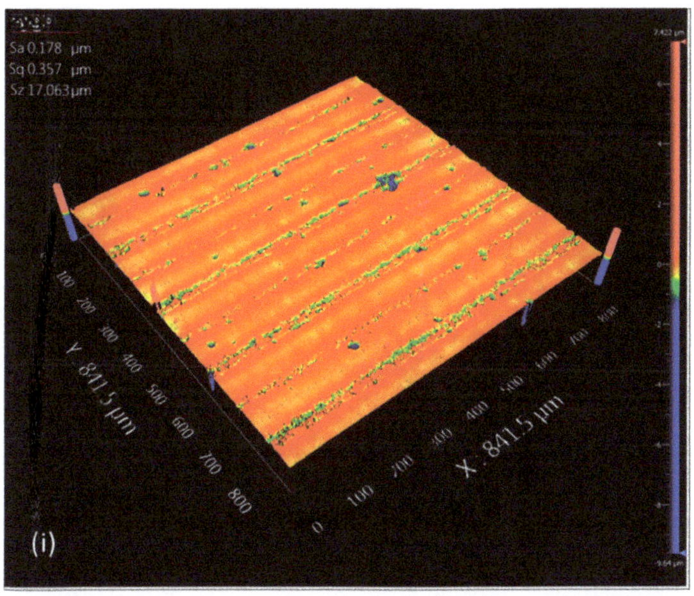
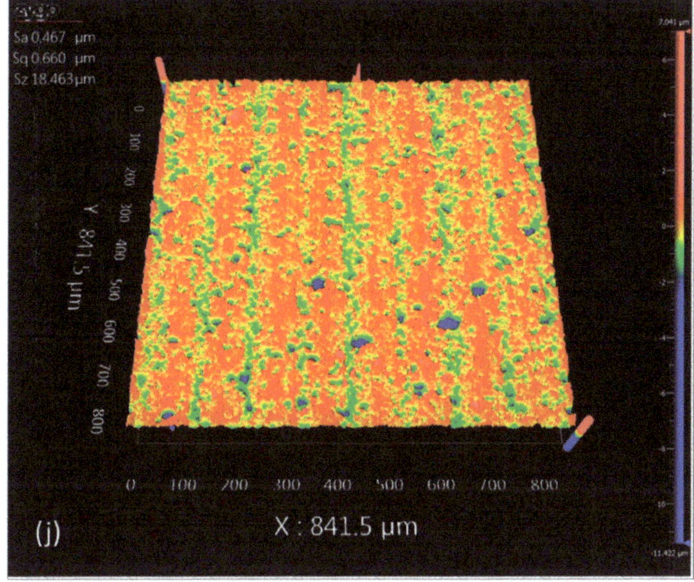

Figure 10. *Cont.*

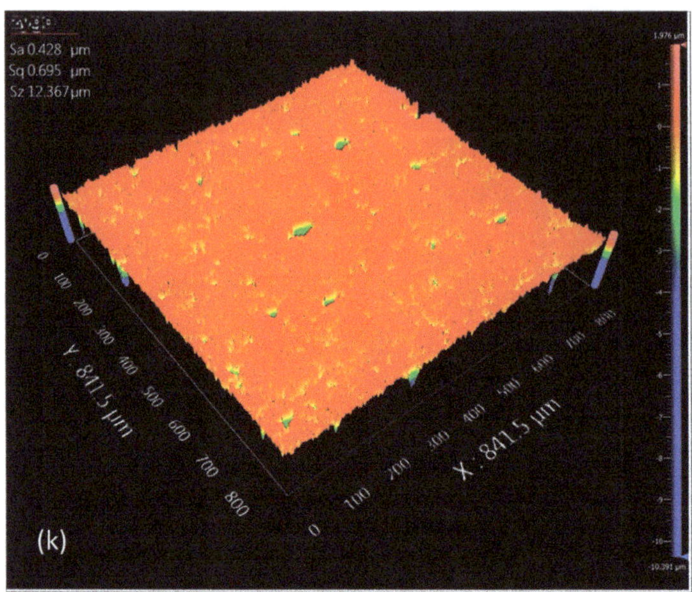

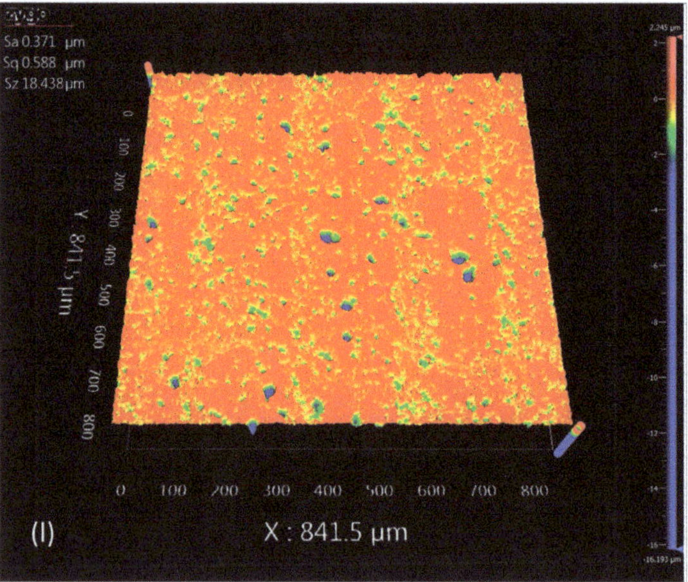

Figure 10. 3D profile images of zirconia surface machined by diamond-coated milling tools from set A, feed rate: 100 mm/min (**a**) cutting time: 200 min, (**c**) cutting time: 400 min, (**e**) cutting time: 800 min; feed rate: 200 mm/min, (**g**) cutting time: 200 min, (**i**) cutting time: 400 min, (**k**) cutting time: 800 min and set B, feed rate: 100 mm/min, (**b**) cutting time: 200 min, (**d**) cutting time: 400 min, (**f**) cutting time: 800 min; feed rate: 200 mm/min, (**h**) cutting time: 200 min, (**j**) cutting time: 400 min, (**l**) cutting time: 400 min.

5. Conclusions

In this work, half-sintered zirconia blocks are milled using tungsten carbide milling tools coated by varied substrate-to-filament distances in HFCVD. The goal is to investigate the influence of substrate-to-filament distance in an HFCVD process on coating grain size, coating thickness, diamond purity, and ZrO_2 machinability. This research implies that milling tools produced with shorter substrate-to-filament distances have a better tool life and milling performance on ZrO_2 ceramics; the findings are summarized below.

a.   Shorter substrate-to-filament distances lead to a smaller grain size and a thinner coating thickness due to excessively high temperatures and secondary nucleation.

b.   Shorter substrate to filament distances lead to a lower diamond purity of the coating.

c.   Tools with coating produced under a shorter substrate-to-filament distance are less prone towards tool breakage.

d.   An appropriately thin coating and fine grain size from a shorter substrate-to-filament distance produces a better machined surface with far less micro-voids and cavities.

Author Contributions: Conceptualization, L.L.F.; Methodology, L.L.F.; Software, L.L.F. and B.Z.; Validation, W.S.Y. and S.T.; Formal analysis, L.L.F.; Investigation, L.L.F. and Z.S.; Resources, Z.S. and S.T.; Data curation, L.L.F., Z.S. and B.Z.; Writing—original draft, L.L.F. and B.Z.; Writing—review & editing, W.S.Y.; Visualization, B.Z.; Supervision, W.S.Y. and S.T.; Project administration, S.T.; Funding acquisition, S.T. All authors have read and agreed to the published version of the manuscript.

Funding: The work described in this paper was partially supported by a grant from the General Research Fund from the Research Grants Council of the Hong Kong Special Administrative Region (Grant number PolyU15221322); the major project of the National Natural Science Foundation of China (Project No.: U19A20104); Shenzhen Science and Technology Program (Project No.: JCYJ20210324131214039). The authors also would like to thank the financial support from the State Key Laboratory of Ultra-precision Machining Technology and the Research Committee of The Hong Kong Polytechnic University (Project code: RK18).

Data Availability Statement: Not applicable.

Conflicts of Interest: The authors declare that they have no known competing financial interests or personal relationships that could have appeared to influence the work reported in this paper.

References

1.   Gautam, C.; Joyner, J.; Gautam, A.; Rao, J.; Vajtai, R. Zirconia based dental ceramics: Structure, mechanical properties, biocompatibility and applications. *Dalton Trans.* **2016**, *45*, 19194–19215. [CrossRef] [PubMed]

2.   Mondal, B.; Chattopadhyay, A.B.; Virkar, A.; Paul, A. Development and performance of zirconia-toughened alumina ceramic tools. *Wear* **1992**, *156*, 365–383. [CrossRef]

3.   Pantea, M.; Antoniac, I.; Trante, O.; Ciocoiu, R.; Fischer, C.A.; Traistaru, T. Correlations between connector geometry and strength of zirconia-based fixed partial dentures. *Mater. Chem. Phys.* **2019**, *222*, 96–109. [CrossRef]

4.   Chen, H.; Ding, C.X. Nanostructured zirconia coating prepared by atmospheric plasma spraying. *Surf. Coat. Technol.* **2002**, *150*, 31–36. [CrossRef]

5.   Pfefferkorn, F.E.; Shin, Y.C.; Tian, Y.; Incropera, F.P. Laser-assisted machining of magnesia-partially-stabilized zirconia. *J. Manuf. Sci. Eng.* **2004**, *126*, 42–51. [CrossRef]

6.   Yin, L.; Jahanmir, S.; Ives, L.K. Abrasive machining of porcelain and zirconia with a dental handpiece. *Wear* **2003**, *255*, 975–989. [CrossRef]

7.   Hocheng, H.; Kuo, K.L.; Lin, J.T. Machinability of Zirconia Ceramics in Ultrasonic Drilling. *Mater. Manuf. Process.* **1999**, *14*, 713–724. [CrossRef]

8.   Juri, A.Z.; Zhang, Y.; Kotousov, A.; Yin, L. Zirconia responses to edge chipping damage induced in conventional and ultrasonic vibration-assisted diamond machining. *J. Mater. Res. Technol.* **2021**, *13*, 573–589. [CrossRef]

9.   Camposilvan, E.; Leone, R.; Gremillard, L.; Sorrentino, R.; Zarone, F.; Ferrari, M.; Chevalier, J. Aging resistance, mechanical properties and translucency of different yttria-stabilized zirconia ceramics for monolithic dental crown applications. *Dent. Mater.* **2018**, *34*, 879–890. [CrossRef]

10.   Tsugawa, K.; Ishihara, M.; Kim, J.; Hasegawa, M.; Koga, Y. Large-Area and Low-Temperature Nanodiamond Coating by Microwave Plasma Chemical Vapor Deposition. 2006. Available online: https://www.researchgate.net/publication/236737100 (accessed on 30 May 2023).

11. Shen, B.; Sun, F. Effect of surface morphology on the frictional behaviour of hot filament chemical vapour deposition diamond films. *Proc. Inst. Mech. Eng. Part J J. Eng. Tribol.* **2009**, *223*, 1049–1058. [CrossRef]

12. Polini, R.; Mantini, F.P.; Braic, M.; Amar, M.; Ahmed, W.; Taylor, H. Effects of Ti- and Zr-based interlayer coatings on the hot filament chemical vapour deposition of diamond on high speed steel. *Thin Solid Films* **2006**, *494*, 116–122. [CrossRef]

13. Zhu, W.; Mccune, R.C.; Devries, J.E.; Tamor, M.A.; Ng, K.Y.S. Characterization of diamond films on binderless W-Mo composite carbide. *Diam. Relat. Mater.* **1994**, *3*, 1270–1276. [CrossRef]

14. Ehrhardt, H. New developments in the field of superhard coatings. *Surf. Coat. Technol.* **1995**, *74*, 29–35. [CrossRef]

15. Liu, H.; Dandy, D.S. Studies on nucleation process in diamond CVD: An overview of recent developments. *Diam. Relat. Mater.* **1995**, *4*, 1173–1188. [CrossRef]

16. Michler, J.; Stiegler, J.; von Kaenel, Y.; Moeckli, P.; Dorsch, W.; Stenkamp, D.; Blank, E. Microstructure evolution and non-diamond carbon incorporation in CVD diamond thin films grown at low substrate temperatures. *J. Cryst. Growth* **1997**, *172*, 404–415. [CrossRef]

17. Rozbicki, R.T.; Sarin, V.K. Nucleation and growth of combustion flame deposited diamond on silicon nitride. *Int. J. Refract. Met. Hard Mater.* **1998**, *16*, 377–388. [CrossRef]

18. Geng, D.; Zhang, D.; Xu, Y.; He, F.; Liu, F. Comparison of drill wear mechanism between rotary ultrasonic elliptical machining and conventional drilling of CFRP. *J. Reinf. Plast. Compos.* **2014**, *33*, 797–809. [CrossRef]

19. Xu, S.; Shimada, K.; Mizutani, M.; Kuriyagawa, T. Fabrication of hybrid micro/nano-textured surfaces using rotary ultrasonic machining with one-point diamond tool. *Int. J. Mach. Tools Manuf.* **2014**, *86*, 12–17. [CrossRef]

20. Singh, R.P.; Singhal, S. Investigation of machining characteristics in rotary ultrasonic machining of alumina ceramic. *Mater. Manuf. Process.* **2017**, *32*, 309–326. [CrossRef]

21. Wojciechowski, S.; Mrozek, K. Mechanical and technological aspects of micro ball end milling with various tool inclinations. *Int. J. Mech. Sci.* **2017**, *134*, 424–435. [CrossRef]

22. Kizaki, T.; Harada, K.; Mitsuishi, M. Efficient and precise cutting of zirconia ceramics using heated cutting tool. *CIRP Ann. Manuf. Technol.* **2014**, *63*, 105–108. [CrossRef]

23. Dong, X.; Zheng, L.; Wei, W.; Sun, X.; Liu, Z.; Sun, Y.; He, Z. Drilling force characteristics and casing drill wear analysis of GFRP low frequency axial vibration hole making. *Diam. Abras. Eng.* **2023**, *43*, 82–89. [CrossRef]

24. Peng, F.; Lei, D.; Wang, W. Diamond grinding quality analysis based on control grinding depth. *Diam. Abras. Eng.* **2023**, *43*, 118–125. [CrossRef]

25. Yang, C.; Zhang, N.; Su, H.; Ding, W. Research on abrasive wear of CBN Hinge honing Tool based on Single Abrasive Cutting. *Diam. Abras. Eng.* **2022**, *42*, 728–737. [CrossRef]

26. Neves, D.; Diniz, A.E.; Lima, M.S.F. Microstructural analyses and wear behavior of the cemented carbide tools after laser surface treatment and PVD coating. *Appl. Surf. Sci.* **2013**, *282*, 680–688. [CrossRef]

27. Osswald, S.; Yushin, G.; Mochalin, V.; Kucheyev, S.O.; Gogotsi, Y. Control of sp2/sp3 carbon ratio and surface chemistry of nanodiamond powders by selective oxidation in air. *J. Am. Chem. Soc.* **2006**, *128*, 11635–11642. [CrossRef] [PubMed]

28. Das, D.; Singh, R.N. A review of nucleation, growth and low temperature synthesis of diamond thin films. *Int. Mater. Rev.* **2007**, *52*, 29–64. [CrossRef]

29. Bian, R.; He, N.; Ding, W.; Liu, S. A study on the tool wear of PCD micro end mills in ductile milling of ZrO_2 ceramics. *Int. J. Adv. Manuf. Technol.* **2017**, *92*, 2197–2206. [CrossRef]

30. Zhang, J.; Yuan, Y.; Zhang, J. Cutting performance of microcrystalline, nanocrystalline and dual-layer composite diamond coated tools in drilling carbon fiber reinforced plastics. *Appl. Sci.* **2018**, *8*, 1642. [CrossRef]

31. Yuan, Z.; Guo, Y.; Li, C.; Liu, L.; Yang, B.; Song, H.; Zhai, Z.; Lu, Z.; Li, H.; Jiang, X.; et al. New multilayered diamond/β-SiC composite architectures for high-performance hard coating. *Mater. Des.* **2020**, *186*, 108207. [CrossRef]

32. Carroll, L.J.; Cabet, C.; Carroll, M.C.; Wright, R.N. The development of microstructural damage during high temperature creep-fatigue of a nickel alloy. *Int. J. Fatigue* **2013**, *47*, 115–125. [CrossRef]

 *processes*

Article

A Novel Approach to Optimizing Grinding Parameters in the Parallel Grinding Process

Tengfei Yin [1], Hanqian Zhang [2,3], Wei Hang [4,*] and Suet To [1,*]

1   State Key Laboratory of Ultra-Precision Machining Technology, Department of Industrial and Systems Engineering, The Hong Kong Polytechnic University, Hung Hom, Kowloon, Hong Kong SAR, China
2   Guangzhou Haozhi Industrial Co., Ltd., Guangzhou 511356, China
3   School of Mechanical Engineering and Automation, Harbin Institute of Technology, Shenzhen 518055, China
4   College of Mechanical Engineering, Zhejiang University of Technology, Hangzhou 310014, China
*   Correspondence: whang@zjut.edu.cn (W.H.); sandy.to@polyu.edu.hk (S.T.)

Abstract: Hard materials have found extensive applications in the fields of electronics, optics, and semiconductors. Parallel grinding is a common method for fabricating high-quality surfaces on hard materials with high efficiency. However, the surface generation mechanism has not been fully understood, resulting in a lack of an optimization approach for parallel grinding. In this study, the surface profile formation processes were analyzed under different grinding conditions. Then, a novel method was proposed to improve surface finish in parallel grinding, and grinding experiments were carried out to validate the proposed approach. It was found that the denominator (b) of the simplest form of the rotational speed ratio of the grinding wheel to the workpiece has a great influence on surface generation. The surface finish can be optimized without sacrificing the machining efficiency by slightly adjusting the rotational speeds of the wheel or the workpiece to make the value of b close to the ratio (p) of the wheel contact width to the cross-feed distance per workpiece revolution. Overall, this study provides a novel approach for optimizing the parallel grinding process, which can be applied to industrial applications.

Keywords: parallel grinding; surface generation; surface roughness; speed ratio; optimization

Citation: Yin, T.; Zhang, H.; Hang, W.; To, S. A Novel Approach to Optimizing Grinding Parameters in the Parallel Grinding Process. *Processes* **2024**, *12*, 493. https://doi.org/10.3390/pr12030493

Academic Editor: Chin-Hyung Lee

Received: 30 January 2024
Revised: 20 February 2024
Accepted: 23 February 2024
Published: 28 February 2024

1. Introduction

High-precision parts made from hard materials, such as ceramics, glasses, and single crystals, have important applications in optics, electronics, aerospace, automotive engineering, and semiconductor sectors [1–3]. The high hardness makes them difficult to machine [4]. Diamond cutting processes, including diamond turning and milling, are normally applied to machine soft materials and can reliably achieve high accuracy and good surface integrity [5–7]. By contrast, severe tool wear will occur and result in poor surface quality during diamond cutting of hard materials [8,9]. Although some field-assisted technologies, such as laser [10–12] or ultrasonic vibration [13–15] assisted technologies, have been proposed to improve the cutting performance in terms of the diamond cutting of hard materials, additional assistive devices are required, increasing the machining cost. To date, grinding is still one of the best choices and the most frequently used methods for processing hard materials, due to its simplicity, high machining quality, and cost efficiency [16]. In grinding, the workpiece material is removed by the protruded grits distributed on the working surface of the grinding wheel, and many factors, including the grinding parameters [17,18], wheel geometry [19,20], grit size [21,22], lubrication [23,24], workpiece material properties [25], dressing condition [26], and vibrations [16,27], influence the surface generation processes, determining the final surface quality. In order to optimize the grinding process to obtain good surface quality, it is important to study the surface generation process in grinding.

During the past few decades, numerous efforts have been directed towards understanding the surface formation mechanism in grinding processes. In fact, it is highly dependent on the grinding modes. Different grinding modes, such as straight surface grinding [28], external cylindrical grinding [29], internal cylindrical grinding [30], self-rotating grinding [31], parallel grinding [32], and cross grinding [33], have different geometrical motion relationships between the grinding wheel and the workpiece. Thus, the surface formation mechanism varies from one grinding mode to another. It is easily found in the literature that, in cases where both the wheel and the workpiece rotate in the grinding operations, the rotational speed ratio (the angular speed ratio) of the grinding wheel to the workpiece plays a significant role in the surface generation in the grinding processes. For instance, Patel et al. [34] conducted cylindrical plunge grinding experiments using both the integer and non-integer speed ratios and found that integer speed ratios produced a poor surface finish, while non-integer ratios yielded a complex pattern of roughness spikes. Finally, they concluded that the speed ratio had a great influence on the roughness and texture of the workpiece surfaces and the power consumption. In a self-rotating grinding process, the rotational axis of a cup wheel is parallel with that of the workpiece but offset by the distance of the wheel radius. During grinding, the cup wheel and workpiece both rotate around their own rotational axes, with the wheel feeding toward the workpiece, which can achieve a highly flat surface and is widely used in wafer grinding. Chidambaram et al. [35] built a mathematical model for predicting the grinding marks on the ground wafers and performed grinding experiments to validate the model predictions. The curvature of the grinding lines and the distance between the neighboring grinding lines were found to be highly dependent on the rotational speed ratio. Huo et al. [36] developed a grinding marks formation model for wafer grinding and conducted a series of grinding tests to verify the model and study the influences of grinding parameters including the rotational speeds of the wafer and the grinding wheel, the infeed rate, and the wheel axial run-out on the surface topography. They found that the wavelength of the surface waviness was governed only by the angular speed ratio of the grinding wheel and the silicon wafer, and choosing a proper ratio value can effectively suppress the grinding marks on the surface. Wang et al. [37] established a three-dimensional surface formation model for small ball-end diamond wheel grinding and investigated the effects of various grinding parameters on the surface characteristics. The spatial wavelength in the circumferential direction was found to only be influenced by the integer part of the rotational speed ratio, while the surface residual height was governed by the integer part and the fractional part of the speed ratio, and other grinding parameters.

Parallel grinding is the most frequently employed machining method for manufacturing complex surfaces on hard and brittle materials. The workpiece and the grinding wheel both rotate around their own rotational axes, controlled by the X-Z slides to machine the required surface form. This method has been widely used to generate flat surfaces [38], SiC Fresnel molds [39], aspheric molds [40], a noncoaxial nonaxisymmetric aspheric lens [41], and freeform surfaces [42]. The rotational speed ratio has also been found by researchers to be an important variable governing surface formation. Chen et al. [43] found that the fractional part of the rotational speed ratio of the grinding wheel to the workpiece had a significant influence on the surface roughness and surface patterns, and a phase shift (fractional part of the ratio) of 0.5 is recommended to minimize the scallop height in order to achieve a better surface finish in the parallel grinding process. Pan et al. [39] built a model for the spatial period and amplitude of the surface waviness in parallel grinding. They pointed out that the phase shift introduced by the fractional part of the non-integer rotational speed ratio was helpful in suppressing the waviness amplitude. However, the optimal values of the fractional part of the speed ratio were different in the above two works. This is because the surface roughness was not only influenced by the speed ratios but also the other grinding conditions like the feed rate. When the other grinding conditions change, the optimal value of the fractional part of the speed ratio will change accordingly; but how to determine the optimal rotational speed ratio value while considering the other grinding conditions remains unclear. In previous works [44,45], we studied the vibration characteristics of the grinding wheel spindle and the surface formation mechanism under the spindle vibration in a parallel grinding process, and proposed a method using the reduced fraction of rotational speed ratio (not the fractional part of the decimal form) to predict the surface patterns. In this work, further analysis will be conducted to study how to optimize the grinding processes by using the reduced fraction.

The remainder of this paper is organized as follows. The surface formation mechanism will be revealed theoretically and the optimization method will be described in Section 2. The experimental details for the parallel grinding will be introduced in Section 3. Section 4 validates the method by experimental results and analysis. Conclusions will be drawn in Section 5.

2. Theoretical Analysis

In this section, the surface formation processes under different grinding parameters in parallel will be analyzed, and then an optimization method for improving the surface quality will be derived accordingly.

The configuration of the parallel grinding in this study is shown in Figure 1. The grinding wheel and the workpiece are held by the wheel spindle and the workpiece spindle, respectively. Both of them rotate in an anticlockwise direction at rotational speeds of n_1 and n_2, respectively. The grinding wheel moves from the edge of the workpiece to the center of it at a constant cross-feed rate of v_f during the grinding process. Thus, the locus of the grinding wheel is an Archimedes spiral, and it can be expressed in the global coordinate system $O(X_w Y_w Z_w)$ of the workpiece as a function of time t:

$$\begin{cases} x_w = (R_2 - v_f t)\cos(\frac{2\pi n_2}{60}t) \\ y_w = -(R_2 - v_f t)\sin(\frac{2\pi n_2}{60}t) \end{cases}' \tag{1}$$

where R_2 is the radius of the workpiece.

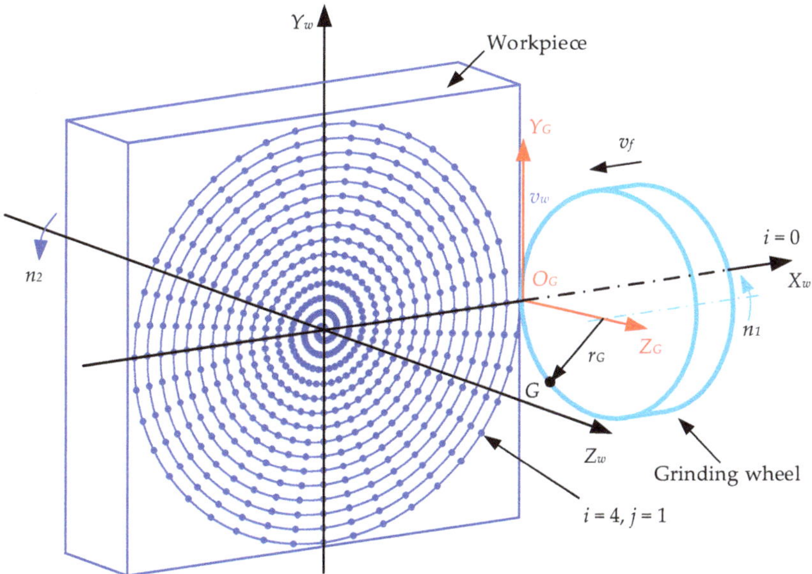

Figure 1. The configuration of the parallel grinding process.

It can also be written as:

$$\begin{cases} x_w = (R_2 - v_f \frac{30\theta}{\pi n_2}) \cos\theta \\ y_w = -(R_2 - v_f \frac{30\theta}{\pi n_2}) \sin\theta' \end{cases} \tag{2}$$

where θ is the rotation angle of the workpiece from $t = 0$ to t.

In the workpiece coordinate system $O(X_w Y_w)$, the workpiece surface is discretized into certain elements in the circumferential direction and the radial direction according to the two-dimensional locus of the workpiece expressed in Equations (1) or (2). The number of sections in the circumferential direction (N_c) and radial direction (N_r) can be calculated as:

$$\begin{cases} N_c = \frac{2\pi}{\Delta\theta} \\ N_r = \frac{R_2}{s}' \end{cases} \tag{3}$$

where $\Delta\theta$ is the angle interval in the circumferential direction, and s is the feeding distance per when the workpiece completes one revolution.

Therefore, the two-dimensional workpiece surface can be expressed in the discrete form as:

$$\begin{cases} x_w(i,j) = \left[R_2 - i\frac{R_2}{N_c N_r} - (j-1)\frac{R_2}{N_r}\right]\cos(i\Delta\theta) \\ y_w(i,j) = -\left[R_2 - i\frac{R_2}{N_c N_r} - (j-1)\frac{R_2}{N_r}\right]\sin(i\Delta\theta) \end{cases} \ for \ \begin{matrix} i = 0,1,2\ldots N_c \\ j = 1,2,3\ldots N_r \end{matrix}. \tag{4}$$

In the grinding process, a series of scallops are often generated due to the wheel run-out, wheel spindle vibration, or the ununiform grit height distribution on the wheel surface. The grinding wheel spindle vibration will produce the same vibration to the center of the grinding wheel, and then the trajectory paths of grits during rotation of the grinding wheel change correspondingly. As a result, the vibration will be copied onto the workpiece, producing a wavy surface [16,46]. Similarly, wheel run-out can also produce ups and downs on the ground surface, and an analysis of the influence of the wheel run-out on the wavy surface formation can be found in [39,47]. The trajectories of ununiform size and the distribution of grains on the grinding wheel surface may form a wavy surface and the corresponding analyses can be referred to in [16,46].

It is accepted in previous studies that the scallop height increases with the radial distance of the workpiece since a larger distance leads to a higher linear velocity when the workpiece spindle is operated at a constant rotational speed. If the scallop height is w_v at a certain radial distance, the surface profile can be simplified as:

$$h(i,j) = w_v(1 - \left|\cos(\frac{n_1}{2n_2}(i\Delta\theta + 2\pi(j-1)))\right|). \tag{5}$$

How to calculate the value of w_v is not the focus of this study, but this can be found in [16,39,45]. It should be noted that the surface profile calculated using Equation (5) is just a simplified form, describing the primary characteristics of the profile. Some secondary characteristics of the final generated surface profile induced by the uneven height and distribution of grits could not be considered in Equation (5).

In our previous research [45], a method using the lowest form of the rotational speed ratio of the grinding wheel to the workpiece has been proposed to reveal the formation mechanisms of the scallop waviness patterns of the workpiece surface in parallel grinding. In this method, b and d are defined as:

$$\frac{d}{b} = \frac{n_1}{n_2}. \tag{6}$$

By reducing the fraction (n_1/n_2) to lowest terms, the corresponding denominator and numerator are the values of b and d, respectively. In other words, b and d are the smallest integers that satisfy Equation (6).

Figure 2 shows the simulated surface profiles during three workpiece revolutions. In the simulation, the scallop height w_v of the wavy surface profile is set to be 1 µm, and the contact width of the grinding wheel is considered to be less than the feeding distance per workpiece revolution (s). $b = 1$ means that the rotational speed of the grinding wheel is an integer (d) times that of the workpiece. Thus, when the workpiece completes one cycle, a wavy surface profile with d periods is generated. In addition, since the rotational speed ratio is an integer, the surface profiles generated during each workpiece revolution are in the same angular position, as shown in Figure 2a. The red point denotes the center of the workpiece. It is clearly seen that the peaks of surface profiles generated during different workpiece revolutions are positioned at the same angular coordinate, as indicated by the green line with an arrow. However, if $b \neq 1$, the rotational speed ratio will not be an integer. Phase shift occurs during every adjacent workpiece revolution, as shown in Figure 2b.

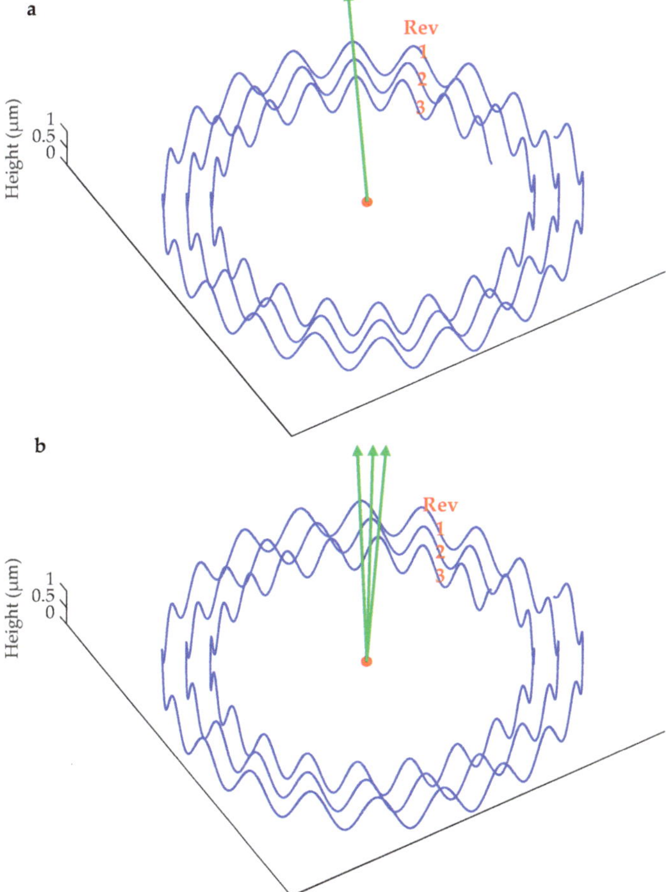

Figure 2. The wave surface profile generation (**a**) $b = 1$; and (**b**) $b \neq 1$.

In the actual grinding process, the cross-feed rate v_f is often selected to be small enough to make sure that the feed distance per workpiece revolution is smaller than the contact width of the workpiece. In this situation, there exists a primary grinding zone and a secondary grinding zone, as shown in Figure 3. The primary zone corresponds to the feed distance per workpiece revolution (s). The secondary grinding zone is introduced since the contact width is larger than s.

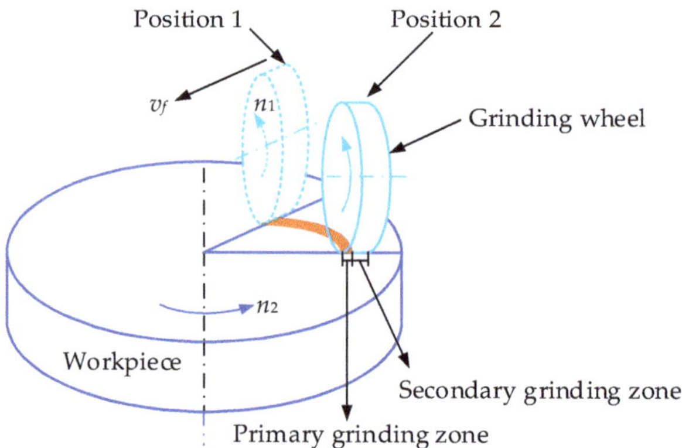

Figure 3. The primary grinding zone and the secondary grinding zone.

Figure 4 illustrates the surface profile generation due to the existence of the secondary grinding zone. It can be clearly seen that the surface profile generated by the primary grinding zone (indicated by the blue line) will be changed by the secondary grinding zone (the red line). If the ratio of the grinding wheel contact width to the feeding distance per workpiece is p, the width ratio of the secondary grinding zone to the primary grinding zone will be $p - 1$, which means that the surface profile generated by the primary grinding zone in the j-th workpiece revolution will be changed by the following $p - 1$ workpiece revolutions if phase shift exists (i.e., $b \neq 1$). This effect is called the overlapping effect.

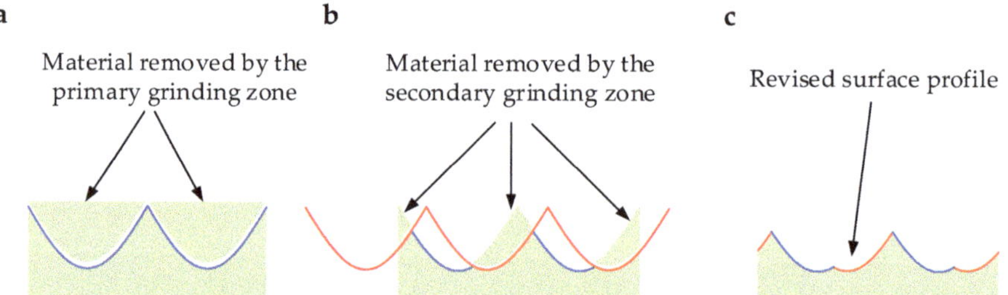

Figure 4. The surface profile generation due to the existence of the secondary grinding zone.

It can be observed that the overlapping effect leads to a reduction of the scallop height of the surface profile, which indicates that $b \neq 1$ can produce a better surface finish in comparison to $b = 1$. The degree of the overlapping effect which is related to the value of b determines how much reduction the scallop height experiences. Thus, how to select the value of b corresponding to the best surface finish is very important, and can be a way to optimize the parallel grinding process.

Figure 5 shows the surface profile generation process at different b values. p is set to be 3, thus the surface profile generated by the primary grinding zone in the first workpiece revolution will be influenced by the following two revolutions. In Figure 5a–d, $b = 5, 4, 3,$ and 2, respectively. It can be seen that the scallop height decreases first and increases as the value of b decreases. More importantly, it achieves the minimum when $b = p$ (3). In Figure 5, the phase shift (fractional part of the rotational speed ratio) is $1/5, 1/4, 1/3,$ and $1/2$. It should be noted that $b = p$ still corresponds to the minimal scallop height, even if the phase shift is changed to $2/5, 3/5, 4/5, 3/4,$ and $2/3$. In Figure 5, $p = 3$ is just an example; if p is changed to other values, the conclusion that $b = p$ produces the minimal scallop height is still valid.

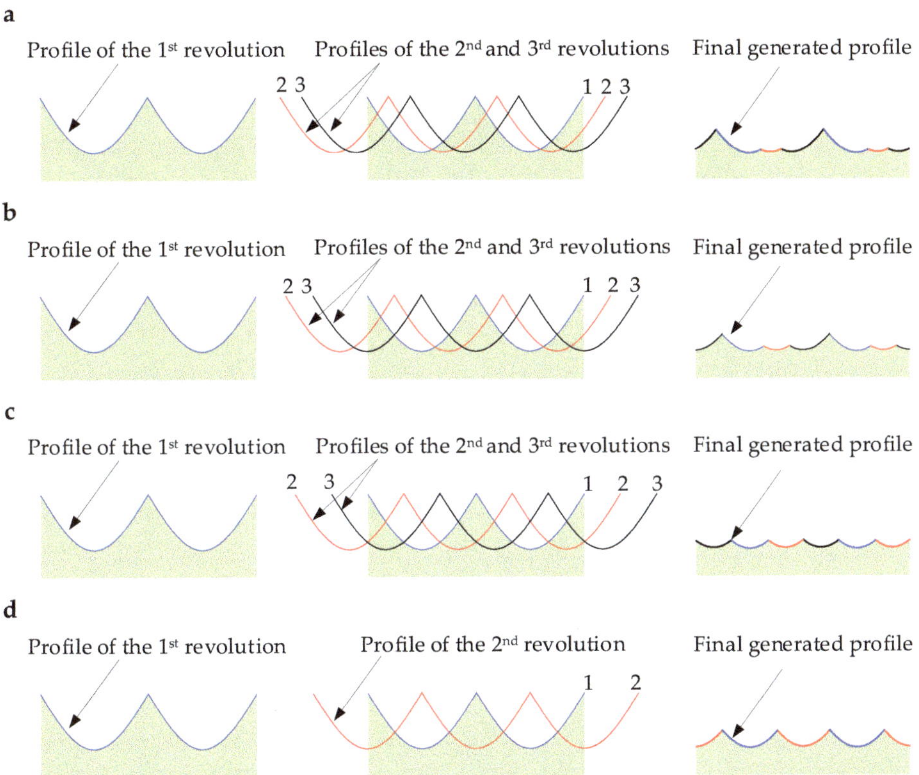

Figure 5. The surface profile generation process at different b values. (**a**) $b = 5$; (**b**) $b = 4$; (**c**) $b = 3$; and (**d**) $b = 2$.

3. Experimental Details

To validate the theoretical analysis in Section 2, a series of grinding experiments were conducted at different rotational speed ratios corresponding to different b values, and the measured ground surface profiles were analyzed.

The experimental setup is shown in Figure 6. Grinding experiments were performed on an ultra-precision grinding machine (Nanotech 450UPL, Moore Nanotechnology Systems, LLC, Swanzey, NH, USA). The workpiece with a fixture was mounted on the workpiece spindle through a vacuum chuck, and the grinding wheel was held by the wheel spindle. Both spindles were supported by ultra-precision aerostatic bearings to provide precise rotation motions. They rotated in an anticlockwise direction in the grinding operation. The grinding depth was controlled by the Z slide and the cross-feed motion was controlled by the X slide. The workpiece material was reaction-bonded silicon carbide sheets with a dimension of 8 mm × 8 mm × 5 mm. A resin-bonded diamond grinding wheel with a mesh number of 325# was employed in the experiments. The diameter and width of the grinding wheel were 20 mm and 5 mm, respectively. The grinding wheel had a sharp edge whose measured radius was approximately 30 μm, as shown in Figure 6. The minimum quantity lubrication was applied to all the grinding tests.

Figure 6. Experimental setup with an enlarged view of the geometrical motion relationship between the workpiece and the grinding wheel.

The values of the grinding parameters are shown in Table 1. The cross-feed rate, the grinding depth, and the workpiece spindle speeds were 8 mm·min^{-1}, 2 μm, and 2000 rpm, respectively. The grinding wheel speeds were set to be 40,000 rpm, 40,200 rpm, 40,400 rpm, 40,667 rpm, and 41,000 rpm. Thus, the corresponding b values are 1, 10, 5, 3, and 2, respectively. For the grinding wheel spindle, the actual rotational speed deviated slightly from the preset one. However, by carefully adjusting the preset values in the experiments, the actual rotational speeds could be ensured to be almost consistent with the values listed in Table 1 within an error of less than 1 rpm.

Table 1. Grinding conditions.

Parameters	Values
Cross-feed rate	8 mm·min^{-1}
Grinding depth	2 μm
Workpiece spindle speed	2000 rpm
Grinding wheel spindle speed	40,000 rpm, 40,200 rpm, 40,400 rpm, 40,667 rpm, 41,000 rpm

Before each grinding test, truing and dressing operations for the grinding wheel were performed on machine to keep the grinding wheel surface fresh and reduce the influences of the grinding wheel wear. Also, the balancing operations were performed carefully for both spindles, because the mass imbalance-induced vibration can have a great influence on surface generation in grinding. After grinding, the workpieces were removed from the fixture and cleaned using alcohol. Then, the surface topographies of the workpieces were measured by an optical surface profiler (Zygo@ Nexview, Zygo Corporation, Middlefield, CT, USA) for analysis.

4. Experimental Results and Discussion

The three-dimensional surface topographies of the workpieces at different grinding wheel spindle speeds, which corresponded to different b values, were obtained by Zygo@ Nexview. The scallop height was different at different radial distances of the workpieces; that was, a larger radial distance produced a higher scallop height due to a higher linear velocity in the parallel grinding processes. Thus, the surface roughness was not uniform along the radial direction of the workpiece. In this study, the surface profiles in the circumferential direction at the radial distance of 1.5 mm were extracted from the three-dimensional surface height data from the measurement results for analyzing the variation of the surface roughness with the value of b.

Figure 7 shows the surface profiles in the circumferential direction for different b values. It can be seen that the scallop height was the highest when $b = 1$ among these five profiles. This is because there was no phase shift when $b = 1$, and the surface profiles generated by the primary grinding zone were not influenced by the secondary grinding zone. If $b \neq 1$, the phase shift occurred between the profiles generated by neighboring workpiece revolutions, resulting in the overlapping effect, which contributed to a reduction of the scallop height. In the experiment, the ratio (p) of the grinding wheel contact width to the feeding distance was close to 3. It can be seen that the scallop height decreased first and then increased with the decreasing value of b. When $b = p$, it reached the minimum.

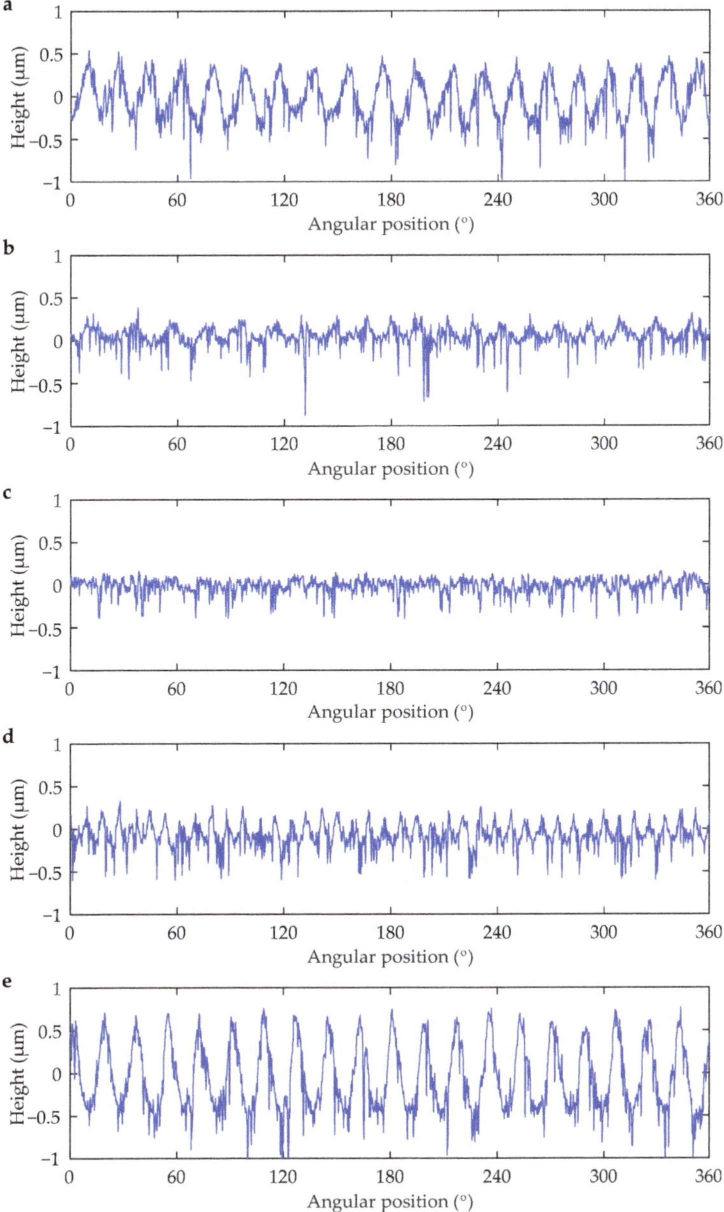

Figure 7. The measured surface profiles of the workpieces at the radial distance of 1.5 mm at different b values (**a**) 10; (**b**) 5; (**c**) 3; (**d**) 2; and (**e**) 1.

In addition, when $b = 1$, the period number of the surface profile was 20, which was equal to the value of d. When $b < p$, the period number would increase due to the overlapping effect (for example, in Figure 7d, the period number increased to 41); however, it was still equal to the value of d. When $b > p$, the period number would change very slightly (less than 1).

The roughness values of the above surface profiles shown in Figure 7 were also calculated for analysis, in which the arithmetic roughness (R_a) and the root-mean-square roughness (R_q) were adopted. Figure 8 shows the variation of the surface roughness including R_a and R_q with the value of b. The measured surface profiles are in the circumferential direction that is perpendicular to the feed direction. The data of the surface profiles are extracted from the three-dimensional surface data measured by the optical surface profiler. The length of the surface profiles is 9.4248 mm and there are 6390 measurement points along the circumferential direction. It can be seen that the value of b has a great influence on the roughness of the surface profiles of the workpiece. When b is close to p, both R_a and R_q are the minimum, though the values of R_q are slightly larger than those of R_a for all the cases.

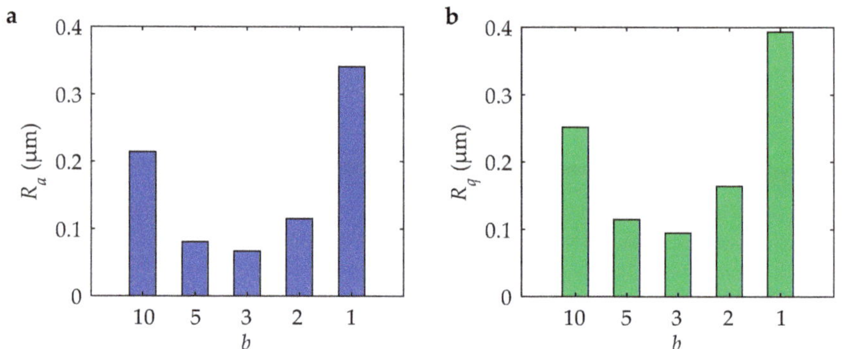

Figure 8. The variation of the roughness of the surface profiles of the workpieces with the value of b. (a) R_a and (b) R_q.

In addition, to evaluate the influence of b values on the whole surfaces, the roughness values of the whole surfaces of the workpieces are obtained, including the arithmetic roughness (S_a) and the root-mean-square roughness (S_q) at different values of b, as shown in Figure 9. Similar to the results of the roughness of the surface profiles in the circumferential direction described in Figure 8, the root-mean-square roughness (S_q) values are slightly higher than the arithmetic roughness (S_a) values for all the b values. It can also be found that the roughness values of the whole surfaces are higher than those of the surface profiles in the circumferential direction at the radial position of 1.5 mm by comparing Figures 8 and 9. More importantly, b values have the same influence on the roughness of the whole surface, that is, the roughness decreases first and increases with b. When b is close to p, the roughness can reach the minimum. This indicates that the grinding process can be optimized in terms of surface roughness by selecting an appropriate b value.

Therefore, adjusting the rotational speed of the workpiece or that of the grinding wheel to regulate the value of b to make it close to the ratio (p) of the grinding wheel contact width to the cross-feed per workpiece revolution can be a convenient way to optimize the parallel grinding process for achieving a better surface finish without compromising the machining efficiency.

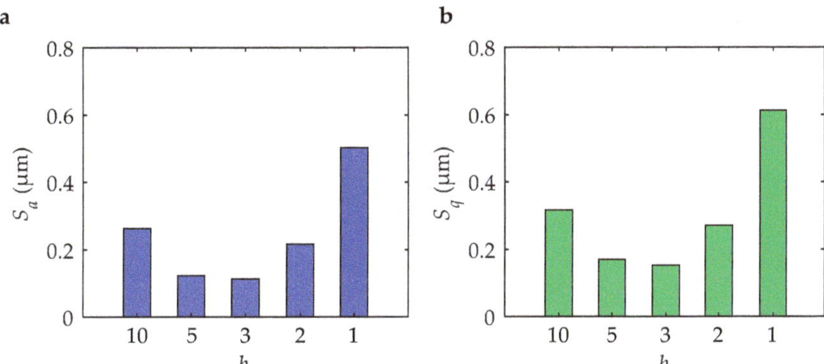

Figure 9. The variation of the roughness of the whole surfaces of the workpieces with the value of b. (**a**) S_a and (**b**) S_q.

In the grinding operation, it is generally accepted that the smaller workpiece speeds and higher grinding wheel speeds can produce a better surface finish if the overlapping effect does not exist. However, it should be noted that the grinding wheel speeds close to the natural frequency of the grinding spindle system may deteriorate the surface roughness due to the resonance [16]. After the rotational speed values of the grinding wheel spindle and the workpiece spindle are roughly determined, the overlapping effect can be introduced to decrease the surface roughness. By carefully adjusting the speed values to make the b value close to the ratio p, the optimal surface roughness can be achieved.

5. Conclusions

In this study, the surface generation processes were analyzed at different grinding conditions for parallel grinding, and a novel approach to optimizing the surface finish was proposed. Finally, grinding experiments were conducted to verify the effectiveness of the approach. The main conclusions are drawn as follows:

(1) There exists a primary grinding zone and a secondary grinding zone when the grinding wheel contact width is larger than the feed per workpiece revolution.

(2) The denominator (b) of the simplest form of the rotational speed ratio of the grinding wheel to the workpiece has a great influence on surface generation. When $b \neq 1$, the phase shift occurs between surface profiles generated by neighboring workpiece revolutions, introducing the overlapping effect. The surface profile generated by the primary grinding zone will be revised by the secondary grinding zone.

(3) The overlapping effect makes a contribution to the reduction of the scallop height on the ground surface. The surface finish can be optimized by adjusting the rotational speeds of the workpiece spindle or the grinding wheel spindle to make the value of b close to the ratio (p) of the wheel contact width to the feed per workpiece revolution.

In summary, the proposed approach for optimizing the parallel grinding process is an effective way to improve the surface quality of workpieces without compromising the machining rate, which can be applied to industrial applications.

Author Contributions: Conceptualization, T.Y.; methodology, T.Y.; software, T.Y.; validation, T.Y., H.Z. and W.H.; formal analysis, T.Y. and H.Z.; investigation, T.Y., H.Z., W.H. and S.T.; resources, S.T.; data curation, T.Y. and S.T.; writing—original draft preparation, T.Y.; writing—review and editing, T.Y. and S.T.; visualization, T.Y.; supervision, S.T.; project administration, S.T.; funding acquisition, S.T. and W.H. All authors have read and agreed to the published version of the manuscript.

Funding: This research was funded by National Key Research and Development Program of China (Grant No. 2023YFE0202900) and International Science and Technology Cooperation Project of Guangzhou Development District/Huangpu District (Project No. 2020GH05). The work was also partially supported by the funding support to the State Key Laboratories in Hong Kong from the Innovation and Technology Commission (ITC) of the Government of the Hong Kong Special Administrative Region (HKSAR), China. The authors also would like to thank the Research Committee of The Hong Kong Polytechnic University.

Data Availability Statement: Data are contained within the article.

Conflicts of Interest: Hanqian Zhang was employed by Guangzhou Haozhi Industrial Co., Ltd., Guangzhou, China. The remaining authors declare that the research was conducted in the absence of any commercial or financial relationships that could be construed as a potential conflict of interest.

References

1. Brinksmeier, E.; Mutlugünes, Y.; Klocke, F.; Aurich, J.; Shore, P.; Ohmori, H. Ultra-precision grinding. *CIRP Ann.* **2010**, *59*, 652–671. [CrossRef]
2. Huang, H.; Li, X.; Mu, D.; Lawn, B.R. Science and art of ductile grinding of brittle solids. *Int. J. Mach. Tools Manuf.* **2021**, *161*, 103675. [CrossRef]
3. Zhao, G.; Zhao, B.; Ding, W.; Xin, L.; Nian, Z.; Peng, J.; He, N.; Xu, J. Nontraditional energy-assisted mechanical machining of difficult-to-cut materials and components in aerospace community: A comparative analysis. *Int. J. Extrem. Manuf.* **2023**, *6*, 022007. [CrossRef]
4. Chen, H.; Wu, Z.; Hong, B.; Hang, W.; Zhang, P.; Cao, X.; Xu, Q.; Chen, P.; Chen, H.; Yuan, J. Study on the affecting factors of material removal mechanism and damage behavior of shear rheological polishing of single crystal silicon carbide. *J. Manuf. Process.* **2024**, *112*, 225–237. [CrossRef]
5. Du, H.; Yin, T.; Yip, W.S.; Zhu, Z.; To, S. Generation of structural colors on pure magnesium surface using the vibration-assisted diamond cutting. *Mater. Lett.* **2021**, *299*, 130041. [CrossRef]
6. Zhao, L.; Zhang, J.; Zhang, J.; Hartmaier, A.; Sun, T. Numerical simulation of materials-oriented ultra-precision diamond cutting: Review and outlook. *Int. J. Extrem. Manuf.* **2023**, *5*, 022001. [CrossRef]
7. Wang, J.; Wang, Y.; Zhang, G.; Xu, B.; Zhao, Z.; Yin, T. Fabrication of Polymethyl Methacrylate (PMMA) Hydrophilic Surfaces Using Combined Offset-Tool-Servo Flycutting and Hot Embossing Methods. *Polymers* **2023**, *15*, 4532. [CrossRef]
8. Zhang, Z.; Yan, J.; Kuriyagawa, T. Study on tool wear characteristics in diamond turning of reaction-bonded silicon carbide. *Int. J. Adv. Manuf. Technol.* **2011**, *57*, 117–125. [CrossRef]
9. Yingfei, G.; Jiuhua, X.; Hui, Y. Diamond tools wear and their applicability when ultra-precision turning of SiCp/2009Al matrix composite. *Wear* **2010**, *269*, 699–708. [CrossRef]
10. Chen, X.; Liu, C.; Ke, J.; Zhang, J.; Shu, X.; Xu, J. Subsurface damage and phase transformation in laser-assisted nanometric cutting of single crystal silicon. *Mater. Des.* **2020**, *190*, 108524. [CrossRef]
11. Zhao, G.; Hu, M.; Li, L.; Zhao, C.; Zhang, J.; Zhang, X. Enhanced machinability of SiCp/Al composites with laser-induced oxidation assisted milling. *Ceram. Int.* **2020**, *46*, 18592–18600. [CrossRef]
12. You, K.; Yan, G.; Luo, X.; Gilchrist, M.D.; Fang, F. Advances in laser assisted machining of hard and brittle materials. *J. Manuf. Process.* **2020**, *58*, 677–692. [CrossRef]
13. Bertolini, R.; Andrea, G.; Alagan, N.T.; Bruschi, S. Tool wear reduction in ultrasonic vibration-assisted turning of SiC-reinforced metal-matrix composite. *Wear* **2023**, *523*, 204785. [CrossRef]
14. Ahmed, F.; Ko, T.J.; Kurniawan, R.; Kwack, Y. Machinability analysis of difficult-to-cut material during ultrasonic vibration-assisted ball end milling. *Mater. Manuf. Process.* **2021**, *36*, 1734–1745. [CrossRef]
15. Xing, Y.; Xue, C.; Liu, Y.; Du, H.; Yip, W.S.; To, S. Freeform surfaces manufacturing of optical glass by ultrasonic vibration-assisted slow tool servo turning. *J. Mater. Process. Technol.* **2023**, *324*, 118271. [CrossRef]
16. Yin, T.; Du, H.; Zhang, G.; Hang, W.; To, S. Theoretical and experimental investigation into the formation mechanism of surface waviness in ultra-precision grinding. *Tribol. Int.* **2023**, *180*, 108269. [CrossRef]
17. Mayer, J.E., Jr.; Fang, G.-P. Effect of grinding parameters on surface finish of ground ceramics. *CIRP Ann.* **1995**, *44*, 279–282. [CrossRef]
18. De Souza Ruzzi, R.; da Silva, R.B.; da Silva, L.R.R.; Machado, Á.R.; Jackson, M.J.; Hassui, A. Influence of grinding parameters on Inconel 625 surface grinding. *J. Manuf. Process.* **2020**, *55*, 174–185. [CrossRef]
19. Nadolny, K.; Kapłonek, W. Design of a device for precision shaping of grinding wheel macro-and micro-geometry. *J. Cent. South Univ.* **2012**, *19*, 135–143. [CrossRef]
20. Wasif, M.; Iqbal, S.A.; Ahmed, A.; Tufail, M.; Rababah, M. Optimization of simplified grinding wheel geometry for the accurate generation of end-mill cutters using the five-axis CNC grinding process. *Int. J. Adv. Manuf. Technol.* **2019**, *105*, 4325–4344. [CrossRef]
21. Gopal, A.V.; Rao, P.V. Selection of optimum conditions for maximum material removal rate with surface finish and damage as constraints in SiC grinding. *Int. J. Mach. Tools Manuf.* **2003**, *43*, 1327–1336. [CrossRef]
22. Demir, H.; Gullu, A.; Ciftci, I.; Seker, U. An investigation into the influences of grain size and grinding parameters on surface roughness and grinding forces when grinding. *J. Mech. Eng.* **2010**, *56*, 447–454.

23. Yang, M.; Kong, M.; Li, C.; Long, Y.; Zhang, Y.; Sharma, S.; Li, R.; Gao, T.; Liu, M.; Cui, X. Temperature field model in surface grinding: A comparative assessment. *Int. J. Extrem. Manuf.* **2023**, *5*, 042011. [CrossRef]
24. Shen, B.; Shih, A.J.; Tung, S.C. Application of nanofluids in minimum quantity lubrication grinding. *Tribol. Trans.* **2008**, *51*, 730–737. [CrossRef]
25. Wang, Y.; Li, C.; Zhang, Y.; Yang, M.; Li, B.; Dong, L.; Wang, J. Processing characteristics of vegetable oil-based nanofluid MQL for grinding different workpiece materials. *Int. J. Precis. Eng. Manuf. Green Technol.* **2018**, *5*, 327–339. [CrossRef]
26. Daneshi, A.; Jandaghi, N.; Tawakoli, T. Effect of dressing on internal cylindrical grinding. *Procedia CIRP* **2014**, *14*, 37–41. [CrossRef]
27. Hassui, A.; Diniz, A.E. Correlating surface roughness and vibration on plunge cylindrical grinding of steel. *Int. J. Mach. Tools Manuf.* **2003**, *43*, 855–862. [CrossRef]
28. Tang, J.; Du, J.; Chen, Y. Modeling and experimental study of grinding forces in surface grinding. *J. Mater. Process. Technol.* **2009**, *209*, 2847–2854. [CrossRef]
29. Deresse, N.C.; Deshpande, V.; Taifa, I.W. Experimental investigation of the effects of process parameters on material removal rate using Taguchi method in external cylindrical grinding operation. *Eng. Sci. Technol. Int. J.* **2020**, *23*, 405–420. [CrossRef]
30. Nadolny, K.; Kieraś, S. Experimental studies on the centrifugal MQL-CCA method of applying coolant during the internal cylindrical grinding process. *Materials* **2020**, *13*, 2383. [CrossRef]
31. Qin, F.; Zhang, L.; Chen, P.; An, T.; Dai, Y.; Gong, Y.; Yi, Z.; Wang, H. In situ wireless measurement of grinding force in silicon wafer self-rotating grinding process. *Mech. Syst. Signal Process.* **2021**, *154*, 107550. [CrossRef]
32. Wang, Z.; Guo, J. Research on an optimized machining method for parallel grinding of f-θ optics. *Int. J. Adv. Manuf. Technol.* **2015**, *80*, 1411–1419. [CrossRef]
33. Xi, J.; Zhao, H.; Li, B.; Ren, D. Profile error compensation in cross-grinding mode for large-diameter aspheric mirrors. *Int. J. Adv. Manuf. Technol.* **2016**, *83*, 1515–1523. [CrossRef]
34. Patel, A.; Bauer, R.; Warkentin, A. Investigation of the effect of speed ratio on workpiece surface topography and grinding power in cylindrical plunge grinding using grooved and non-grooved grinding wheels. *Int. J. Adv. Manuf. Technol.* **2019**, *105*, 2977–2987. [CrossRef]
35. Chidambaram, S.; Pei, Z.; Kassir, S. Fine grinding of silicon wafers: A mathematical model for grinding marks. *Int. J. Mach. Tools Manuf.* **2003**, *43*, 1595–1602. [CrossRef]
36. Huo, F.; Kang, R.; Li, Z.; Guo, D. Origin, modeling and suppression of grinding marks in ultra precision grinding of silicon wafers. *Int. J. Mach. Tools Manuf.* **2013**, *66*, 54–65. [CrossRef]
37. Wang, T.; Liu, H.; Wu, C.; Cheng, J.; Chen, M. Three-dimensional modeling and theoretical investigation of grinding marks on the surface in small ball-end diamond wheel grinding. *Int. J. Mech. Sci.* **2020**, *173*, 105467. [CrossRef]
38. Sun, X.; Stephenson, D.; Ohnishi, O.; Baldwin, A. An investigation into parallel and cross grinding of BK7 glass. *Precis. Eng.* **2006**, *30*, 145–153. [CrossRef]
39. Pan, Y.; Zhao, Q.; Guo, B.; Chen, B.; Wang, J.; Wu, X. An investigation of the surface waviness features of ground surface in parallel grinding process. *Int. J. Mech. Sci.* **2020**, *170*, 105351. [CrossRef]
40. Chen, B.; Guo, B.; Zhao, Q. An investigation into parallel and cross grinding of aspheric surface on monocrystal silicon. *Int. J. Adv. Manuf. Technol.* **2015**, *80*, 737–746. [CrossRef]
41. Jiang, C.; Guo, Y.; Li, H. Parallel grinding error for a noncoaxial nonaxisymmetric aspheric lens using a fixture with adjustable gradient. *Int. J. Adv. Manuf. Technol.* **2013**, *66*, 537–545. [CrossRef]
42. Chen, S.; Cheung, C.F.; Zhang, F.; Liu, M. Optimization of tool path for uniform scallop-height in ultra-precision grinding of free-form surfaces. *Nanomanuf. Metrol.* **2019**, *2*, 215–224. [CrossRef]
43. Chen, S.; Cheung, C.; Zhao, C.; Zhang, F. Simulated and measured surface roughness in high-speed grinding of silicon carbide wafers. *Int. J. Adv. Manuf. Technol.* **2017**, *91*, 719–730. [CrossRef]
44. Yin, T.; Zhang, G.; Du, J.; To, S. Nonlinear analysis of stability and rotational accuracy of an unbalanced rotor supported by aerostatic journal bearings. *IEEE Access* **2021**, *9*, 61887–61900. [CrossRef]
45. Yin, T.; To, S.; Du, H.; Zhang, G. Effects of wheel spindle error motion on surface generation in grinding. *Int. J. Mech. Sci.* **2022**, *218*, 107046. [CrossRef]
46. Cao, Y.; Guan, J.; Li, B.; Chen, X.; Yang, J.; Gan, C. Modeling and simulation of grinding surface topography considering wheel vibration. *Int. J. Adv. Manuf. Technol.* **2013**, *66*, 937–945. [CrossRef]
47. Badger, J.; Murphy, S.; O'Donnell, G. The effect of wheel eccentricity and run-out on grinding forces, waviness, wheel wear and chatter. *Int. J. Mach. Tools Manuf.* **2011**, *51*, 766–774. [CrossRef]

Article

Improving the Quality of Tantalum Cylindrical Deep-Drawn Part Formation Using Different Lubricating Media-Coated Dies

Teng Xu [1], Shihao Dou [1], Mingwu Su [1], Jianbin Huang [1], Ningyuan Zhu [2], Shangpang Yu [3] and Likuan Zhu [1,*]

[1] Shenzhen Key Laboratory of High Performance Nontraditional Manufacturing, College of Mechatronics and Control Engineering, Shenzhen University, Shenzhen 518060, China; tengxu@szu.edu.cn (T.X.); 2100291010@email.szu.edu.cn (S.D.); 15361914178@163.com (M.S.); hdzshjb2021@163.com (J.H.)

[2] School of Mechanical and Electrical Engineering, Jiangxi University of Science and Technology, Ganzhou 341000, China; zhuningyuan@126.com

[3] Department of Mechanical Engineering, National Taipei University of Technology, No. 1, Sec. 3, Zhongxiao E. Rd., Taipei 10608, Taiwan; ysp@ntut.edu.tw

* Correspondence: zhulikuan@yeah.net; Tel.: +86-18588960496

Abstract: Lubrication is one of the key factors to improve metal-forming quality. In the process of deep drawing, seizing tumors easily occur on the contact surfaces between the tantalum metal and the mold, which greatly affects the forming quality of the deep-drawn parts. Quality-forming quality problems that occur during the deep drawing of tantalum metal are studied from the perspective of lubrication in this paper. Three lubrication media, caster oil, PE (polyethylene) film, and DLC (Diamond Like Carbon) film, were adopted in the deep drawing of tantalum cylindrical cups. A universal testing machine and microscope were used to investigate the effect of lubrication media on the limit-drawing ratio, maximum forming force, and surface topography quality during the deep drawing process of the tantalum sheet. The results reveal that the lubrication of the PE film and DLC film can greatly improve the forming quality of the tantalum metal sheet, in which the DLC film has higher wear resistance and lower friction coefficient and can be used as the lubricating medium in the industrial forming process of tantalum deep-drawn parts.

Keywords: tantalum metal; lubrication media; DLC; limit drawing ratio; surface topography quality

Citation: Xu, T.; Dou, S.; Su, M.; Huang, J.; Zhu, N.; Yu, S.; Zhu, L. Improving the Quality of Tantalum Cylindrical Deep-Drawn Part Formation Using Different Lubricating Media-Coated Dies. *Processes* **2024**, *12*, 210. https://doi.org/10.3390/pr12010210

Academic Editor: Hideki KITA

Received: 1 December 2023
Revised: 8 January 2024
Accepted: 9 January 2024
Published: 18 January 2024

1. Introduction

The tantalum and tantalum alloy are some of the most important high-temperature and corrosion-resistant functional materials in modern industry [1], which have a high melting point, good electrical and thermal conductivity, chemical stability, high-temperature strength, and workability [2]. Due to the excellent properties of the tantalum material, it has been used in aerospace [3], weapons and equipment [4], chemical industry [5], electronics [6] and medical and health fields [7]. At present, tantalum is widely produced in tantalum capacitors. Compared to other capacitors in the field of capacitors, tantalum capacitors have outstanding advantages, such as a high specific capacity, high reliability, long life, small size, lightweight, and wide operating temperature range, which makes it one of the indispensable ideal electronic components in high-tech fields such as aerospace, aviation, military engineering, electronic computers, automatic control equipment, communication equipment, and medical electronic devices. Among the many branches of tantalum capacitors, liquid tantalum capacitors [8], which are known as "never fail" capacitors in the industry, have excellent performance and are mainly made of a deep-drawn tantalum alloy shell containing electrolytes and sealed by glass [9].

Deep drawing is an important process for forming thin-walled tantalum alloy containers and other shell parts [10]. The problem of adhesive wear is one of the key problems in the deep drawing of thin wall tantalum alloy members. In addition to common cracking and wrinkling defects, during the deep drawing process of the tantalum alloy sheet, the material is transferred from the sheet to the mold due to factors such as low deformation resistance, low hardness as well as weak scratch resistance, resulting in adhesion, the surface wear of the sheet and the formation of a seizing tumor on the mold surface [11]. The growth and drop out of the seizing tumor are periodic, which can eventually lead to the severe adhesive wear of the die and sheet [12] and further cause a decrease in the service performance of the deep-drawn parts and result in great safety risks when the deep-drawn parts experience extreme service conditions. Taking the case of the tantalum alloy electrolytic capacitor as an example, the surface wear of the deep-drawn part could lead to the leakage of the embedded electrolyte.

To solve the seizing tumor phenomenon during the deep drawing process, numerous scholars have carried out in-depth-related studies. In the study of Freiße [13], a laser-generated tool surface with a supporting plateau of hard particles is presented. Spherical fused tungsten carbides were injected into the surface via a laser melt injection. The metallic matrix of the composite was rejected by applying laser ablation. As a consequence, the hard particles stood out of the matrix and were in direct contact with the sheet material. Dry and lubricated forming experiments were carried out by strip drawing with the bending and deep drawing of cups. Within this work, the feasibility of dry metal forming high-alloy steel was demonstrated by applying the MMC surface, whereby adhesive wear could be reduced. Marchin and Ashrafifizadeh [14] investigated the effect of carbon addition on the tribological behavior of multilayered TiSiN coating, and the performance of TiSiCN and TiSiN coatings on cold-forming steel dies were compared. Tribological tests were conducted on a ball-on-disc wear tester using a zirconia ball, indicating the lower coefficient of friction of the coatings compared to the steel substrate; the lowest value of 0.2 was experienced by the TiSiCN-coated surface. A study of wear mechanisms proved that the application of coatings changed the mode of prevailing wear from adhesive to abrasive, decreasing the wear rate to one-third. Adding carbon to TiSiN changed the tribological behavior of the coating, similar to a self-lubricant film. Field tests of TiSiCN coating on forming dies for deep drawing operation of steel tubes and the results revealed that the service life of the die increased by 10 times.

Ke Xugui and Zhang Rongqing [15] believed that the effective prevention of seizing tumors was closely related to the mold material, mold surface treatment, and lubricating medium. High-speed steel SKH2 was chosen instead of cemented carbide as the mold material, and then oxygen and nitrogen treatment was carried out on the mold surface to make its hardness reach more than 60 HRC. Tantalum cylindrical cups with smooth surfaces which meet the quality requirements have been successfully formed. The research results showed that effective measures to prevent the adhesion wear of pure tantalum sheets during deep drawing should be considered from many aspects, such as mold material, mold surface quality, and the lubricating medium. Liu Hao et al. [16] designed a mold using Si_3N_4 ceramic as the mold material and analyzed the influence of the lubricating medium on the surface quality of pure tantalum cylindrical cups and the flowability of the material in specific areas. The results showed that the use of ceramic molds can significantly reduce the surface seizing tumor of pure tantalum cylindrical cups, proving that a suitable lubricating medium can reduce the surface roughness of pure tantalum cylindrical cups and improve the material flowability of the pure tantalum sheet formed.

B. Sresomroeng and V. Premanond et al. [17] evaluated the anti-adhesion performance of commercial nitride and DLC films coated on cold work tool steel against HSS in the forming operation. The anti-adhesion performance of biofilm-coated tools in the metal stamping process was also investigated by performing a U-bending experiment. The results displayed how, for high-strength steel (HSS), the adhesion wear of workpiece material on a non-coated die surface was detected after 49 strokes, whereas adhesion wear could not be found in the case of stamping SPCC sheets up to 500 strokes. This indicates that the type of workpiece material is one of the major effects that influences the occurrence of adhesion wear. All types of coating selected on the die surface can be used to improve anti-adhesion properties when HSS is formed. Andreas Wank, Guido Reisel, and Bernhard Wielage [18] investigated the behavior of DLC coatings in the lubricant-free, cold, massive formation of aluminum. The capability of diamond-like carbon coatings to permit the formation of massive lubricant-free cold aluminum is tested by forced-in tests featuring a high deformation degree. The frictional behavior during penetration into AA6016 material is tested for PVD a-C as well as PECVD a-C: H and a-C:H: Si coatings in comparison to an uncoated 1.3343 tool of steel. It is shown that the use of diamond-like carbon coatings does not generally result in a reduction in aluminum workpiece material transfer to respective tool surfaces in cold, massive forming processes with high deformation degrees. Despite increased friction and material take-up, a-C: H-coated punches show important advantages concerning the formation of the product surface quality improvement. Both in axial and circumferential directions, the lower roughness of produced bores compared to uncoated punches and silicon-doped or hydrogen-free diamond-like carbon coatings is achieved. Also, in contrast to uncoated punches, no change in surface quality is observed for repeated testing. Steiner. J, Andreas. K, Merkleini. M [19] investigated the potential of carbon-based coatings for dry deep drawing and friction, and the wear behavior of different coating compositions was evaluated in strip drawing tests. The tribological behavior of tetrahedral amorphous (ta-C) and hydrogenated amorphous carbon coatings with and without tungsten modification (a-C:H: W, a-C: H) was investigated. The influence of tool topography is analyzed by applying different finishes to the surface. The results show reduced friction with a decreased roughness for coated tools. Besides tool topography, the coating type determines the tribological conditions. Smooth tools with ta-C and a-C: H coatings reveal low friction and prevent adhesive wear. By contrast, smooth a-C:H: W-coated tools only lead to a slight improvement compared to rough, uncoated specimens.

Although the above studies achieved a reduction in the generation rate of seizing tumors, the processing costs are high, the processing processes are complicated, and the research reports on pure tantalum materials are few. In this paper, different lubrication media are utilized to compare the results of deep drawing with this lubrication condition as well as that with dry friction to explore the effect of the lubrication media on the limit drawing ratio and surface topography quality of the tantalum cylindrical cups and determine the suitable lubrication medium.

2. Experimental Section

2.1. Schematic of Deep Drawing

A schematic of the deep drawing process in this paper is shown in Figure 1. The whole working process is as follows: First of all, the blank was placed into the circular groove in the center of the die (the depth of the circular groove is the same as the thickness of the blank, 0.2 mm. It locates the position of the blank and provides a constant blank holding force during the drawing process). Then, the blank holder was installed and tightened with bolts. Then, the punch was placed in the right position through the positioning hole of the blank holder. The function of the positioning hole was to align the punch with the die. In the end, the height of the universal testing machine was adjusted so that the punch and the surface of the plate were tight. After the universal test machine started up. With the punch down, the blank began to deform and was slowly pulled into the mold until the universal testing machine completed the set stroke. After the automatic return of the punch, the blank holder was withdrawn, taking the forming cup out and placing it in the ultrasonic cleaning machine to wash the surface impurities away; it stayed under natural air-drying conditions for further experimental observation.

When the gap between the punch and die is too small, the friction resistance, as well as the extrusion stress between the sheet and the die, increase, resulting in an increase in deep drawing force and a reduction in the limit drawing ratio, which can easily cause the fracture defect. The excessive gap between the punch and die can reduce the straightening effect of the die on the sheet, resulting in edge wrinkle defects during the deep drawing. To observe the surface morphology quality of tantalum cylindrical cups and explore the effect of different lubrication conditions on the forming quality of tantalum cylindrical cups, the gap was set as 1.1 t, where t = 0.2 mm was the thickness of the tantalum sheet.

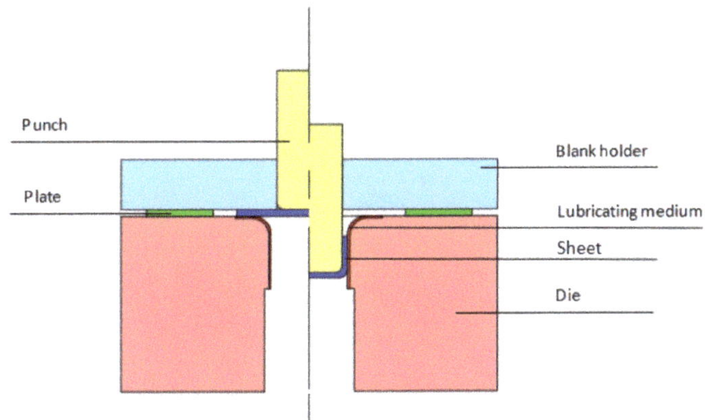

Figure 1. Schematic of deep drawing.

In the process of deep drawing, the blank has a large bending deformation when sliding in the round corner of the punch, and it is straightened again when entering the straight wall from the round corner of the die as it passes the gap between the punch and die.

The diameter of the punch was 6 mm, and the corner radius of the punch was 0.4 mm (Figure 2). The influence of the corner radius of the punch on the process of deep drawing cis described as follows: If the corner radius of the punch is too small, the blank is subjected to excessive bending deformation in this part, the strength of the dangerous section of the blank is reduced, and the limit drawing coefficient is increased. In addition, even if the blank is not fractured in the dangerous section, if it is too small, the corner radius of the punch causes the local thinning of the thickness of the blank near the dangerous section, and this local thinning and bending trace is left on the side wall of the finished part after the deep drawing process, ultimately affecting the surface quality of the forming part.

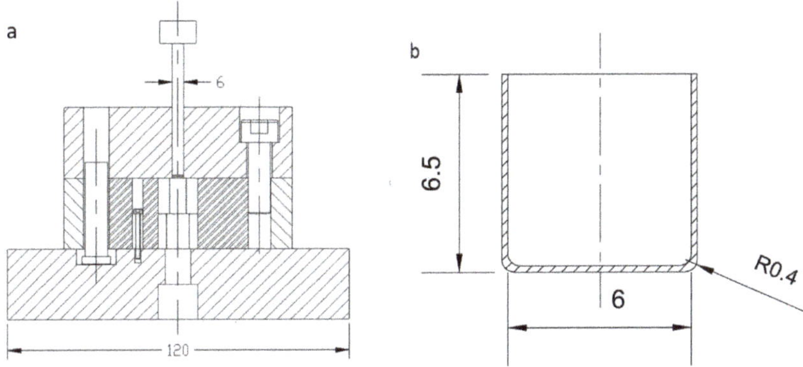

Figure 2. Schematic (a) deep-drawing die and (b) tantalum cup.

If the corner radius of the die is too small, the deformation resistance of the blank when passing through the rounded corner increases, and the frictional resistance when passing through the gap of the punch and die also increases, increasing the total deep drawing force. And the life of the mold also reduces because of the increase in the bending force and friction. Therefore, when the corner radius of the die is too small, a larger limit drawing coefficient must be used. If the corner radius of the die is too large, it increases the width of the blank that is not in contact with the die surface at the initial stage of deep drawing reducing the blanking effect during deep drawing, and making this part of the blank very easy to wrinkle. In the later period of deep drawing, too large a radius of rounded corners also causes the outer edge of the wool to break away from the role of the blank holder prematurely and wrinkle, especially when the relative thickness of the blank is small. Therefore, when the thickness of blank $t \leq 2$ mm, the corner radius of the die R_A is 3 t~6 t. In this paper, the corner radius of the die was 0.8 mm.

2.2. Experimental Scheme

The high-purity tantalum metal with a thickness of 0.2 mm was adopted as the workpiece for deep drawing. Tool steel alloy SKD11 with a hardness of HRC60 after quenching was selected as the material for the blank holders, punches, dies, and plate, as shown in Figure 1. Steel was widely used in punch and die materials in sheet forming.

A series of tantalum sheets with a diameter of 13.2 mm, 13.4 mm, 13.6 mm, 13.8 mm, 14.0 mm, 14.2 mm, 14.4 mm, 14.6 mm, and 14.8 mm were processed using wire cutting, respectively. In this paper, the deep drawing experiment was carried out on the universal MTS-SANS CMT6104 testing machine, where the stamping speed was 20 mm/min, and the maximum forming force was 10,000 N. The forming stroke was set to 7 mm, and the forming force was obtained directly from the machine. Four experimental observation groups (DLC, PE, caster oil, and Unlubrication) were set up to complete the deep drawing experiment. The experimental data of maximum forming force, limit drawing ratio, and surface topography were collected, and the group with the best gain effect was selected after analysis and summary.

In this paper, the influence of different lubrication media on the depth of the tantalum plate is discussed; caster oil, PE film, and DLC were selected from the perspective of oil lubrication and solid lubrication film lubrication, respectively. The caster oil had a density of 0.93×103 kg/m^3 and a viscosity of 0.61 Pa·s. PE film selected a thickness of 0.03 mm below the low-density polyethylene film. The caster oil and PE film were purchased from a company in Guangdong, China. The manufacture of DLC film coated on the punch and mold was made by a company through the PIID equipment in city of Dongguan, China.

3. Results and Discussion

3.1. Maximum Forming Force

The relationship curve between the drawing stroke and forming force during the deep drawing of tantalum cylindrical cups with different lubrication media is shown in Figure 3. The trends of the curves are roughly the same under different lubrication conditions. With the move downward of the deep drawing punch, the forming force increased and reached the maximum peak force. After that, the flange area of the tantalum sheet decreased, and the forming force began to gradually decrease until the flange completely disappeared and the cup was successfully formed. Good lubrication conditions were maintained throughout the deep drawing process with the lubrication media of caster oil and DLC. As seen in Figure 4, the PE film broke when the tantalum sheet reached a certain depth height, resulting in a second peak of the forming force; then, the forming force decreased, and the cup shape was finally formed successfully.

Compared to the deep drawing results without lubrication, the maximum forming forces were reduced by 14.4% and 13.6% under the lubrication conditions of DLC and PE film. Great improvements were created compared with the reduction of 5.6% under the condition of caster oil. This conclusion shows that the lubrication performance of DLC and PE film is much better than caster oil lubrication, and the solid lubrication DLC film is more suitable for industrial production compared with the shortcomings that the PE film might fracture during the experiment.

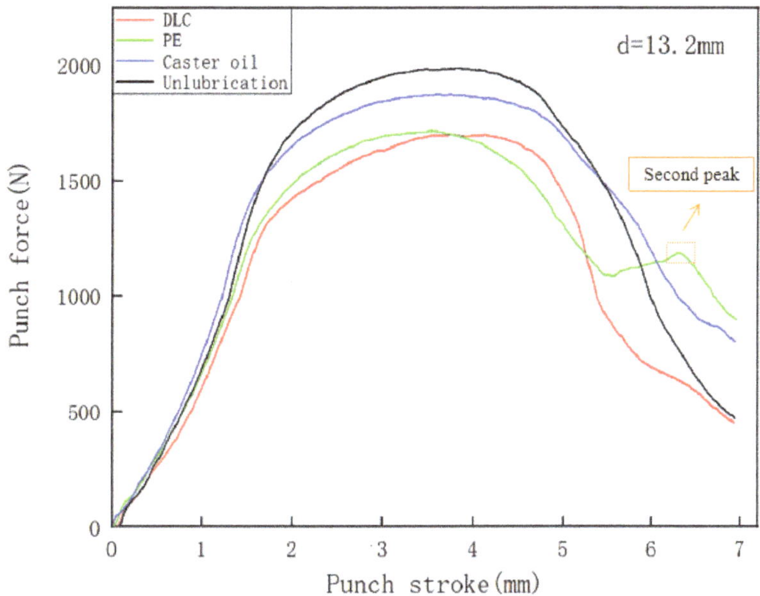

Figure 3. Punch stroke–forming force curve.

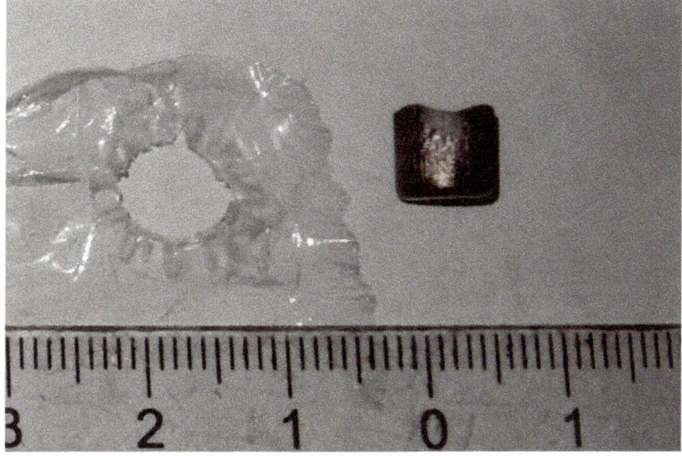

Figure 4. Deep drawing under the PE film.

3.2. Friction Coefficient

The friction curve of the DLC film achieved by the ball-on-disk shown in Figure 5. The test parameters were set as follows: the rotation speed was 100 r/min, the load was 500 g, the rotation radius was 3 mm, and the test time was 15 min. In this experiment, the surface of the steel ball with a diameter of 5.556 mm was coated with the lubricating medium of DLC, which was clamped at the test head. The tantalum sheet with a 15 mm diameter and 0.4 mm thickness was clamped on the platform to make the ball contact the tantalum sheet and perform continuous circumference movement until the end of the test time.

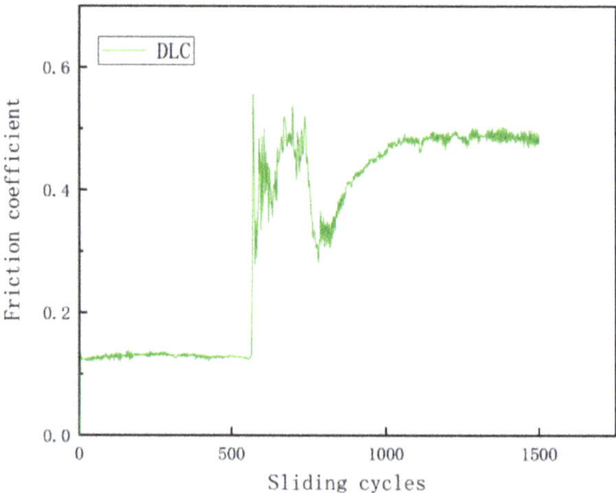

Figure 5. Friction curve of DLC film.

The friction coefficient value of the DLC film rose in a straight line after 565 revolutions, which indicates that the DLC film coated on the surface of the ball was worn. After that, the material of the ball itself began to contact with the tantalum sheet, resulting in a sharp increase in the friction coefficient value. The average friction coefficient value of the DLC film was calculated as 0.128. And this was much better than liquid lubricants, whose friction coefficient was larger than 0.2 [20]. The friction coefficient of PE film was 0.15–0.2 [21]. The wear resistance of PE film is far less than the DLC film, which is easy to fracture. From an industrial point of view, the DLC film with low friction performance and high wear resistance is more suitable for actual production demand. Frictional experiments have proved the excellent wear resistance and low friction coefficient of DLC. However, the PE film was easily damaged with poor wear resistance because it slid only a few laps before breaking.

3.3. Limit-Drawing Ratio

The limit drawing ratio image of tantalum sheets under different lubrication media is shown in Figure 6. To study the effect of different lubrication media on the limited deep-drawing ratio of the tantalum sheet, the formula is used as follows: $K = D/R$. Here, D is the outer round diameter of the tantalum sheet, and R is the diameter of the punch head.

Figure 6. Formed cups under different lubrication media (**a**) DLC (**b**) PE (**c**) caster oil.

The results show that the K with PE membrane can reach 2.43 and that with the DLC film can also reach 2.37, which is higher than 2.27 without lubrication and 2.3 under caster oil. In the forming process of cups, the larger the area size of the tantalum sheet, the larger the area it contacts with the pressure circle, which could cause an increase in friction resistance. When the friction resistance exceeds a certain limit, the cup can fracture in the forming process. It is clear that PE film provides an effective method to reduce friction resistance, and the DLC film can also effectively reduce friction resistance, which is greatly improved compared to the results under the condition of dry friction and caster oil lubrication.

3.4. The Surface Topography

To study the effect of different lubricating mediums on the surface quality of tantalum sheets during deep drawing, 13.2 mm diameter tantalum sheets were selected for deep drawing experiments. Tantalum cup-shaped parts obtained from the deep drawing were cleaned and air dried using an ultrasonic vibration cleaning machine and then placed under the super depth of a field microscope to observe the surface morphology of each workpiece. The observed surface morphology is shown in Figure 7.

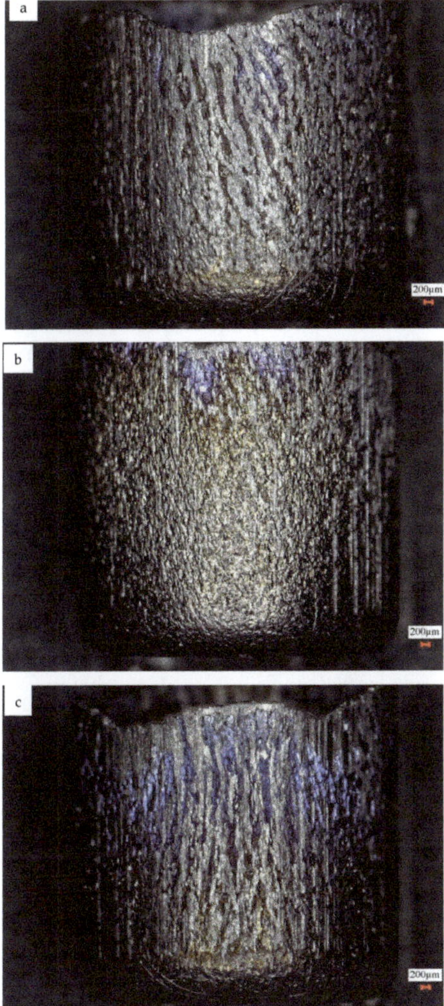

Figure 7. *Cont.*

Figure 7. Deep-drawn parts under *d* = 13.2 mm (**a**) DLC, (**b**) PE, (**c**) caster oil (**d**) Not lubricated.

DLC is a kind of amorphous film material, which is very similar to diamond in terms of its hardness and electrochemical properties. The internal valence bond structure of DLC is relatively complex, and there are both the sp3 hybrid orbitals of diamond and the sp2 hybrid orbitals of graphite between the carbon atoms. Because of the intermingling of this valence bond between the carbon atoms, DLC forms its unique three-dimensional network structure [22]. This 3D network structure feature makes the frictional behavior of DLC films significantly different from that of diamonds. The DLC film not only has excellent wear resistance and high hardness characteristics at low-temperature conditions but also has an excellent friction performance because of its sp2 hybrid structure. As seen in Figure 7a, the outer wall of the tantalum cup has the best morphology quality under the lubrication condition of the DLC film, there is no clear evidence of a seizing tumor on the surface, and the scratch depth produced by deep drawing is much shallower. The surface roughness obtained using the keyence VK-X series confocal microscopy is 6.121 μm, as seen in Figure 8a [23].

Caster oil, as a lubricating oil, is widely used in the industrial field. It has a composition hydroxyl group, and this polar group can make the oiliness additive to produce chemical or physical adsorption, forming a solid lubrication film on the metal surface so that the lubrication performance can be improved. However, when the temperature of the friction surface rises to a certain temperature, the molecular arrangement of the oiliness additive adsorption film is destroyed and loses the lubrication effect. At this time, the boundary lubrication film covering the two metal surfaces disappears, creating direct contact between the metals. As the direct contact area increases, the friction energy increases accordingly, resulting in a temperature rise, and this result is prone to the sintering phenomenon. As shown in Figure 7c, the outer wall of the tantalum cup under caster oil lubrication has a clear adhesion tumor because the caster oil lubrication film is destroyed, causing direct contact between the metals. It has been proved that caster oil is not enough to protect the outer wall of the tantalum sheet in contact with the mold surface during the deep drawing process. The surface roughness obtained after the measurement is 6.387 μm, as seen in Figure 8c.

Due to insufficient wear resistance and ductility, the PE membrane was damaged in the back half of the deep pull, resulting in direct contact between the workpiece and the inner surface of the mold, and, finally, adhesion to the tumor on the upper end of the cup, as shown in Figure 7b. The surface roughness obtained after the measurement was 7.488 μm, as seen in Figure 8b.

To sum up, DLC has good wear resistance and plays a good role in lubrication during deep drawing, which can avoid the sticking of the tantalum sheet and is the most significant for improving the quality of forming deep-drawn cylindrical cups among the three lubrication media.

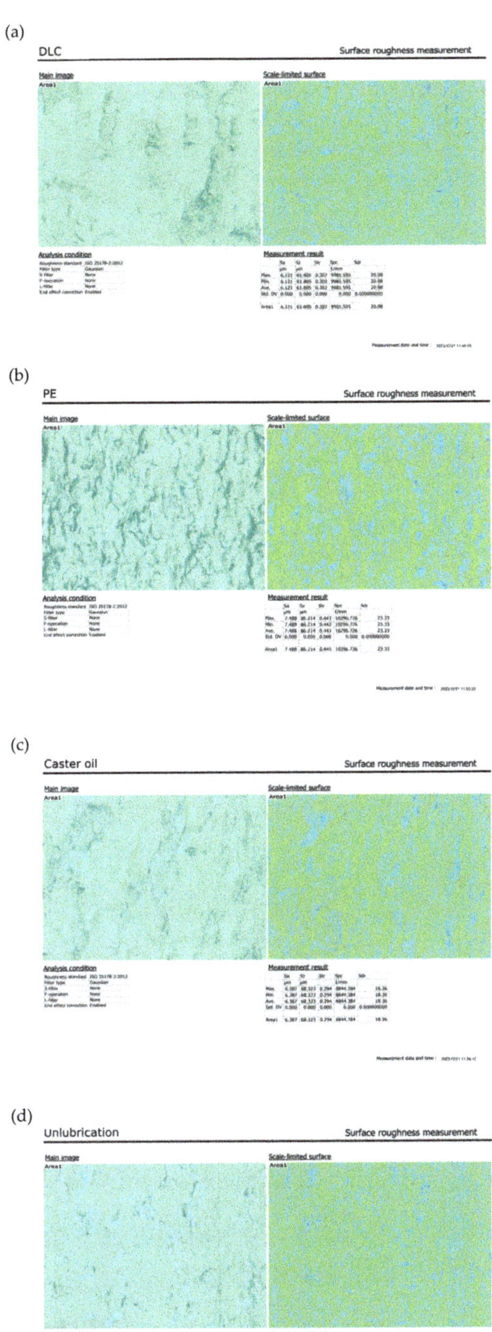

Figure 8. Surface roughness measurement of cups (**a**) DLC, (**b**) PE, (**c**) Caster oil, (**d**) Not lubricated.

4. Conclusions

In this paper, deep drawing experiments of tantalum cylindrical cups under different lubrication conditions were carried out. Four groups of experiments were set up as follows: DLC, PE, caster oil, and no lubrication. The maximum forming force under different lubrication conditions was compared, and the surface morphology quality of the drawn parts was observed using the super depth of a field optical microscope; the following results can be concluded:

(1) All the lubrication media of the DLC, PE membrane, and caster oil can effectively reduce the maximum forming force in the deep drawing of tantalum cylindrical cups, and the maximum forming force could be reduced by 13.6% and 14.4% through the utilization of the PE membrane and the DLC membrane, respectively, which are particularly effective for the deep drawing of tantalum cylindrical cups.

(2) The DLC film is not destroyed after more than 500 revolving cycles, and the friction coefficient is 0.128. The wear resistance of the DLC film is much higher than that of the PE film.

(3) The tantalum sheet with a deep drawing ratio of $K = 2.37$ can be successfully formed under the lubrication of the DLC film, and the K under the lubrication of the PE film reaches 2.43. Due to the lower sliding friction of the PE film, the limited deep drawing ratio obtained with the PE film is better than that of the DLC film.

(4) The DLC film can effectively reduce the forming force of tantalum sheets and improve the limit drawing ratio and the surface morphology quality of the forming parts. It has good wear resistance, which can be used as a lubrication medium in the tantalum metal-forming production process.

Author Contributions: All authors have contributed to the creation of this manuscript for important intellectual content. T.X.: methodology; data curation; formal analysis; writing, original draft, review, and editing; S.D.: experiment; validation; writing, review, and provide method; M.S. and J.H.: review, validation, and editing; N.Z. and S.Y.: experiment and analyze the data; L.Z.: review and editing, supervise and guide. All authors have read and agreed to the published version of the manuscript.

Funding: This research is funded by the National Natural Science Foundation of China (Grant No.: 52005341; 51905241), the Basic and Applied Basic Research Fund of Guangdong Province (2019A1515111115), and the National Taipei University of Technology-Shenzhen University Joint Research Program (Project no. 2023010).

Data Availability Statement: All data generated or analyzed during this study are included in this published article.

Conflicts of Interest: The authors declare no conflicts of interest.

References

1. Mitchell, T.E.; Raffo, P.L. Mechanical properties of some tantalum alloys. *Can. J. Phys.* **2011**, *45*, 1047–1062.
2. Robin, A.; Gomes, E.A.; Nunes, C.A.; Coelho, G.C.; Baldan, C.A. Preparation and characterization of Zircaloy 4-Tantalum alloys for application in corrosive media. *Int. J. Refract. Met. Hard Mater.* **2012**, *35*, 90–95. [CrossRef]
3. Vishay Intertechnology. Liquid tantalum capacitors bring high capacity value and flexibility to aerospace applications. *Electron. Prod.* **2017**, *65*.
4. Wang, H.; Zhang, X.M.; Li, L.P.; Wang, F.; Cai, X.M. Industry Application of tantalum and tantalum alloys. *Equip. Manuf. Technol.* **2013**, *8*, 115–117.
5. Ren, X.; Zhang, Y.J. Vortex detection study of chemical anticorrosive tantalum alloy pipe material. *China Met. Bull.* **2018**, *3*, 100–101.
6. Tian, C. The Invention Relates to a Method for Improving the Voltage Resistance of Chip Tantalum Capacitor. Ph.D. Thesis, Xidian University, Xi'An, China, 2019.
7. Liu, S.Y. Application of tantalum in high-tech. *Rare Met. Cem. Carbides* **1998**, *2*, 55–57+26.
8. Prokhorova, T.; Orlov, V.; Miroshnichenko, M. Effect of tantalum capacitor powder preparation conditions on the dielectric loss tangent of anodes. *Inorg. Mater.* **2014**, *50*, 145–149. [CrossRef]
9. Zhang, Y.J.; Gong, Z.J.; Wang, K. Processing Technology Research on Deep drawing EB Tantalum Ribbon for Liquid Tantalum Capacitor Shell. *Dev. Appl. Mater.* **2014**, *29*, 58–61.
10. Kardan, M.; Parvizi, A.; Askari, A. Influence of process parameters on residual stresses in deep-drawing process with FEM and experimental evaluations. *J. Braz. Soc. Mech. Sci. Eng.* **2018**, *40*, 157–169.

11. Dong, W.; Xu, L.; Lin, Q.; Wang, Z. Experimental and numerical investigation on galling behavior in sheet metal forming process. *Int. J. Adv. Manuf. Technol.* **2017**, *88*, 1101–1109. [CrossRef]
12. Dong, W.Z.; Lin, Q.Q.; Wang, Z.G. On the galling behavior in HSS sheet metal forming process by FEM-Archard model. *Mater. Sci. Technol.* **2015**, *3*, 42–45.
13. Freiße, H.; Ditsche, A.; Seefeld, T. Reducing adhesive wear in dry deep drawing of high-alloy steels by using MMC tool. *Manuf. Rev.* **2019**, *6*, 12. [CrossRef]
14. Marchin, N.; Ashrafifizadeh, F. Effect of carbon addition on tribological performance of TiSiN coatings produced by cathodic arc physical vapour deposition. *Surf. Coat. Technol.* **2021**, *407*, 126781. [CrossRef]
15. Ke, X.G.; Zhang, R.Q. Study of drawing key techniques and design of tranfer die for shell of wet all-tantalum capacitor. *FST* **2008**, *33*, 82–86.
16. Liu, H.; Chen, Z.L.; Wang, C.R. Research on cupping test and drawing test of tantalum sheet. *FST* **2019**, *44*, 150–153.
17. Sresomroeng, B.; Premanond, V.; Kaewtatip, P.; Khantachawana, A.; Koga, N. Anti-adhesion performance of various nitride and DLC films against high strength steel in metal forming operation. *Diam. Relat. Mater.* **2010**, *19*, 833–836.
18. Wank, A.; Reisel, G.; Wielage, B. Behavior of DLC coatings in lubricant free cold massive forming of aluminum. *Surf. Coat. Technol.* **2006**, *210*, 822–827. [CrossRef]
19. Steriner, J.; Andreas, K.; Merkleini, M. Investigation of surface finishing of carbon based coated tools for dry deep drawing of aluminium alloys. In *IOP Conference Series: Materials Science and Engineering*; IOP Publishing: Bristol, UK, 2016.
20. Vollertsen, F.; Hu, Z. Tribological size effects in sheet metal forming measured by a strip drawing test. *Cirp Ann. Manuf. Technol.* **2006**, *55*, 291–294.
21. Sufen, Z.; Mengliang, L.; Xiaoyan, L. Analysis the Friction Coefficient Performance and Application of the PE Film for Solvent Free Composite. *Plastics* **2019**, *48*, 16–18.
22. Wang, L.; Huang, L.; Wang, Y.; Xie, Z.; Wang, X. Duplex DLC coatings fabricated on the inner surface of a tube using plasma immersion ion implantation and deposition. *Diam. Relat. Mater.* **2008**, *17*, 43–47. [CrossRef]
23. *ISO 25178-2:2012*; Geometrical Product Specifications (GPS)—Surface Texture: Areal, Part 1: Terms, definitions and Surface Texture Parameters. ISO: Geneva, Switzerland, 2012.

Article

Growth Substrate Geometry Optimization for the Productive Mechanical Dry Transfer of Carbon Nanotubes

Andre Butzerin [1,*], Sascha Weikert [2] and Konrad Wegener [2]

[1] Institute of Machine Tools and Manufacturing, ETH Zürich, 8092 Zürich, Switzerland
[2] inspire AG, 8005 Zürich, Switzerland
* Correspondence: butzerin@iwf.mavt.ethz.ch

Abstract: The selection of growth substrate geometries for the mechanical dry transfer of carbon nanotubes to device substrates depends on the precision of the assembly equipment. Since these geometries play a decisive role in the overall efficiency of the process, an investigation of the most important geometry parameters is carried out. The substrate geometry affects the number of carbon nanotubes suspended during the growth process and the speed of mechanical assembly at the same time. Since those two criteria are interlinked and affect productivity, a meta-model for the growth and selection of the nanotubes is simulated and a time study of the resulting assembly motions is subsequently performed. The geometry parameters are then evaluated based on the total number of suspended carbon nanotubes and the throughput rate, measured in transfers per hour. The accuracy specifications are then taken into account. Depending on the overall accuracy that can be achieved, different offset angles and overlaps between the growth and receiving substrate can be reached, which affect productivity differently for different substrate geometries. To increase the overall productivity, growth substrate designs are adapted to allow fully automated operation. This measure also reduces the frequency of substrate exchanges once all carbon nanotubes have been harvested. The introduction of substrates with multiple, polygonally arranged edges increases the total number of nanotubes that can be harvested. The inclusion of polygonally arranged edges in the initial analysis shows a significant increase in overall productivity.

Keywords: optimization; productivity; mechanical dry transfer; substrate geometry; suspended carbon nanotube

Citation: Butzerin, A.; Weikert, S.; Wegener, K. Growth Substrate Geometry Optimization for the Productive Mechanical Dry Transfer of Carbon Nanotubes. *Processes* **2024**, *12*, 928. https://doi.org/10.3390/pr12050928

Academic Editors: Guoqing Zhang, Zejia Zhao and Wai Sze YIP

Received: 1 April 2024
Revised: 22 April 2024
Accepted: 25 April 2024
Published: 1 May 2024

1. Introduction

The mechanical dry transfer method of placing carbon nanotubes onto substrates is known to be very clean and therefore results in exceptional device performance. With this type of production process, a distinction can be made between deterministic and non-deterministic placement. With the non-deterministic placement, as reported in [1], carbon nanotubes are scattered randomly over a device substrate. With this method, a very high number of transfers per hour (TPH) can be achieved, but there is no control over the position and orientation of the individual carbon nanotubes. The deterministic mechanical dry transfer manufacturing technique of carbon nanotube devices, on the other hand, utilizes a growth substrate as a carbon nanotube donor to integrate those into a device substrate. The growth substrate has cantilever pairs that form trenches in which a carbon nanotube can be suspended. In [2–9], nanotubes are suspended on such cantilever pairs and are transferred to the device substrate by mechanically breaking them off their support structure. Van der Waals forces guarantee adhesion to the device electrodes. This procedure is schematically depicted in Figure 1. As, however, each single carbon nanotube has to be transferred individually, the achievable production rate is a severe disadvantage. Moreover, as it is necessary to use a motion system for the positioning of carbon nanotubes, travel distances and ranges need to be considered as well.

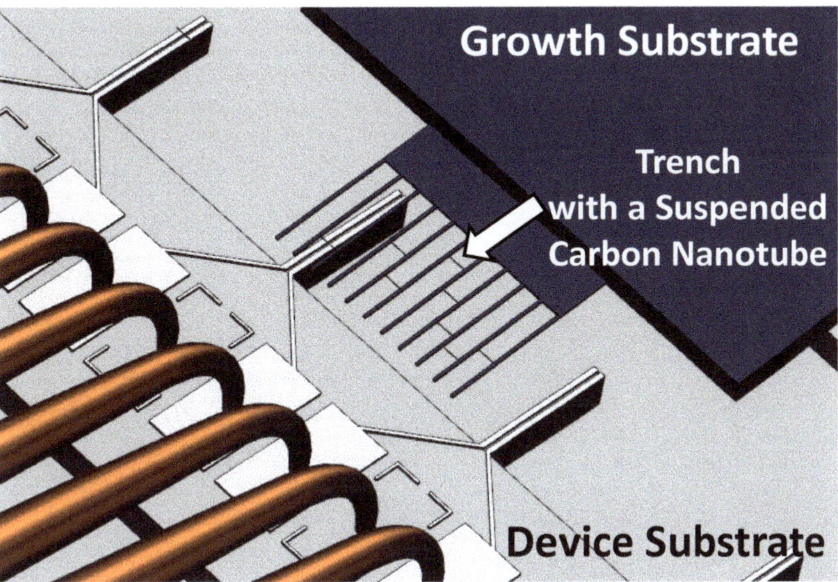

Figure 1. Visualization of growth and device substrate interaction during mechanical dry assembly. Carbon nanotubes are transferred by breaking them off from the cantilevers. They are held in place on the device substrate by Van der Waals forces. After each transfer, the next carbon nanotube on the growth substrate is moved to the next electrodes on the device substrate and the process is repeated.

The growth substrate's geometry also affects the carbon nanotube synthesis. During the synthesis procedure, the number of nanotubes which are suspended across the cantilevers is determined. In addition, their relative angle to the substrate and their actual length are defined as well. Depending on the application, it can be desired to have multiple carbon nanotubes on a single device. As the number of tubes which are suspended on the same pair of cantilevers varies, a selection regarding the desired number of tubes per trench (TPT) is required too. Hence, substrate geometries must be optimized to increase the production rate while taking effectiveness of growth, selection and assembly into account.

Growth substrates are already commercially available and do not have to be manufactured in-house. However, their use does not address large batch fabrication. As shown in [8], only eight devices are fabricated with a growth substrate of 48 cantilevers and a cantilever pitch of 60 μm.

The selection of the optimal type of growth substrate geometry can be based on a multicriteria decision-making model [10,11]. The approach presented in this work, however, focuses on how the growth substrate geometry parameters affect both productivity and accuracy throughout the whole process chain, from synthesis to assembly, using simulation.

2. Productivity

The geometry of a substrate has a significant impact on the overall process efficiency. In addition to the growth parameters, the geometry of the part where the nanotubes are suspended on the substrate dictates the density of harvestable tubes. Hence, the achievable productivity is mainly characterized by these features besides the growth process requirements. Therefore, the process scheme shown in Figure 2 is carried out. A simple simulation of nanotube growth is introduced which considers the cantilever geometries and the growth density of the CVD process. The output of this simulation yields a list of the coordinates of each transferable nanotube. Experimentally grown carbon nanotube substrates validate the proportion of harvestable tubes determined by simulations and the parameters used. The resulting tube list serves, together with the substrate geometry and trajectory parameters of the manipulator, as input for a consecutive time study. The final result is a transfer rate in transfers per hour (TPH) for a given substrate geometry that is achievable with a manipulator's positioning performance. Furthermore, hints about the replacement rate of the growth substrate are given by the number of transferable trenches.

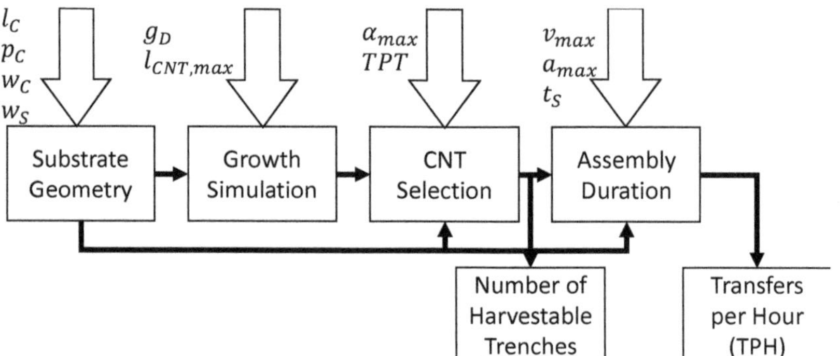

Figure 2. Analysis scheme of growth substrate geometry parameters for the simulation of productivity.

2.1. Simulation of Nanotube Growth and Selection

The first step that is required for the growth simulation is the substrate geometry. The simulation only considers a two-dimensional substrate geometry in order to keep the level of complexity low. The input parameters for the creation of the substrate geometry are the length l_C and width w_C of the cantilevers and their pitch p_C as shown in Figure 3. With these parameters given and additional knowledge of the total substrate edge width w_e, sufficient information is given to create the desired substrate. As shown in Figure 3, the resulting number of cantilevers n_C for carbon nanotube growth can be calculated as

$$n_C = \frac{w_e - w_C}{p_C} + 1 \tag{1}$$

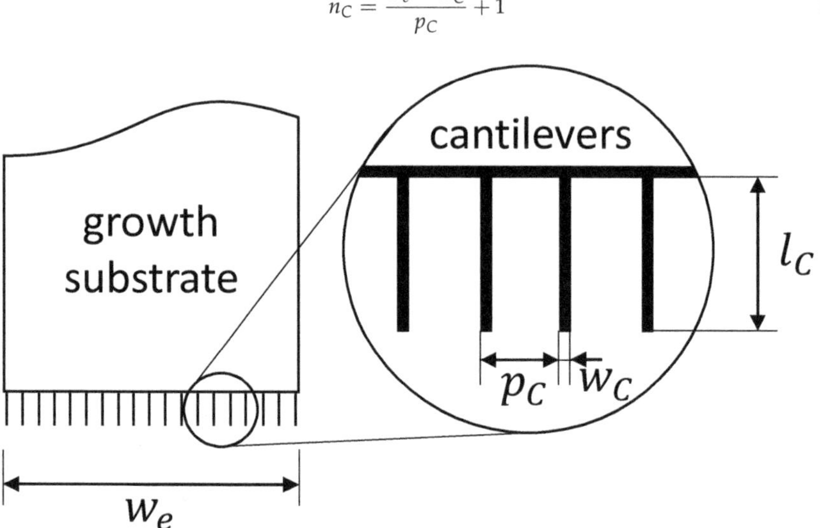

Figure 3. The crucial cantilever geometry of a comb-like growth substrate used for mechanical dry transfer.

After the determination of the substrate geometry the growth can be simulated. Additional required parameters for this are the maximum length $l_{CNT,max}$ of carbon nanotubes to be grown and the growth density g_D with which they are distributed across the substrate. The growth density g_D therefore denotes the density of the carbon nanotubes of the desired type, such as semiconducting and/or metallic.

For the sake of simplicity, it is assumed that grown carbon nanotubes are always straight and that they are defect-free. Only their length, position and orientation is required to determine the location of the cantilever pair between which they are suspended. In order to reduce computational load, the growth area of simulated nanotube synthesis is constrained to twice the cantilever length l_C times the substrate edge width w_e as depicted in Figure 4 in light green.

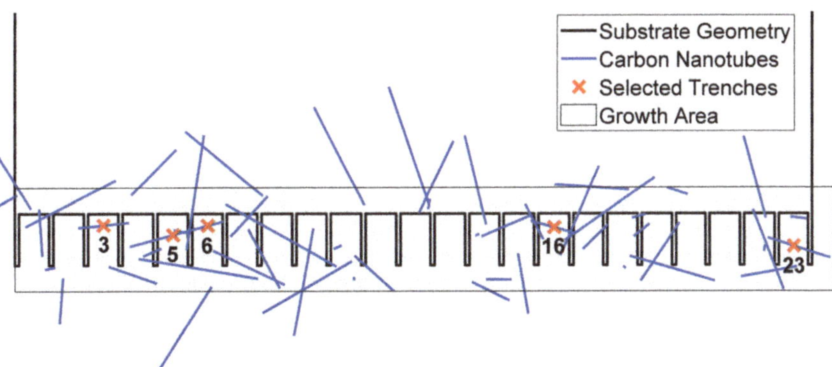

Figure 4. Example visualization of random carbon nanotube growth with a nominal growth density of $g_D = 3000\frac{tubes}{mm^2}$ on one single substrate edge. The chosen substrate geometry is $n_C = 24$, $p_C = 14$ µm, $w_C = 2$ µm, $l_C = 24$ µm. The tube angle tolerance α_{max} for trench selection is $\pm 20°$ and the desired number of suspended tubes per trench $TPT = 1$.

The simulation considers small variations in growth density via uniform distribution and over a sample size of n_S in total. Those affect the position and orientation of the suspended nanotubes. Larger variations in growth density resulting from experimental imperfections, however, have not been considered.

The tubes designated for transfer can be determined once there is a given substrate geometry with carbon nanotubes distributed across its cantilevers. At this step a distinction between tubes that are suspended and tubes that are not suspended is made. A tube is considered suspended if it has at least two intersections with the contour of the substrate geometry that are at least one nominal trench width $(p_C - w_C)$ apart along the tube axis. All other tubes are disregarded.

Since multiple carbon nanotubes can be suspended on a single pair of cantilevers, it is important to categorize each trench by the count of tubes per trench TPT. This is required in order to be able to filter the trenches later by this number, as it may be necessary to assemble devices with a specific number of carbon nanotubes.

In Figure 4, the trenches chosen for transfer are indicated with a red "x". It is the outcome of a filter criterion of one carbon nanotube per trench $TPT = 1$ with a maximum relative angle $\alpha_{max} = \pm 20°$. The resulting list of tubes contains an address table of all trenches to be transferred.

2.2. Time Study

With the tube list as information it is feasible to derive an expected productivity of the assembly. The result is quantified in transfers per hour (TPH) and is a measure of the achievable productivity with the parameters applied. It can be calculated based on the substrate geometry and trajectory parameters. Assuming an acceleration limited trajectory over a distance D and a maximum acceleration a_{max}, the duration Δt for a positioning step is calculated depending on whether maximum velocity v_{max} can be reached or not. If

$$D \geq \frac{v_{max}^2}{a_{max}} \tag{2}$$

is true, maximum velocity v_{max} is reached. For this case the duration is calculated as

$$\Delta t = 2 \cdot \frac{v_{max}}{a_{max}} + \frac{D - \frac{v_{max}^2}{a_{max}}}{v_{max}} \tag{3}$$

and for the case where maximum velocity v_{max} is not reached the duration is calculated as

$$\Delta t = 2 \cdot \sqrt{\frac{D}{a_{max}}} \tag{4}$$

The sequence of X, Y and Z motions for one assembly cycle are taken from [12] where the required travel distances are calculated according to the addresses in the tube list and the substrate geometry. In order to have sufficient statistical significance, the simulation is repeated n_S times for one type of substrate geometry. Each type of substrate geometry has n_S of differently distributed sets of carbon nanotubes that are suspended between the substrate's cantilevers. From these simulation samples, an average transfer rate is calculated from the ratio of the total number of assembly cycles that have been performed and the sum of their duration. The number of trenches with the specified amount of tubes per trench TPT is tracked as well. This value can be associated with the frequency of growth substrate replacements.

In Figure 5, the productivity P in terms of TPH and number of harvestable trenches n_{harv} with $TPT = 1$ is shown over the geometry parameters of the cantilever pitch p_C and the cantilever length l_C. While the productivity in form of TPH only takes manipulator movements into account, the replacement rate of the substrate itself is considered by the number of harvestable trenches n_{harv}. The substrates need to be replaced less frequently if the number of harvestable trenches n_{harv} is high.

Conclusions from Figure 5a,c show that substrate geometries with narrow cantilevers are preferable. The overall productivity is higher since substrates with such a cantilever geometry offer a higher trench density. This leads to a higher number of harvestable trenches n_{harv} while at the same time the travel distance is shortened. In contrast, and as it can be seen in Figure 5b,d, the cantilever length l_C is limited to the lower and upper end. Short cantilevers cannot suspend as many carbon nanotubes and long cantilevers suspend too many carbon nanotubes. Hence, the optimum cantilever length lies somewhere in between and depends on TPT and the growth parameters chosen.

The results from Section 2 stipulate to minimize the cantilever pitch p_C for a maximum productivity, while adjusting the cantilever length l_C to a desired carbon nanotube growth. However, due to the finite accuracy of the manipulator, which has to move the growth substrate for the nanotube transfer, the cantilever pitch p_C is limited on the lower end.

The simulation results of the cantilever geometry used in [8] show a median productivity of 128 TPH for the same simulation parameters only with an adapted maximum nanotube length $l_{CNT,max} = 120$ µm. This clearly shows the potential for optimizing growth substrate geometries for large-scale production.

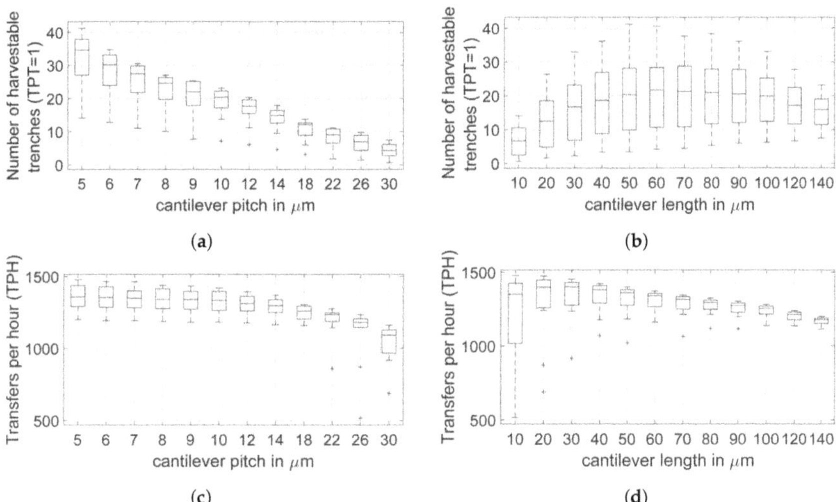

Figure 5. Simulation results for 144 substrate geometry variations with the according number of harvestable trenches n_{harv} where $TPT = 1$ and the TPH over the cantilever length l_C and cantilever pitch p_C. Each p_C is evaluated for all l_C and vice versa. Boxes represent 25th and 75th percentiles, whiskers extreme values and outliers are plotted individually. Each geometry has a cantilever width $w_C = 2$ μm and was simulated $n_S = 50$ times. Used parameters for growth are $g_D = 10,000\frac{tubes}{mm^2}$ and $l_{CNT,max} = 40$ μm. The angle tolerance α_{max} for tube selection was set to $\pm10°$. Results presented are the outcome of a polygonal substrate with a number of edges $n_{edges} = 2$, the effects of which are discussed in Section 3. (**a**) As one nanotube can be suspended across multiple trenches, a reduction in cantilever pitch p_C increases the number of harvestable trenches n_{harv}. (**b**) The optimal cantilever length l_C, here 60 μm, is impacted by the maximum nanotube length $l_{CNT,max}$ used in the growth simulation. (**c**) With decreasing cantilever pitch p_C, the number of TPH increases as the travel distance is shortened and the trench density becomes higher. (**d**) The longer the cantilevers are, the higher the required travel distance becomes, causing a lower TPH.

2.3. Clearance Due to Geometry and Accuracy

Figure 6 depicts the schematic trench and device geometry during transfer. The clearance c is the horizontal minimal distance between both substrates, w_D is the device width and the errors Δx and Δy are the total positioning errors along the x and y coordinate axes. Those include errors from the manipulator and measurements errors, but also the error due to imprecise substrate fabrication. Depending on the application it may be desirable to manufacture devices with aligned carbon nanotubes. If this is the case the inclination angle Θ has to correct for the angle of the tube. The error $\Delta\varepsilon$ of Θ must then also be taken into account. Based on these considerations, the available clearance c can be defined as

$$c = \frac{p_C - w_C - w_D}{2} - (y_O + \Delta y) \cdot |\tan(\Theta + \Delta\varepsilon)| - |\Delta x| \tag{5}$$

If the positioning and preparation of the growth substrate is inaccurate, the transfer of carbon nanotubes becomes less efficient and less reliable, as the probability of collisions between the two substrates is higher. With a comparison of coefficients in Equation (5), it can be seen that Δx impacts clearance the most. The angular deviation $\Delta\varepsilon$ and the deviation in Δy are rather uncritical. However, if those deviations are underestimated, it becomes apparent that the two substrates are likely to crash with another. Depending on the magnitude of error, it can happen such that only individual cantilevers break off or even the whole substrate.

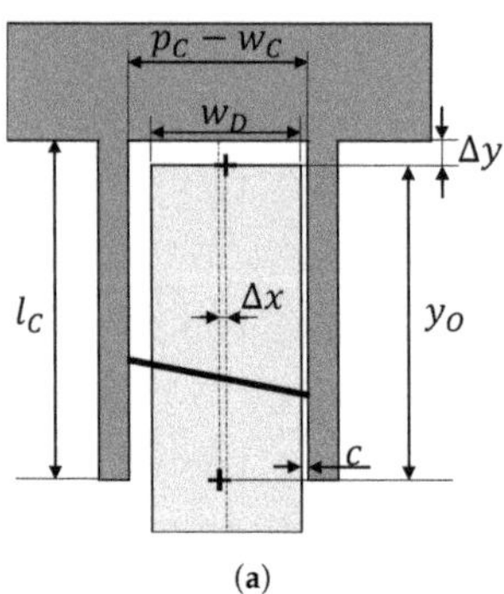
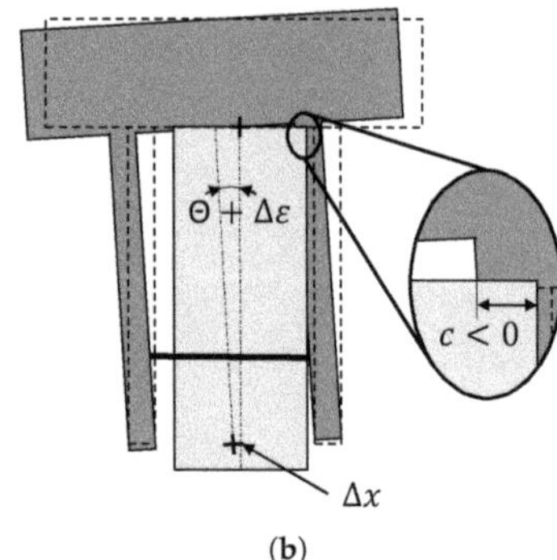

Figure 6. Geometries for the growth substrate in dark gray and the device in light gray. A carbon nanotube, depicted between growth substrate cantilevers in black is transferred to the device by moving perpendicular to the image plane. The offset is shown by the magnitudes of Δx, Δy and $\Delta\varepsilon$. (**a**) The existing clearance c depending on substrate geometries, overlap y_O and position errors Δx and Δy. (**b**) Negative example with collision of substrates. The alignment error $\Delta\varepsilon$ and the overlap y_O are too large to avoid collision, respectively.

As collision must be avoided at all times, c must in any case be greater than zero. It is not required to have an overlap y_O equal to the full length l_C for a successful transfer. Hence, the clearance along the y direction is chosen such that it is equal to the maximum expected error Δy. However, the number of harvestable trenches n_{harv} decreases proportionally with y_O as carbon nanotubes located deeper in the trench cannot be reached. The consequence is a productivity reduction (TPH) by a factor of η_O.

$$\eta_O = \frac{y_O}{l_C} \tag{6}$$

With a more precise manipulator and accurate substrate fabrication, y_O can be increased because of a smaller error margin Δy.

For the case of automatized carbon nanotube assembly, where device after device is approached like in Figure 6, the overshoots have to be taken into account. In [12], the average overshoots $\bar{x}_{os}$ and $\bar{y}_{os}$ for a parallel kinematic micromanipulator are 0.3 µm for the x axis and 0.8 µm for the y axis, respectively. Assuming perfect substrate fabrication and negligibly small measurements errors, it can be concluded that $\Delta x \approx \bar{x}_{os}$ and $\Delta y \approx \bar{y}_{os}$. Following this assumption, a family of curves, as in Figure 7, shows the transfer parameters overlap y_O over Θ where the clearance $c = 0$.

2.4. Implications to Productivity

Analogous to Figure 7, the accuracy and the substrate geometry affect the maximum possible overlap y_O. The maximum possible overlap for a given accuracy and geometry can be calculated by inserting $c = 0$ in

$$y_O = \frac{\frac{p_C - w_C - w_D}{2} - |\Delta x| - \Delta y \cdot |\tan(\Theta + \Delta\varepsilon)| - c}{\tan(\Theta + \Delta\varepsilon)} \tag{7}$$

As it can be seen from Equation (6), η_O is also a function of y_O and the cantilever length l_C. Since the different geometry affects productivity (TPH) and the overlap factor η_O alike, the cantilever geometries need to by chosen according to Θ in Figure 8 for the desired overlap to reach maximum productivity.

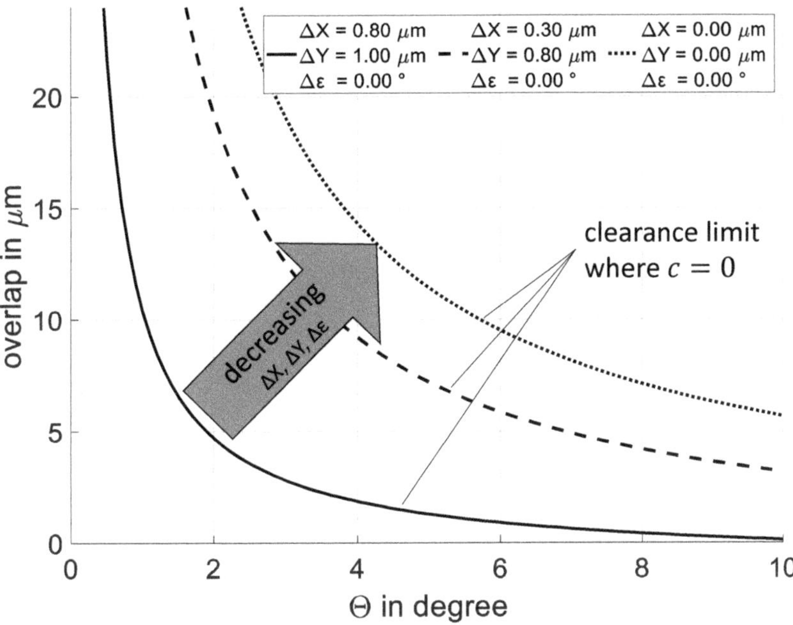

Figure 7. Three examples according to Figure 6 with different errors Δx and Δy where clearance $c = 0$. Parameter sets of overlap and Θ below each curve maintain $c > 0$. The alignment error $\Delta\varepsilon$ shifts graphs to the left along the x axis. The dotted curve highlights the maximum possible overlap y_O with cantilever geometry $p_C = 14$ μm, $w_C = 2$ μm, $l_C = 24$ μm and device width $w_D = 10$ μm. The solid and dashed curves show the maximum possible overlap y_o with the same substrate geometry and error values reported in [12].

Considering the geometric accuracy constraints from Equation (5), it is apparent that for $|\Theta| \approx 0$, the total error $|\Delta x|$ shows the highest sensitivity for a given substrate geometry. Therefore, $|\Delta x|$, which is the sum of the individual errors of overshoot x_o, static positioning of the axis x_p, substrate fabrication x_f and re-referencing x_{ref}, must be kept low by machine design or tight fabrication tolerances. Otherwise, re-referencing of relative substrate positions or mapping is required. With re-referencing, the relative distance between the two substrates can be measured very close to the tool center point by optical means and thus allowing an appropriate compensation. However, this approach can also lead to an error x_{ref}, which must be taken into account.

As the accuracy demands to the x axis is more crucial, the overshoot error x_o should be targeted to be minimized first. This means damping plays an important role, but also the moved masses, which affect the eigenfrequencies, take substantial influence. To reduce the moved mass of the x axis, it is placed on top of the axes stack.

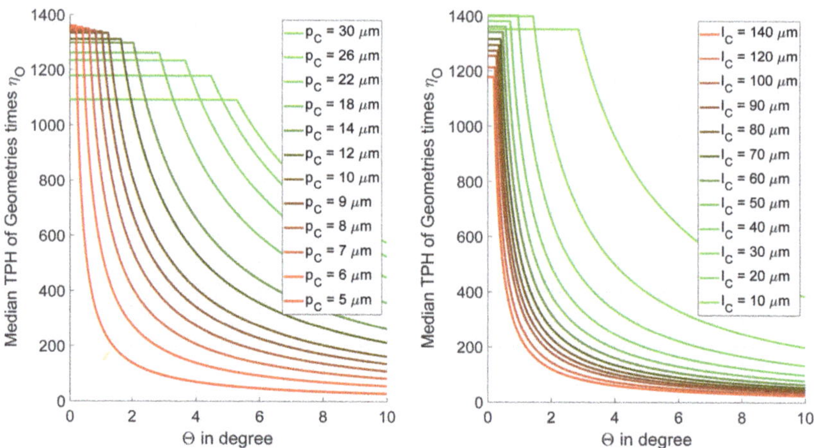

Figure 8. Effective productivity (TPH) of each substrate geometry parameter from the median transfers per hour of Section 2 with overlap factor η_o taken into account. Geometry parameters w_C and w_D are constant for both plots. Left plot shows the effect of cantilever pitch p_C on the effective productivity for a constant cantilever length $l_C = 140$ μm. The right plot shows the effect of cantilever length l_C on the effective productivity for a constant cantilever pitch $p_C = 5$ μm. Both evaluations assume that errors $\Delta x = \Delta y = 0$ μm and $\Delta\varepsilon = 0°$.

3. Design for Automation

With the introduction of fully automated assembly, it is either required to choose a substrate geometry which offers sufficient tolerance or add a re-referencing mechanism to compensate for position errors. Beyond that, the number of harvestable trenches n_{harv} is the decisive parameter for the frequency of substrate exchanges. The higher their number, the less frequent substrate exchanges are required. On top of cantilever geometry optimization, substrates with multiple edges can be introduced to decrease the required time until an exchange. This enhances the time a substrate can be used but also requires an additional rotation axis. Therefore, the total duration of the positioning motion has to be considered for an additional axis with its jerk, acceleration, velocity, travel distance and settling time. Based on this consideration, the number of substrate edges n_e has to be optimized for a corresponding rotation axis. Since the substrates are of polygonal shape, the circumcircle diameter d can be chosen as their size factor. Increasing the number of harvestable trenches n_{harv} by increasing the diameter d of the growth substrate would reduce the total number of substrates that can be produced in one batch. Furthermore, as the diameter of the growth substrate increases, the distance between the axis of rotation and the tool center point increases, resulting in a longer settling time and worse accuracy.

The number of substrate edges n_e influences the edge width w_e and the angular travel distance to bring the next edge into position.

$$w_e = d \cdot \sin \frac{\pi}{n_e} \tag{8}$$

The value of w_e can also be calculated after Equation (1) with the number of cantilevers per edge n_C and their respective geometry parameters p_C and w_C.

$$w_e = (n_C - 1) \cdot p_C + w_C \tag{9}$$

As shown in Figure 9, the diameter d of the circumcircle is selected as a constant in order to make substrates of different shapes comparable to another. Continuing from this assumption, the procedure described in Section 2 is carried out for all substrate edges. Additionally, trajectory parameters and the settling time of a rotation axis are added to the assessment. The parameters for nanotube growth are not changed to be able to compare it to the results from Section 2.

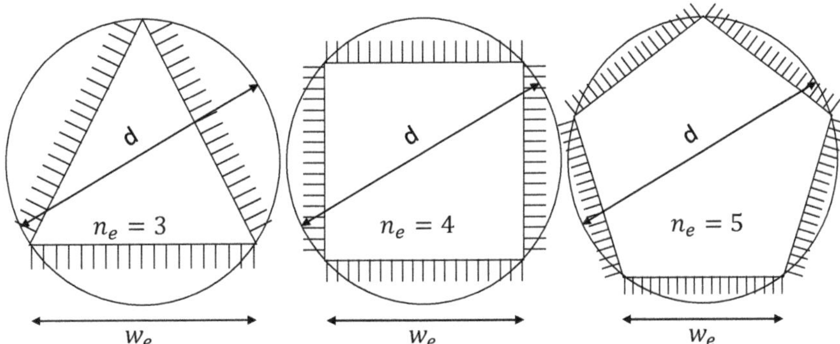

Figure 9. Different substrate geometries with polygonal shapes and the same circumcircle diameter d characterized by the number of edges n_e. The width of each edge w_e and the circumcircle diameter d determine their size.

If Equations (8) and (9) are combined, the relationship between the number of edges n_e and the number of cantilevers n_C can be written as

$$d \cdot \sin \frac{\pi}{n_e} = (n_C - 1) \cdot p_C + w_C \tag{10}$$

Equation (10) points out the inverse relationship between the number of edges n_e and the number of cantilevers per edge n_C. At one point while increasing the number of substrate edges n_e the number of trenches, which is the number of cantilevers $n_C - 1$, will become zero and nanotubes could only be suspended across substrate edges. However, for practical reasons, like maintaining parallelism between cantilevers, this case is avoided in consequent investigations. With the condition that the number of cantilevers per substrate edge $n_C > 1$, the upper limit of the number of substrate edges n_C can be calculated for a certain geometry.

According to Equation (10), the number of cantilevers per edge n_C scale linearly with the circumcircle diameter d. Hence, the number of harvestable trenches n_{harv} also scales with this diameter, since the nanotube distribution is uniform. This allows one to approximate the average number of harvestable trenches n_{harv} for one edge and to assume that each substrate edge has a similar number. This effectively eliminates the time-consuming simulation of growth and selection for each individual edge.

Figure 10 exemplarily shows the influence of the different number of substrate edges n_e and the trajectory parameters of the rotation axis onto productivity. The substrate design in terms of number of edges n_e can be assessed by its required median exchange frequency $\tilde{f}_e$. This value is the quotient of the median transfers per hour $\overline{P}$ and the median number of trenches $\tilde{n}_{harv}$.

$$\tilde{f}_e = \frac{\overline{P}}{\tilde{n}_{harv}} \tag{11}$$

In Table 1, the median exchange frequency $\tilde{f}_e$ indicates how many times the substrate must be exchanged per hour for continuous production. Taking the average required exchange time $\bar{t}_{ex}$ per whole growth substrate into account, the median effective productivity $\overline{P}_e$ can be calculated as

$$\overline{P}_e = \overline{P} \cdot (1 - \tilde{f}_e \cdot \bar{t}_{ex}) \tag{12}$$

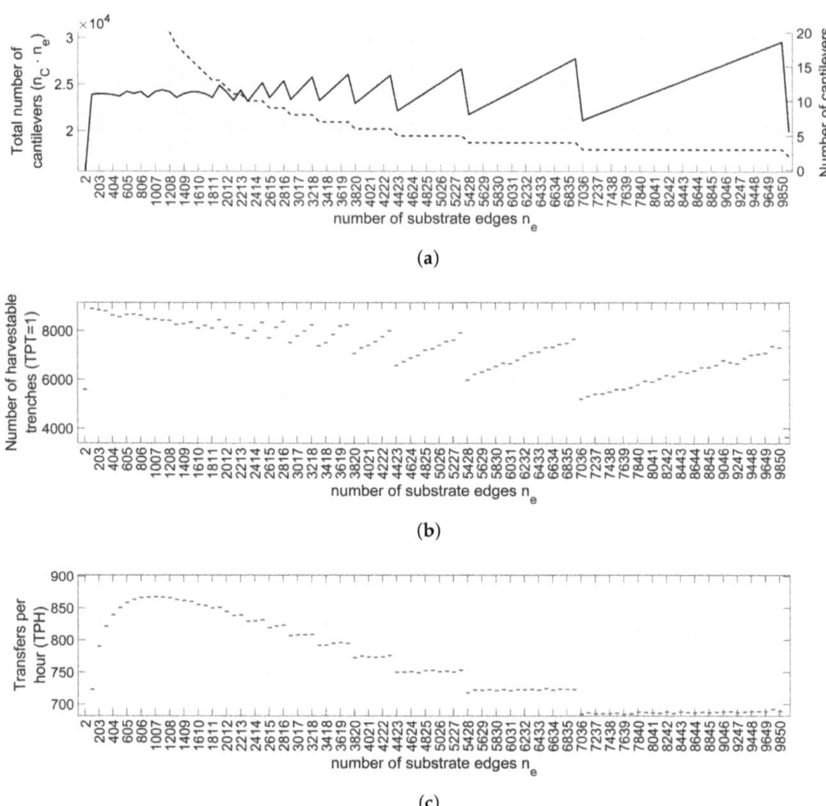

Figure 10. Simulation of varying polygonal substrate edges n_e for a growth substrate with constant cantilever geometry $p_C = 5$ μm and $l_C = 60$ μm and a circumcircle diameter $d = 38$ mm. (**a**) The total number of cantilevers on a substrate $n_C \cdot n_e$ as solid line with its vertical axis on the left side and the number of cantilevers per substrate edge n_C as a function of the number of substrate edges as dashed line and with its vertical axis on the right side. (**b**) Depending on the number of substrate edges, the count of harvestable trenches differs. This is significantly impacted by growth parameters and the number of cantilevers as shown in Figure 10a. (**c**) The frequency of substrate rotation changes with the number of substrate edges. This affects the assembly speed and is dependent on trajectory parameters and settling time of the rotation axis.

Taking an average required exchange time $\bar{t}_{ex}$ of 2 min into account, the preferable number of substrate edges n_e would be 1007 according to the values from Table 1. This example is, however, only valid for the rotation axis with its chosen trajectory parameters and settling time. As depicted in Figure 11, a change in these parameters could cause a shift of the optimum substrate geometry with another number of edges n_e. A prototype of an assembly machine verifies the movement parameters used for the simulation and leads to a comparable effective productivity if overhead time is excluded. The proposed growth substrate design for automation already considers positioning-related factors, and a proof of concept has successfully been demonstrated. However, for future industrial utilization it is necessary to also consider the Mean Time Between Failure (MTBF) of various components to assess effective productivity.

Table 1. The required exchange frequencies and the median effective productivity $\overline{P_e}$ in TPH of substrates with different numbers of edges n_e after Equation (11) and some values from Figure 10 with an average exchange duration $t_{ex} = 120$ s.

n_e	2	1007	2012	3017	4021	5026	6031	7036	8041	9046
$\overline{P}$	181	867	844	808	774	751	721	685	687	689
$\bar{n}_{harv}$	5584	8488	8132	7788	7400	7407	6646	5204	5912	6784
$\bar{f}_e$	0.03	0.1	0.1	0.1	0.1	0.1	0.11	0.13	0.12	0.1
$\overline{P_e}$	181	864	842	805	771	749	719	682	684	686

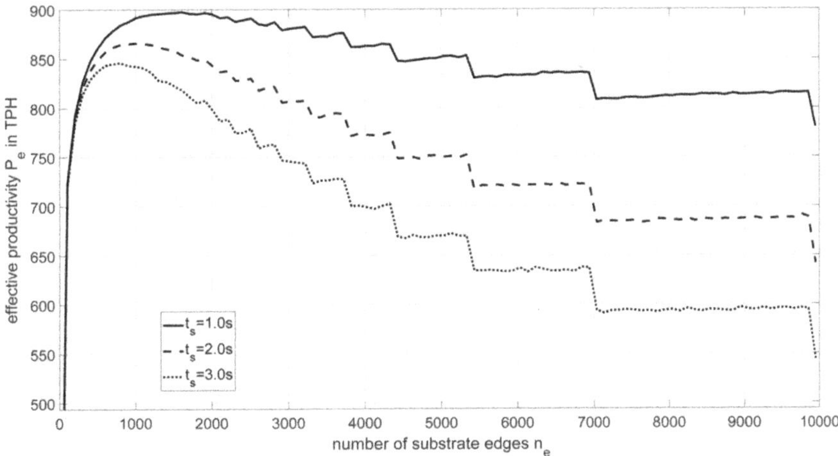

Figure 11. The effective productivity over the number of substrate edges for different settling times t_s of the rotation axis.

4. Conclusions

The productivity of the mechanical dry transfer of carbon nanotubes from a growth substrate to a device substrate relies on the cantilever geometry. While the length of the cantilevers must be adjusted to the growth of the nanotubes and the desired number of tubes per trench, the cantilever pitch must be reduced to achieve the maximum productivity.

However, as positioning errors are present at all times, the cantilever pitch is limited on the lower end. With the total errors $\Delta X, \Delta Y, \Delta \varepsilon$ decreasing, higher device-trench-overlaps at higher angles can be achieved. This leads to an increase in the number of harvestable tubes and thus to a less frequent exchange of substrates. Furthermore, it is shown how the overlap factor η_O affects the median transfer speed for various angles and cantilever pitches and lengths.

By adding a rotation axis, it becomes feasible to use substrates with multiple edges. This approach of increasing the total number of harvestable trenches leads to less frequent substrate exchanges. The analysis of the number of these edges shows that, depending on the trajectory parameters and the settling time of the rotation axis, the optimum number of substrate edges can be selected in order to reduce the substrate exchange frequency. Consequently, productivity is significantly amplified and also cost-effectiveness of the assembly process is improved. However, with each additional axis, system complexity increases. Therefore, the machine design with its configuration of axes must be considered already at early design stages when the required assembly motions are known.

Author Contributions: Conceptualization, A.B.; methodology, A.B.; software, A.B.; validation, A.B., S.W. and K.W.; formal analysis, A.B.; investigation, A.B. and S.W.; resources, A.B. and K.W.; data curation, A.B.; writing—original draft preparation, A.B.; writing—review and editing, A.B., S.W. and K.W.; visualization, A.B.; supervision, S.W. and K.W.; project administration, S.W. and K.W.; funding acquisition, K.W. All authors have read and agreed to the published version of the manuscript.

Funding: This research received no external funding.

Data Availability Statement: Data can be made available on request.

Conflicts of Interest: Authors Sascha Weikert and Konrad Wegener were employed by the company inspire AG. The remaining authors declare that the research was conducted in the absence of any commercial or financial relationships that could be construed as a potential conflict of interest.

References

1. Wei, N.; Laiho, P.; Khan, A.T.; Hussain, A.; Lyuleeva, A.; Ahmed, S.; Zhang, Q.; Liao, Y.; Tian, Y.; Ding, E.X.; et al. Fast and Ultraclean Approach for Measuring the Transport Properties of Carbon Nanotubes. *Adv. Funct. Mater.* **2020**, *30*, 1907150. [CrossRef]
2. Wu, C.C.; Liu, C.H.; Zhong, Z. One-Step Direct Transfer of Pristine Single-Walled Carbon Nanotubes for Functional Nanoelectronics. *Nano Lett.* **2010**, *10*, 1032–1036. [CrossRef]
3. Muoth, M.; Hierold, C. Transfer of carbon nanotubes onto microactuators for hysteresis-free transistors at low thermal budget. In Proceedings of the 2012 IEEE 25th International Conference on Micro Electro Mechanical Systems (MEMS), Paris, France, 29 January–2 February 2012; pp. 1352–1355. [CrossRef]
4. Pei, F.; Laird, E.A.; Steele, G.A.; Kouwenhoven, L.P. Valley–spin blockade and spin resonance in carbon nanotubes. *Nat. Nanotechnol.* **2012**, *7*, 630–634. [CrossRef] [PubMed]
5. Waissman, J.; Honig, M.; Pecker, S.; Benyamini, A.; Hamo, A.; Ilani, S. Realization of pristine and locally tunable one-dimensional electron systems in carbon nanotubes. *Nat. Nanotechnol.* **2013**, *8*, 569–574. [CrossRef]
6. Ranjan, V.; Puebla-Hellmann, G.; Jung, M.; Hasler, T.; Nunnenkamp, A.; Muoth, M.; Hierold, C.; Wallraff, A.; Schönenberger, C. Clean carbon nanotubes coupled to superconducting impedance-matching circuits. *Nat. Commun.* **2015**, *6*, 7165. [CrossRef] [PubMed]
7. Blien, S.; Steger, P.; Albang, A.; Paradiso, N.; Hüttel, A.K. Quartz Tuning-Fork Based Carbon Nanotube Transfer into Quantum Device Geometries. *Phys. Status Solidi B* **2018**, *255*, 1800118. [CrossRef]
8. Cubaynes, T.; Contamin, L.C.; Dartiailh, M.C.; Desjardins, M.M.; Cottet, A.; Delbecq, M.R.; Kontos, T. Nanoassembly technique of carbon nanotubes for hybrid circuit-QED. *Appl. Phys. Lett.* **2020**, *117*, 114001. [CrossRef]
9. Althuon, T.; Cubaynes, T.; Auer, A.; Sürgers, C.; Wernsdorfer, W. Nano-assembled open quantum dot nanotube devices. *Commun. Mater.* **2024**, *5*, 5. [CrossRef]
10. Ordu, M.; Der, O. Polymeric Materials Selection for Flexible Pulsating Heat Pipe Manufacturing Using a Comparative Hybrid MCDM Approach. *Polymers* **2023**, *15*, 2933. [CrossRef] [PubMed]
11. Emovon, I.; Oghenenyerovwho, O.S. Application of MCDM method in material selection for optimal design: A review. *Results Mater.* **2020**, *7*, 100115. [CrossRef]
12. Butzerin, A.; Lanz, N.; Weikert, S.; Wegener, K. Design and Performance Evaluation of a Novel Parallel Kinematic Micromanipulator. *Precis. Eng.* **2024**, *under review*.

MDPI AG
Grosspeteranlage 5
4052 Basel
Switzerland
Tel.: +41 61 683 77 34
www.mdpi.com

Processes Editorial Office
E-mail: processes@mdpi.com
www.mdpi.com/journal/processes